Texts in Applied Mathematics **36**

Texts in Applied Mathematics

(continued after index)

Geir E. Dullerud   Fernando Paganini

A Course in
Robust Control Theory

A Convex Approach

With 36 Illustrations

 Springer

Geir E. Dullerud
Department of Mechanical and
    Industrial Engineering
University of Illinois
Urbana, IL 61801
USA
dullerud@legend.me.uiuc.edu

Fernando Paganini
Department of Electrical Engineering
University of California
Los Angeles, CA 90095-1594
USA
paganini@ee.ucla.edu

Series Editors

J.E. Marsden
Control and Dynamical Systems, 107-81
California Institute of Technology
Pasadena, CA 91125
USA
marsden@cds.caltech.edu

L. Sirovich
Division of Applied Mathematics
Brown University
Providence, RI 02912
USA
chico@camelot.mssm.edu

S.S. Antman
Department of Mathematics
and
Institute for Physical Science
    and Technology
University of Maryland
College Park, MD 20742-4015
USA
ssa@math.umd.edu

Mathematics Subject Classification (2000): 93-01, 49-01

Library of Congress Cataloging-in-Publication Data
Dullerud, Geir E., 1966–
    A course in robust control theory: a convex approach/Geir E. Dullerud,
Fernando G. Paganini.
        p. cm. – (Texts in applied mathematics)
    Includes bibliographical references and index.
    ISBN 978-1-4419-3189-4
        1. Robust control. 2. Control theory. I. Paganini, Fernando G. II. Title.
III. Series.
TJ217.2 D85 1999
629.8′312–dc21   99-046358

Printed in the United States of America.   (IBT)

9 8 7 6 5 4 3 2 (Corrected printing, 2005)

springeronline.com

To our Families

Series Preface

Mathematics is playing an ever more important role in the physical and biological sciences, provoking a blurring of boundaries between scientific disciplines and a resurgence of interest in the modern as well as the classical techniques of applied mathematics. This renewal of interest, both in research and teaching, has led to the establishment of the series: *Texts in Applied Mathematics (TAM)*.

The development of new courses is a natural consequence of a high level of excitement on the research frontier as newer techniques, such as numerical and symbolic computer systems, dynamical systems, and chaos, mix with and reinforce the traditional methods of applied mathematics. Thus, the purpose of this textbook series is to meet the current and future needs of these advances and encourage the teaching of new courses.

TAM will publish textbooks suitable for use in advanced undergraduate and beginning graduate courses, and will complement the *Applied Mathematical Sciences (AMS)* series, which will focus on advanced textbooks and research level monographs.

Preface

Robust control theory has been one of the most active areas of mainstream systems theory since the late 1970s. This research activity has been at the confluence of dynamical systems theory, functional analysis, matrix analysis, numerical methods, complexity theory, and control engineering applications. Remarkably, the discipline has exhibited deep theoretical aspects of interest to pure and applied mathematicians, and at the same time has remained in close connection with tangible engineering applications. By now this research effort has produced a rather extensive set of approaches using a wide variety of mathematical techniques, and applications of robust control theory are spreading to diverse areas such as aerospace systems, chemical processes, power networks, and control of fluids. Some major milestones of the theory were developed in the 1980s, and by now have been reported in a number of textbooks. However, during the 1990s the theory has seen major advances and achieved a new maturity, hinging around the notion of convexity. This emphasis is two-fold. On one hand, a close, and mutually beneficial interaction has developed between control theory and the active area of convex optimization; in particular the methods of convex programming have proven relevant to a wide class of control problems. Simultaneously a new understanding has developed on the computational complexity implications of uncertainty modeling; in particular it has become clear that one must go beyond a time invariant structure to describe uncertainty in terms amenable to convex robustness analysis.

Our broad goal in this book is to give a graduate-level course on robust control theory that emphasizes these new developments, but at the same time conveys the main principles and ubiquitous tools at the heart of the subject. We thus aim at a coherent and unified introduction to robust control which starts at the "beginning" of basic systems theory, covers some of the more established landscape of the field, and leads up to the more recent themes and to topics of current research. The effect of this intention on the book is that as it progresses the new topics are increasingly emphasized, whereas more conventional techniques appear less frequently. While many aspects of the vast literature have been left out, we have selected those that we believe are central and most effectively form a launching point for further study of the field. We therefore hope the book will be of value to graduate students planning to do research in the area, to mathematical researchers and computer scientists wishing to learn about robust control theory, and engineers interested in the basis of advanced control techniques.

The text is written to comprise a two-quarter or two-semester graduate course in applied mathematics or engineering. Alternatively, for students with background in state space methods, a serious approach at a significant portion of the material can be achieved in one semester. The material has been successfully taught in this capacity during the past few years by the authors at Caltech, University of Waterloo, University of Illinois, and UCLA. Students are assumed to have some familiarity with linear algebra, and otherwise only advanced calculus and basic complex analysis are strictly required. The presentation style assumes, however, a mathematically inclined reader, since we focus on a complete theoretical foundation rather than on application examples.

After a conceptual introduction in Chapter 0, the next three chapters cover basic tools to be used throughout the course. Chapter 1 reviews some basic elements of linear algebra and matrix theory, including the Jordan form and the singular value decomposition. It then presents some elementary concepts of convex analysis in finite dimensional space, and ends by introducing the important concept of a linear matrix inequality (LMI). Chapter 2 provides a compact, yet thorough treatment of state space systems theory, including controllability and observability, eigenvalue assignment and minimal realizations. In Chapter 3 we introduce the functional analysis and operator theory used in the sequel, beginning with normed and inner product spaces, and following with operators on Hilbert space. The focus of the text is on L_2 signal norms, so a number of associated function spaces are presented, including the H_2 and H_∞ spaces which are related to the concepts of time invariance and causality in systems.

Chapter 4 revisits the topic of open-loop systems and realizations by using the newly introduced concept of a norm. Controllability and observability can now be quantified by means of gramians, and balanced realizations are introduced. Hankel operators and singular values are then discussed, leading up to model reduction using balanced truncation.

In Chapters 5 through 7 three feedback design problems are discussed: stabilization, H_2 optimization, and H_∞ control. We take advantage of this progression to present three main tools for controller synthesis: the parametrization of stabilizing controllers via coprime factorizations, Riccati equation techniques for optimization, and the most recent LMI techniques. The latter play a relatively minor role in Chapters 5 and 6, but are essentially the exclusive tool employed in Chapter 7. In this way we endeavor to exhibit the power of the new methods while simultaneously providing an in depth covering of important, more classical techniques.

In Chapters 8 and 9 we discuss uncertain system models and techniques for analysis and synthesis in the presence of uncertainty. Chapter 8, which focuses exclusively on analysis, adopts an abstract operator viewpoint which leads to the study of structural properties of uncertainty and the corresponding robustness analysis tools, based on generalizations of the Small gain theorem. The more recent convex necessary and sufficient conditions which apply to uncertainty that is not time invariant are discussed first, and are then followed by the presentation of the more classical structured singular value methods. The operator perspective allows one to develop the theory in isolation from some technical details which arise when dealing with potentially unstable systems; these are later tackled in Chapter 9, and subsequently techniques for robust controller synthesis are presented.

The final two chapters are devoted to the presentation of four topics of recent research, in a more descriptive manner. Chapter 10 covers extensions of robustness analysis. First the method of integral quadratic constraints (IQCs) is discussed, which provides generalized uncertainty descriptions and is closely tied to the tools of Chapter 8. Secondly, the analysis of robust H_2 performance is covered, which aims at reconciling the desirable performance characteristics of the H_2 norm with the advantages of L_2-induced norms to measure uncertainty. Chapter 11 provides some generalizations with an emphasis on synthesis. The first is a consideration of multidimensional state space techniques, aimed at uncertain system realization, distributed systems, and parameter-varying systems. Linear parameter-varying control synthesis and multidimensional model reduction are discussed. Finally, time-varying systems are studied using an operator theoretic framework, which shows how these systems can be treated formally as though they were time invariant, leading to analogs of many of the earlier results of the book in a more general setting.

The core chapters of the book are accompanied by exercises, ranging from simple application examples to more significant extensions of the exposed theory.

We are indebted to a number of people as we complete this book. Much of what we report here we have learned from our Ph.D. advisors, Keith Glover and John Doyle, who have also had a considerable influence on our careers and our views of research and control. John also provided significant encouragement for this project, and has tried out our draft in recent courses at Caltech. Raff D'Andrea at Cornell also used early drafts and provided very useful feedback. Bruce Francis provided his notes on introductory state space theory, which were used as a basis for the succinct treatment in Chapter 2. We also thank Carolyn Beck and Sanjay Lall for their in depth comments on the manuscript, and the anonymous reviewers for valuable suggestions. This text was fine-tuned with the help of many students who studied from earlier versions and pointed out numerous mistakes; we particularly thank Carol Pirie and Federico Najson for very thorough revisions. In addition, we are grateful to Carolyn Jackson for her assistance in typing a preliminary version of the manuscript.

Finally, we wish to thank our wives, Carolyn and Malena for their support and patience, and our children, Natalie, Rafael and Fernando for the time taken away from them during this effort.

Geir E. Dullerud   Fernando Paganini
Urbana   Los Angeles

November, 1999

Contents

Figures

0
Introduction

In this course we will explore and study a mathematical approach aimed directly at dealing with complex physical systems that are coupled in feedback. The general methodology we study has analytical applications to both human-engineered systems and systems that arise in nature, and the context of our course will be its use for feedback control.

The direction we will take is based on two related observations about models for complex physical systems. The first is that analytical or computational models which closely describe physical systems are difficult or impossible to precisely characterize and simulate. The second is that a model, no matter how detailed, is *never* a completely accurate representation of a real physical system. The first observation means that we are forced to use simplified system models for reasons of tractability; the latter simply states that models are innately inaccurate. In this course both aspects will be termed model *uncertainty*, and our main objective is to develop systematic techniques and tools for the design and analysis of systems which are uncertain. The predominant idea that is used to contend with such uncertainty or unpredictability is *feedback* compensation.

There are several ways in which systems can be uncertain, and in this course we will target the main three:

- The initial conditions of a system may not be accurately specified or completely known.

- Systems experience disturbances from their environment, in fact the separation between system and environment is always an idealization. Also system commands are typically not known a priori.

- Uncertainty in the accuracy of a system model itself is a central source. Any dynamical model of a system will neglect some physical phenomena, and this means that any analytical control approach based solely on this model will neglect some regimes of operation.

In short: *the major objective of feedback control is to minimize the effects of unknown initial conditions and external influences on system behavior, subject to the constraint of not having a complete representation of the system.* This is a formidable challenge in that predictable behavior is expected from a controlled system, and yet the strategies used to achieve this must do so using an inexact system model. The term *robust* in the title of this course refers to the fact that the methods we pursue will be expected to operate in an uncertain environment with respect to the system dynamics. The mathematical tools and models we use will be primarily linear, motivated mainly by the requirement of computability of our methods; however, the theory we develop is aimed at the control of complex, nonlinear systems. In this introductory chapter we will devote some space to discuss, at an informal level, the interplay between linear and nonlinear aspects in this approach.

The purpose of this chapter is to provide some context and motivation for the mathematical work and problems we will encounter in the course. For this reason we do not provide many technical details here; however, it might be informative to refer back to this chapter periodically during the course.

0.1   System representations

We will now introduce the diagrams and models used in this course.

0.1.1   Block diagrams

We will often view physical or mathematical systems as mappings. From this perspective a system maps an *input* to an *output*; for *dynamical* systems these are regarded as functions of time. This is not the only or most primitive way to view systems, although we will find this viewpoint to be very attractive both mathematically and for guiding and building intuition. In this section we introduce the notion of a *block diagram* for representing systems, and most importantly for specifying their interconnections.

We use the symbol P to denote a system that maps an input function $u(t)$ to an output function $y(t)$. This relationship is denoted by

$$y = P(u).$$

Figure 0.1 illustrates this relationship. The direction of the arrows indicate whether a function is an input or an output of the system P. The details

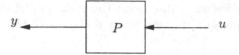

Figure 0.1. Basic block diagram

of how P constructs y from the input u are not depicted in the diagram; instead the benefit of using such block diagrams is that interconnections of systems can be readily visualized.

Consider the so-called cascade interconnection of the two subsystems. This interconnection represents the equations

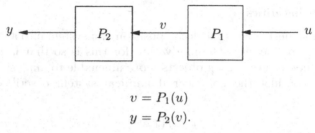

$$v = P_1(u)$$
$$y = P_2(v).$$

We see that this interconnection takes the two subsystems P_1 and P_2 to form a system P defined by $P(u) = P_2(P_1(u))$. Thus this diagram simply depicts a composition of maps. Notice that the input to P_2 is the output of P_1.

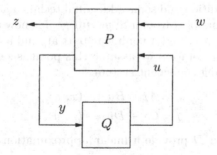

Another type of interconnection involves feedback. In the figure above we have such an arrangement. Here P has inputs given by the ordered pair (w, u) and the outputs (z, y). The system Q has input y and output u. This block diagram therefore pictorially represents the equations

$$(z, y) = P(w, u)$$
$$u = Q(y).$$

Since part of the output of P is an input to Q, and conversely the output of Q is an input to P, these systems are coupled in feedback.

We will now move on to discussing the basic modeling concept of this course and in doing so will immediately make use of block diagrams.

0.1.2   Nonlinear equations and linear decompositions

We have just introduced the idea of representing a system as an input–output mapping, and did not concern ourselves with how such a mapping might be defined. We will now outline the main idea behind the modeling framework used in this course, which is to represent a complex system as a combination of a perturbation and a simpler system. We will illustrate this by studying two important cases.

Isolating nonlinearities

The first case considered is the decomposition of a system into a linear part and a static nonlinearity. The motivation for this is so that later we can replace the nonlinearity using objects more amenable to analysis.

To start consider the nonlinear dynamical system described by the equations

$$\dot{x} = f(x, u) \tag{0.1}$$
$$y = h(x, u),$$

with the initial condition $x(0)$. Here $x(t)$, $y(t)$, and $u(t)$ are vector valued functions, and f and h are smooth vector valued functions. The first of these equations is a differential equation and the second is purely algebraic. Given an initial condition and some additional technical assumptions, these equations define a mapping from u to y. Our goal is now to decompose this system into a linear part and a nonlinear part around a specified point; to reduce clutter in the notation we assume this point is zero.

Define the following equivalent system:

$$\dot{x} = Ax + Bu + g(x, u) \tag{0.2}$$
$$y = Cx + Du + r(x, u),$$

where A, B, C, and D provide a linear approximation to the dynamics, and

$$g(x, u) = f(x, u) - Ax - Bu$$
$$r(x, u) = h(x, u) - Cx - Du.$$

For instance, one could take the Jacobian linearization

$$A = d_1 f(0, 0), \ B = d_2 f(0, 0),$$
$$C = d_1 h(0, 0), \text{ and } D = d_2 h(0, 0),$$

where d_1 and d_2 denote vector differentiation by the first and second vector variables, respectively. The following discussion, however, does not require this assumption. The system in (0.2) consists of linear functions and the possibly nonlinear functions g and r. It is clear that the solutions to this system have a one-to-one correspondence with the solutions of (0.1), since we have simply rewritten the functions. Further, let us write these equations in the equivalent form

$$\dot{x} = Ax + Bu + w_1 \tag{0.3}$$
$$y = Cx + Du + w_2 \tag{0.4}$$
$$(w_1, w_2) = (g(x, u), r(x, u)). \tag{0.5}$$

Now let G be the mapping described by (0.3) and (0.4), which satisfies

$$G : (w_1, w_2, u) \mapsto (x, u, y),$$

given an initial condition $x(0)$. Further let Q be the mapping which takes $(x, u) \mapsto (w_1, w_2)$ as described by (0.5). Thus the system of equations defined by (0.3 – 0.5) has the block diagram below. The system G is totally

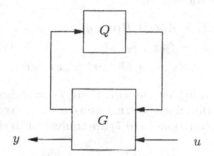

Figure 0.2. System decomposition

described by linear differential equations, and Q is a *static* nonlinear mapping. By static we mean that the output of Q at any point in time depends only on the input at that particular time, or equivalently that Q has no memory. Thus all of the nonlinear behavior of the initial system (0.1) is captured in Q and the feedback interconnection.

We will almost exclusively work with the case where the point $(0, 0)$, around which this decomposition is taken, is an *equilibrium point* of (0.1). Namely,

$$f(0, 0) = 0.$$

In this case the function g satisfies $g(0, 0) = 0$, and therefore $Q(0, 0) = (0, r(0, 0))$. Also where the linear system described by

$$\dot{x} = Ax + Bu$$
$$y = Cx + Du$$

is a *linearization* of (0.1) around the equilibrium point. The linear system G is thus an augmented version of the linearization.

Higher-order dynamics

In the construction just considered we were able to isolate nonlinear system aspects in the mapping Q, and our motivation for this was so that later we will be able to replace Q with an alternative description which is more easily analyzed. For the same reason we will sometimes wish to do this not only with the nonlinear part of a system, but also with some of its dynamics. Let us now move on to consider this more complex scenario. We have the equations

$$\begin{bmatrix} \dot{x}_1 \\ \dot{x}_2 \end{bmatrix} = \begin{bmatrix} f_1(x_1,\, x_2,\, u) \\ f_2(x_1,\, x_2,\, u) \end{bmatrix}, \tag{0.6}$$

$$y = h(x_1,\, x_2,\, u).$$

Following a similar procedure to the one we just carried out on the system in (0.1), we can decompose the system described in (0.6) to arrive at the equivalent set of equations:

$$\dot{x}_1 = A_1 x_1 + B_1 u + g_1(x_1,\, x_2,\, u), \tag{0.7}$$
$$\dot{x}_2 = f_2(x_1,\, x_2,\, u),$$
$$y = C_1 x_1 + Du + r(x_1,\, x_2,\, u).$$

This is done by focusing on the equations $\dot{x}_1 = f_1(x_1,\, x_2,\, u)$ and $y = h(x_1,\, x_2,\, u)$, and performing the same steps as before treating both x_2 and u as the inputs. The equations in (0.7) are equivalent to the linear equations

$$\dot{x}_1 = A_1 x_1 + B_1 u + w_1 \tag{0.8}$$
$$y = C_1 x_1 + Du + w_2,$$

coupled with the nonlinear equations

$$\dot{x}_2 = f_2(x_1,\, x_2,\, u) \tag{0.9}$$
$$(w_1,\, w_2) = (g_1(x_1,\, x_2,\, u),\, r(x_1,\, x_2,\, u)).$$

Now as before we set G to be the linear system

$$G : (w_1,\, w_2,\, u) \mapsto (x_1,\, u,\, y)$$

which satisfies the equations in (0.8). Also define Q to be the system described by (0.9) where

$$Q : (x_1,\, u) \mapsto (w_1,\, w_2).$$

With these new definitions of P and Q we see that Figure 0.2 depicts the system described in (0.6). Furthermore part of the system dynamics and all of the system nonlinearity is isolated in the mapping Q. Notice that the decomposition we performed on (0.1) is a special case of the current one.

Modeling Q

In each of the two decompositions just considered we split the initial systems, given by (0.1) or (0.6), into a G-part and a Q-part. The system G was described by linear differential equations, whereas nonlinearities were confined to Q. This decomposing of systems into a linear low dimensional part, and a potentially nonlinear and high dimensional part, is at the heart of this course. The main approach adopted here to deal with the Q-part will be to capture its behavior with a *set of mappings* $\boldsymbol{\Delta}$.

Formally stated we require the set $\boldsymbol{\Delta}$ to have the following property: if $q = Q(p)$, for some input p, then there should exist a mapping Δ in the set $\boldsymbol{\Delta}$ such that

$$q = \Delta(p). \tag{0.10}$$

The key idea here is that the elements of the set $\boldsymbol{\Delta}$ can be much simpler dynamically than Q. In particular, while Q is in general nonlinear, we will use *linear* mappings Δ, which, when combined in a set are actually able to generate all of the possible input–output pairs (p, q) which satisfy $q = Q(p)$. Therein lies the power of introducing $\boldsymbol{\Delta}$: one complex object can be replaced by a set of simpler ones. The motivation for doing this is that the resulting analysis can frequently be made tractable.

We now discuss how this idea can be used for analysis of the system depicted in Figure 0.2. Let

$\bar{S}(G, Q)$ denote the mapping $u \mapsto y$ in Figure 0.2.

Now replace this map with the set of maps

$$\bar{S}(G, \boldsymbol{\Delta})$$

generated by choosing Δ from the set $\boldsymbol{\Delta}$. Then we see that if the input–output behaviors associated with all the mappings $\bar{S}(G, \Delta)$ satisfy a given property, then so must any input–output behavior of $\bar{S}(G, Q)$. Thus any property which holds over the set $\boldsymbol{\Delta}$ is guaranteed to hold for the system $\bar{S}(G, Q)$. However, the converse is not true and so analysis using $\boldsymbol{\Delta}$ can in general be conservative. Let us consider this issue.

If a set $\boldsymbol{\Delta}$ has the property described in (0.10), then providing that it has more than one element, it will necessarily generate more input–output pairs than Q. Specifically

$$\{(p, q): \ q = Q(p)\} \subset \{(p, q): \ \text{there exists } \Delta \in \boldsymbol{\Delta}, \text{ such that } q = \Delta(p)\}.$$

Clearly the set on the left defines a function, whereas the input–output pairs generated by $\boldsymbol{\Delta}$ is in general only a relation. The degree of closeness of these sets determines the level of conservatism introduced by using $\boldsymbol{\Delta}$ in place of Q.

We now illustrate how the behavior of Q can be captured by $\boldsymbol{\Delta}$ with two simple examples.

Examples:

We begin with the decomposition for (0.1). For simplicity assume that x, y, and u are all scalar valued functions. Now suppose that the functions r and g, which define Q, are known to satisfy the sector or Lipschitz bounds

$$|w_1(t)| \leq k_{11}|x(t)| + k_{12}|u(t)| \qquad (0.11)$$
$$|w_2(t)| \leq k_{21}|x(t)| + k_{22}|u(t)|,$$

for some positive constants k_{ij}. It follows that if for particular signals $(w_1, w_2) = Q(x, u)$, then there exist scalar functions of time $\delta_{11}(t), \delta_{12}(t), \delta_{21}(t)$ and $\delta_{22}(t)$, each satisfying $\delta_{ij}(t) \in [-k_{ij}, \, k_{ij}]$, such that

$$w_1(t) = \delta_{11}(t)x(t) + \delta_{12}(t)u(t) \qquad (0.12)$$
$$w_2(t) = \delta_{21}(t)x(t) + \delta_{22}(t)u(t),$$

Define the set $\boldsymbol{\Delta}$ to consist of all 2×2 matrix functions Δ which satisfy

$$\Delta = \begin{bmatrix} \delta_{11}(t) & \delta_{12}(t) \\ \delta_{21}(t) & \delta_{22}(t) \end{bmatrix}, \text{ where } |\delta_{ij}(t)| \leq k_{ij} \text{ for each time } t \geq 0.$$

From the above discussion it is clear the set $\boldsymbol{\Delta}$ has the property that given any inputs and outputs satisfying $(w_1, w_2) = Q(x, u)$, there exists $\Delta \in \boldsymbol{\Delta}$ satisfying (0.12).

Let us turn to an analogous construction associated with the decomposition of the system governed by (0.6), recalling that Q is now dynamic. Assume x_1 and u are scalar, and suppose it is known that if $(w_1, w_2) = Q(x_1, u)$, then the following energy inequalities hold:

$$\int_0^\infty |w_1(t)|^2 dt \leq k_1 \left(\int_0^\infty |x_1(t)|^2 dt + \int_0^\infty |u(t)|^2 dt \right) \qquad (0.13)$$

$$\int_0^\infty |w_2(t)|^2 dt \leq k_2 \left(\int_0^\infty |x_1(t)|^2 dt + \int_0^\infty |u(t)|^2 dt \right),$$

when the right-hand side integrals are finite. Define $\boldsymbol{\Delta}$ to consist of all *linear* mappings $\Delta : (x_1, u) \mapsto (w_1, w_2)$ which satisfy the above inequalities for all functions x_1 and u from a suitably defined class. It is possible to show that if $(w_1, w_2) = Q(x_1, u)$, for some bounded energy functions x_1 and u, then there exists a mapping Δ in $\boldsymbol{\Delta}$ such that $\Delta(x_1, u) = (w_1, w_2)$. In this sense $\boldsymbol{\Delta}$ can generate any behavior of Q. As a remark, inequalities such as the above assume implicitly that initial conditions in the state x_2 can be neglected; in the language of the next section, there is a requirement of stability in high-order dynamics that can be isolated in this way. $\qquad \square$

Using a set $\boldsymbol{\Delta}$ instead of the mapping Q has another purpose. As already pointed out physical systems will never be exactly represented by models of the form (0.1) or (0.6). Thus the introduction of the set $\boldsymbol{\Delta}$ affords a way to account for potential system behaviors without explicitly modeling them. For example the inequalities in (0.13) may be all that is known about some higher-order dynamics of a system; note that given these bounds we would not even need to know the order of these dynamics to account for them using Δ. Therefore this provides a way explicitly to incorporate knowledge about the unpredictability of a physical system into a formal model. Thus the introduction of the Δ serves two related but distinct purposes:

- provides a technique for simplifying a given model;

- can be used to model and account for uncertain dynamics.

In the course we will study analysis and synthesis using these types of models, particularly when systems are formed by the interconnection of many such subsystems. We call these types of models *uncertain* systems.

0.2 Robust control problems and uncertainty

In this section we outline three of the basic control scenarios pursued in this course. Our discussion is informal and is intended to provide motivation for the mathematical analysis in the sequel.

0.2.1 Stabilization

One of the most basic goals of a feedback control system is stabilization, which renders a system insensitive to uncertainty surrounding its initial conditions. Before explaining this in more detail we review some basic concepts.

Consider the autonomous system

$$\dot{x} = f(x), \text{ with some initial condition } x(0). \tag{0.14}$$

We will be concerned with equilibrium points x_e of this system, namely, points where $f(x_e) = 0$ is satisfied. Without loss of generality in this discussion we shall assume that $x_e = 0$, since this can always be arranged by redefining f appropriately. We say that the equilibrium point zero is stable if for any initial condition $x(0)$ sufficiently near to zero, the time trajectory $x(t)$ remains near to zero. The equilibrium is *exponentially stable* if it is stable and furthermore the function $x(t)$ tends to zero at an exponential rate when $x(0)$ is chosen sufficiently small. Stability is an important property because it is unlikely that a physical system is ever exactly at an equilibrium point. It says that if the initial state of the system is slightly perturbed away from the equilibrium, the resulting state trajectory will not diverge. Exponential stability goes further to say that if such initial deviations are small then the system trajectory will tend quickly back to the equilibrium point. Thus stable systems are insensitive to uncertainty about their initial conditions.

We now review a test for exponential stability. Suppose

$$\dot{x} = Ax$$

is the linearization of (0.14) at the point zero. It is possible to show that: *the zero point of (0.14) is exponentially stable if and only if the linearization is exponentially stable at zero.*[1] The linearization is exponentially stable exactly when all the eigenvalues of the matrix A have negative real part. Thus exponential stability can be checked directly by calculating A, the Jacobian matrix of f at zero. Further it can be shown that if any of the eigenvalues have positive real part, then the equilibrium point zero is not even stable.

We now move to the issue of stabilization. Below is a controlled nonlinear system

$$\dot{x} = f(x, u), \tag{0.15}$$

where the input is the function u. Suppose that $f(0, 0) = 0$ and thus if u is fixed at zero the point $x = 0$ is an equilibrium. This equilibrium may, however, be unstable; the stabilization problem involves using the control input u to turn it into an exponentially stable one.

We discuss first a special type of control strategy called a state feedback. In this scenario we seek a control feedback law of the form

$$u(t) = p(x(t)),$$

where p is a smooth function, such that the closed-loop system

$$\dot{x} = f(x, p(x(t)))$$

is exponentially stable around zero. That is, we want to find a function p which maps the state of the system x to a control action u. Let us assume that such a p exists and examine some of its properties. First notice that in order for zero to be an equilibrium point of the closed loop we may assume

$$p(0) = 0.$$

Given this the linearization of the closed loop is

$$\dot{x} = (A + BF)x,$$

where $A = d_1 f(0, 0)$, $B = d_2 f(0, 0)$ and $F = dp(0)$. Thus we see that the closed-loop system is exponentially stable if and only if all the eigenvalues of $A + BF$ are strictly in the left half of the complex plane. Conversely notice that if a matrix F exists such that $A + BF$ has the desired stability property, then the state feedback law $p(x) = Fx$ will stabilize the closed loop.

In the scenario just discussed the state x was directly available to use in the feedback law. A more general control situation occurs when only

---

[1] This is under the assumption that f is sufficiently smooth.

an observation $y = h(x, u)$ is available for feedback, or when a dynamic control law is employed. For these the analysis is more complicated and we defer the study to subsequent chapters. We now illustrate these concepts with an example.

Stabilization of a double pendulum

Shown below in Figure 0.3 is a double pendulum. In the figure two rigid links are connected by a hinge joint, so that they can rotate with respect to each other. The first link is also constrained to rotate about a point which is fixed in space. The control input to this rigid body system is a torque

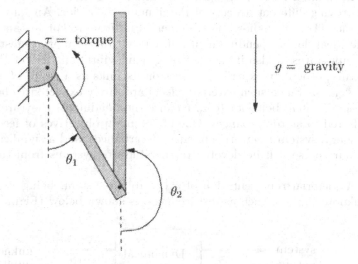

Figure 0.3. Double pendulum with torque input

τ which is applied to the first link. The configuration of this system is completely specified by θ_1 and θ_2, which are the respective angles that the first and second links make with the vertical. Since we have assumed that the links are ideal rigid bodies, this system can be described by a differential equation of the form (0.15), where $x = (\theta_1, \theta_2, \dot{\theta}_1, \dot{\theta}_2)$ and $u = \tau$. It is routine to show that for each fixed angle θ_{1e} of the first link, there exists a torque τ_e, such that

$$((\theta_{1e}, \theta_{2e}, 0, 0), \tau_e) \text{ is an equilibrium point,}$$

where θ_{2e} is equal to either zero or π. That is, for any value of θ_{1e} we have two equilibrium points of the system; both occur when the second link is vertical. When $\theta_{2e} = 0$ the equilibrium point is stable. The $\theta_{2e} = \pi$ equilibrium point is unstable.

We may wish to stabilize (i.e. balance) the pendulum about its upright position at such an equilibrium point. To apply a state feedback control law as discussed above we require the ability to measure x; namely, we base our control law on the link angles and their velocities. In the more general scheme, also described above, we would only have access to some function of these four measurements; a typical situation is that the observation is the position of the tip of the second pendulum.

0.2.2   Disturbances and commands

Dynamic stability is usually a basic requirement of a controlled system; however, typically much more is demanded, and in fact often feedback is introduced in systems which are already stable, with the objective of improving different aspects of the dynamic behavior. An important issue is the effect of unknown environmental influences. For instance, consider the ideal double pendulum just discussed above; more realistically such an apparatus is also influenced by ground vibrations felt at the point of fixture, or it may experience forces on its links as a result of air currents or "gusts." Since such external effects are rarely expected to help achieve desired system behavior (e.g., balance the pendulum) they are commonly referred to as *disturbances*. One of the main objectives of feedback is to render a system insensitive to such disturbances, and a significant portion of our course will be devoted to the study of systems from this point of view.

A pictorial representation of a controlled system, being influenced by unknown inputs, such as disturbances, is shown below (Figure 0.4). Here

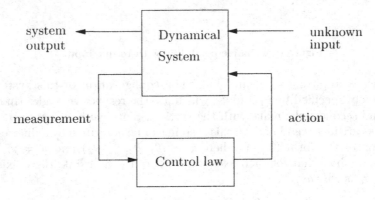

Figure 0.4. Controlled system

we have a dynamical system in feedback with a control law, which takes some actions based on measurement information. However, there are other environmental influences acting on the systems, which are unknown at the time of control design, and could be disturbances, or also external commands. The behavior of the system is characterized by the outputs. To bring sharper focus to our discussion we consider a second example.

Position control of an electric motor

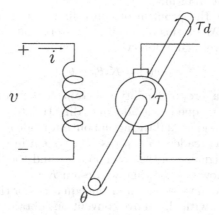

The figure depicts a schematic diagram of an electric motor controlled by excitation: a voltage applied to the motor windings results in a torque applied to the motor shaft. While physical details are not central to our discussion, we will find it useful to write down an elementary dynamical model. The key variables are

- v, applied voltage;
- i, current in the field windings;
- τ, motor torque and τ_d opposing torque from the environment;
- θ, angle of the shaft and $\omega = \dot{\theta}$ angular velocity.

The objective of the control system is for the angular position θ to follow a reference command θ_r, despite the effect of an unknown resisting torque τ_d. This so-called servomechanism problem is common in many applications; for instance, we could think of moving a robot arm in an uncertain environment.

We begin by writing down a differential equation model for the motor:

$$v = Ri + L\frac{di}{dt} \tag{0.16}$$

$$\tau = \gamma i \tag{0.17}$$

$$J\frac{d\omega}{dt} = \tau - \tau_d - B\omega \tag{0.18}$$

Here (0.16) models the electrical circuit in terms of its resistance R and inductance L; (0.17) says the torque is a linear function of the current; and (0.18) is the rotational dynamics of the shaft, where J is the moment of inertia and B is mechanical damping. Since the electrical transients of (0.16) are typically much faster than the mechanical dynamics, it seems reasonable to neglect the former, setting $L = 0$. Also in what follows we normalize, for simplicity, the remaining constants (R, γ, J, B) to unity in an appropriate system of units.

We now address the problem of controlling our motor to achieve the desired objective. We do this by a control system that measures the output θ and acts on the voltage v, following the law

$$v(t) = K(\theta_r - \theta),$$

where $K > 0$ is a proportionality constant to be designed. Intuitively, the system applies torque in the adequate direction to counteract the error between the command θ_r and the actual output angle θ. It is an instructive exercise, left for the reader, to express this system in terms of Figure 0.4. Here the driving signals are the command θ_r and the disturbance torque τ_d, which are unknown at the time we design K.

Given this control law, we can find the equations of the resulting *closed-loop* dynamics, and with the above conventions obtain the following.

$$\frac{d}{dt} \begin{bmatrix} \theta \\ \omega \end{bmatrix} = \begin{bmatrix} 0 & 1 \\ -K & -1 \end{bmatrix} \begin{bmatrix} \theta \\ \omega \end{bmatrix} + \begin{bmatrix} 0 & 0 \\ K & -1 \end{bmatrix} \begin{bmatrix} \theta_r \\ \tau_d \end{bmatrix}$$

Thus our resulting dynamics have the form

$$\dot{x} = Ax + Bw$$

encountered in the previous section.

We first discuss system stability. It is straightforward to verify that the eigenvalues of the above A matrix have negative real part whenever $K > 0$; therefore in the absence of external inputs the system is exponentially stable. In particular, initial conditions will have asymptotically no effect.

Now suppose θ_r and τ_d are constant over time; then by solving the differential equation it follows that the states (θ, ω) converge asymptotically to

$$\theta(\infty) = \theta_r - \frac{\tau_d}{K} \quad \text{and} \quad \omega(\infty) = 0.$$

Thus the motor achieves an asymptotic position which has an error of τ_d/K with respect to the command. We make the following remarks:

- Clearly if we make the constant K very large we will have accurate tracking of θ_r despite the effect of τ_d. This highlights the central role of feedback in achieving system reliability in the presence the uncertainty about the environment. We will revisit this issue in the following section.

- The success in this case depends strongly on the fact that θ_r and τ_d are known to be constant; in other words, while their value was unknown, we had some a priori information about their characteristics.

The last observation is in fact general to any control design question. That is, some information about the unknown inputs is required for us to be able to assess system performance. In the above example the signals were specified except for a parameter. More generally the information available is not so strongly specified; for example, one may know something about the energy or spectral properties of commands and disturbances. The available information is typically expressed in one of two forms:

(i) We may specify a *set* $\mathcal{D}$ and impose that the disturbance should lie in $\mathcal{D}$.

(ii) We may give a statistical description of the disturbance signals.

The first alternative typically leads to questions about the *worst* possible system behavior caused by any element of $\mathcal{D}$. In the second case, one usually is concerned with statistically typical behavior. Which is more appropriate is application dependent, and we will provide methods for both alternatives in this course.

We are now ready to discuss a more complicated instance of uncertainty in the next section.

0.2.3 Unmodeled dynamics

When modeling a physical system we always approximate some aspects of the physical phenomena, owing to an incomplete theory of physics. Also we frequently further simplify our models by choice so as to make analysis more tractable, when we believe such a simplification is innocuous. Now we immediately wonder about the effect of such approximations when we apply *feedback* to a system.

Take, for instance, our previous example of the electric motor, where we deliberately neglected the effects of inductance in the electric circuit which was deemed irrelevant to a study at the time scale of mechanical motion. And the conclusion of our study was that we should make the constant K as large as possible in order to achieve good tracking performance.

However, if we keep the inductance L in our model, and repeat the analysis, it is not difficult to see that the resulting third-order system becomes *unstable* for a sufficiently high value of K. Thus we see that our seemingly benign modeling error can be *amplified* by feedback to the point of making the system unusable. Thus feedback is a double-edged sword: it can render a system insensitive to uncertainty (e.g., the torque disturbance τ_d), but it can also increase sensitivity to it, as was just seen. Feedback design always involves a judicious balance of this fundamental tradeoff.

At this point the reader may be thinking that this difficulty was due exclusively to careless modeling, we should have worked with the full, third-order model. Note, however, that there are many other dynamical aspects which have been neglected. For instance, the bending dynamics of the motor shaft could also be described by additional state equations, and so on. We could go to the level of spatially distributed, infinite dimensional dynamics,

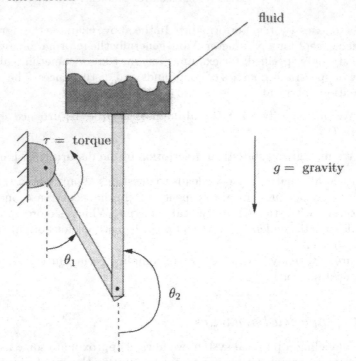

Figure 0.5. Double pendulum with attached fluid vessel

and there would still be neglected effects. No matter where one stops in modeling, the reliability of the conclusions depends strongly on the fact that whatever has been neglected will not become crucial later. In the presence of feedback, this assessment is particularly challenging and is really a central design question.

To emphasize this point in another example, consider the modified double pendulum shown in Figure 0.5. In this new setup a vessel containing fluid has been rigidly attached to the end of the second link. Suppose following our discussion of the previous two sections, that we wish to stabilize this system about one of its equilibria. The addition of the fluid vessel to this system significantly complicates the modeling of this system, the most detailed models being infinite dimensional and highly intractable, to the point where they are even beyond the scope of accurate computer simulation.

However, to balance this system an infinite dimensional model is probably not required, and perhaps a low dimensional one will suffice. An extreme model in this latter category would be one that modeled the fluid vessel system as a point mass; it may well be that a feedback design which renders our system insensitive to the value of this mass would perform well in the real system. But possibly the oscillations of the fluid inside the vessel may compromise performance. In this case a more refined, but still tractable model could consist of an oscillatory mass-spring-type dynamical model.

These modeling issues become even more central if we interconnect many uncertain or complex systems to form a "system of systems." A very simple example is the coupled system formed when the electric motor above is used to generate the control torque for the fluid-pendulum system.

The main conclusion we make is that there is no such thing as the "correct" model for control. A useful model is one in which the remaining uncertainty or unpredictability of the system can be adequately compensated by feedback. Thus we have set the stage for this course. The key players are feedback, stability, performance, uncertainty, and interconnection of systems. The mathematical theory to follow is motivated by the challenging interplay between these aspects of designed dynamical systems.

Notes and references

The basis for the approach described in §0.1.2 has a long history and can be traced back to references [99, 100], where a type of closed-loop stability is considered; a historical account of the early subsequent development of these ideas, which occurred primarily in the Soviet research literature, can be found in reference [126].

For a precise definition of stability and theorems on linearization see any standard text on dynamical systems theory; for instance [69]. For specific results on exponential stability and additional stability results in the context of control theory see [88]. The double pendulum control example of §0.2.1 originates in [152], where it is named the "pendubot."

There are a number of other books on robust control. Excellent references for the developments of the theory until the early 1990s are [28, 56, 67, 188], and a more introductory treatment is given in [44]. Also see [107] for a process control context. More recent books with a similar perspective are [187] which is more compact version of [188], [143] which also covers connections to system identification, and [150] with focus on more practical aspects of design. The recent trends in convex methods and linear matrix inequalities were first summarized in the research monograph [21], and a more recent reference is [149]; both contain solutions to a number of problems focusing on these methods.

1
Preliminaries in Finite Dimensional Space

This chapter is centered around finite dimension vector spaces, mappings on them, and the convexity property.

Much of the material is standard linear algebra, with which the reader is assumed to have some familiarity; correspondingly, our emphasis here is to provide a survey of the key ideas and tools, setting a common notation and presenting some results for future reference. We provide few proofs, but the reader can gain practice with the results and machinery presented by completing some of the exercises at the end of the chapter.

We also cover some of the basic ideas and results from convex analysis in finite dimensional space, which play a key role in this course. Having completed these fundamentals we introduce a new object, *linear matrix inequalities*, or LMIs, which we will use throughout the course as a major theoretical and computational tool.

1.1   Linear spaces and mappings

In this section we will introduce some of the basic ideas in linear algebra. Our treatment is primarily intended as a review for the reader's convenience, with some additional focus on the geometric aspects of the subject. References are given at the end of the chapter for more details at both introductory and advanced levels.

1.1.1 Vector spaces

The structure introduced now will pervade our course, that of a *vector space*, also called a linear space. This is a set that has a natural addition operation defined on it, together with scalar multiplication. Because this is such an important concept, and arises in a number of different ways, it is worth defining it precisely below.

Before proceeding we set some basic notation. The real numbers will be denoted by $\mathbb{R}$, and the complex numbers by $\mathbb{C}$; we will use $j := \sqrt{-1}$ for the imaginary unit. Also, given a complex number $z = x + jy$ with $x, y \in \mathbb{R}$:

- $z^* = x - jy$ is the complex conjugate;

- $|z| = \sqrt{x^2 + y^2}$ is the complex magnitude;

- $x = \operatorname{Re}(z)$ is the real part.

We use $\mathbb{C}^+$ to denote the open right half-plane formed from the complex numbers with positive real part; $\bar{\mathbb{C}}^+$ is the corresponding closed half-plane, and the left half-planes $\mathbb{C}^-$ and $\bar{\mathbb{C}}^-$ are analogously defined. Finally, $j\mathbb{R}$ denotes the imaginary axis.

We now define a vector space. In the definition, the field $\mathbb{F}$ can be taken here to be the real numbers $\mathbb{R}$, or the complex numbers $\mathbb{C}$. The terminology *real* vector space, or *complex* vector space is used to specify these alternatives.

Definition 1.1. *Suppose V is a nonempty set and $\mathbb{F}$ is a field, and that operations of vector addition and scalar multiplication are defined in the following way.*

(a) For every pair u, $v \in V$ a unique element $u + v \in V$ is assigned called their sum;

(b) for each $\alpha \in \mathbb{F}$ and $v \in V$, there is a unique element $\alpha v \in V$ called their product.

Then V is a vector space if the following properties hold for all u, v, $w \in V$, and all α, $\beta \in \mathbb{F}$:

(i) There exists a zero element in V, denoted by 0, such that $v + 0 = v$;

(ii) there exists a vector $-v$ in V, such that $v + (-v) = 0$;

(iii) the association $u + (v + w) = (u + v) + w$ is satisfied;

(iv) the commutativity relationship $u + v = v + u$ holds;

(v) scalar distributivity $\alpha(u + v) = \alpha u + \alpha v$ holds;

(vi) vector distributivity $(\alpha + \beta)v = \alpha v + \beta v$ is satisfied;

(vii) the associative rule $(\alpha\beta)v = \alpha(\beta v)$ for scalar multiplication holds;

(viii) for the unit scalar $1 \in \mathbb{F}$ the equality $1v = v$ holds.

Formally, a vector space is an additive group together with a scalar multiplication operation defined over a field $\mathbb{F}$, which must satisfy the usual rules (v–viii) of distributivity and associativity. Notice that both $\mathcal{V}$ and $\mathbb{F}$ contain the zero element, which we will denote by "0" regardless of the instance.

Given two vector spaces $\mathcal{V}_1$ and $\mathcal{V}_2$, with the same associated scalar field, we use $\mathcal{V}_1 \times \mathcal{V}_2$ to denote the vector space formed by their Cartesian product. Thus every element of $\mathcal{V}_1 \times \mathcal{V}_2$ is of the form

$$(v_1, v_2) \quad \text{where } v_1 \in \mathcal{V}_1 \text{ and } v_2 \in \mathcal{V}_2.$$

Having defined a vector space we now consider a number of examples.

Examples:

Both $\mathbb{R}$ and $\mathbb{C}$ can be considered as real vector spaces, although $\mathbb{C}$ is more commonly regarded as a complex vector space. The most common example of a real vector space is $\mathbb{R}^n = \mathbb{R} \times \cdots \times \mathbb{R}$; namely, n copies of $\mathbb{R}$. We represent elements of $\mathbb{R}^n$ in a column vector notation

$$x = \begin{bmatrix} x_1 \\ \vdots \\ x_n \end{bmatrix} \in \mathbb{R}^n, \quad \text{where each } x_k \in \mathbb{R}.$$

Addition and scalar multiplication in $\mathbb{R}^n$ are defined componentwise:

$$x + y = \begin{bmatrix} x_1 + y_1 \\ x_2 + y_2 \\ \vdots \\ x_n + y_n \end{bmatrix}, \quad \alpha x = \begin{bmatrix} \alpha x_1 \\ \alpha x_2 \\ \vdots \\ \alpha x_n \end{bmatrix}, \quad \text{for} \quad \alpha \in \mathbb{R}, \quad x, y \in \mathbb{R}^n.$$

Identical definitions apply to the complex space $\mathbb{C}^n$. As a further step, consider the space $\mathbb{C}^{m \times n}$ of complex $m \times n$ matrices of the form

$$A = \begin{bmatrix} a_{11} & \cdots & a_{1n} \\ \vdots & \ddots & \vdots \\ a_{m1} & \cdots & a_{mn} \end{bmatrix}.$$

Using once again componentwise addition and scalar multiplication, $\mathbb{C}^{m \times n}$ is a (real or complex) vector space.

We now define two vector spaces of matrices which will be central in our course. First, we define the *transpose* of the above matrix $A \in \mathbb{C}^{m \times n}$ by

$$A' = \begin{bmatrix} a_{11} & \cdots & a_{m1} \\ \vdots & \ddots & \vdots \\ a_{1n} & \cdots & a_{mn} \end{bmatrix} \in \mathbb{C}^{n \times m},$$

and its *Hermitian conjugate* or *adjoint* by

$$A^* = \begin{bmatrix} a_{11}^* & \cdots & a_{m1}^* \\ \vdots & \ddots & \vdots \\ a_{1n}^* & \cdots & a_{mn}^* \end{bmatrix} \in \mathbb{C}^{n \times m}.$$

In both cases the indices have been transposed, but in the latter we also take the complex conjugate of each element. Clearly both operations coincide if the matrix is real; we thus favor the notation A^*, which will serve to indicate both the transpose of a real matrix, and the adjoint of a complex matrix.[1]

The square matrix $A \in \mathbb{C}^{n \times n}$ is *Hermitian* or *self-adjoint* if

$$A = A^*.$$

The space of Hermitian matrices is denoted $\mathbb{H}^n$, and is a *real* vector space. If a Hermitian matrix A is in $\mathbb{R}^{n \times n}$ it is more specifically referred to as *symmetric*. The set of symmetric matrices is also a real vector space and will be written $\mathbb{S}^n$.

The set $\mathcal{F}(\mathbb{R}^m, \mathbb{R}^n)$ of functions mapping m real variables to $\mathbb{R}^n$ is a vector space. Addition between two functions f_1 and f_2 is defined by

$$(f_1 + f_2)(x_1, \ldots, x_m) = f_1(x_1, \ldots, x_m) + f_2(x_1, \ldots, x_m)$$

for any variables $x_1, \ldots, x_m$; this is called pointwise addition. Scalar multiplication by a real number α is defined by

$$(\alpha f)(x_1, \ldots, x_m) = \alpha f(x_1, \ldots, x_m).$$

An example of a less standard vector space is given by the set composed of multinomials in m variables, that have homogeneous order n. We denote this set by $P_m^{[n]}$. To illustrate the elements of this set consider

$$p_1(x_1, x_2, x_3) = x_1^2 x_2 x_3, \quad p_2(x_1, x_2, x_3) = x_1^3 x_2, \quad p_3(x_1, x_2, x_3) = x_1 x_2 x_3.$$

Each of these is a multinomial in three variables; however, p_1 and p_2 have order four, whereas the order of p_3 is three. Thus only p_1 and p_2 are in $P_3^{[4]}$. Similarly of

$$p_4(x_1, x_2, x_3) = x_1^4 + x_2 x_3^3 \quad \text{and} \quad p_5(x_1, x_2, x_3) = x_1^2 x_2 x_3 + x_1$$

only p_4 is in $P_3^{[4]}$, whereas p_5 is not in any $P_3^{[n]}$ space since its terms are not homogeneous in order. Some thought will convince you that $P_m^{[n]}$ is a vector space under pointwise addition. $\square$

---

[1] The transpose, without conjugation, of a complex matrix A will be denoted by A'; however, it is seldom required.

1.1.2 Subspaces

A *subspace* of a vector space $\mathcal{V}$ is a subset of $\mathcal{V}$ which is also a vector space
with respect to the same field and operations; equivalently, it is a subset
which is closed under the operations on $\mathcal{V}$.

Examples:

A vector space can have many subspaces, and the simplest of these is the
zero subspace, denoted by $\{0\}$. This is a subspace of any vector space and
contains only the zero element. Excepting the zero subspace and the entire
space, the simplest type of subspace in $\mathcal{V}$ is of the form

$$\mathcal{S}_v = \{s \in \mathcal{V} : s = \alpha v, \text{ for some } \alpha \in \mathbb{R}\},$$

given v in $\mathcal{V}$. That is, each element in $\mathcal{V}$ generates a subspace by multiplying
it by all possible scalars. In $\mathbb{R}^2$ or $\mathbb{R}^3$, such subspaces correspond to lines
going through the origin.

Going back to our earlier examples of vector spaces we see that the
multinomials $P_m^{[n]}$ are subspaces of $\mathcal{F}(\mathbb{R}^m, \mathbb{R})$, for any n.

Now $\mathbb{R}^n$ has many subspaces and an important set is those associated
with the natural insertion of $\mathbb{R}^m$ into $\mathbb{R}^n$, when $m < n$. Elements of these
subspaces are of the form

$$x = \begin{bmatrix} \bar{x} \\ 0 \end{bmatrix},$$

where $\bar{x} \in \mathbb{R}^m$ and $0 \in \mathbb{R}^{n-m}$. □

Given two subspaces $\mathcal{S}_1$ and $\mathcal{S}_2$ we can define the addition

$$\mathcal{S}_1 + \mathcal{S}_2 = \{s \in \mathcal{V} : s = s_1 + s_2 \text{ for some } s_1 \in \mathcal{S}_1 \text{ and } s_2 \in \mathcal{S}_2\}$$

which is easily verified to be a subspace.

1.1.3 Bases, spans, and linear independence

We now define some key vector space concepts. Given elements $v_1, \ldots, v_m$
in a vector space we denote their *span* by

$$\text{span}\{v_1, \ldots, v_m\},$$

which is the set of all vectors v that can be written as

$$v = \alpha_1 v_1 + \cdots + \alpha_m v_m$$

for some scalars $\alpha_k \in \mathbb{F}$; the above expression is called a *linear combination*
of the vectors $v_1, \ldots, v_m$. It is straightforward to verify that the span always
defines a subspace. If for some vectors we have

$$\text{span}\{v_1, \ldots, v_m\} = \mathcal{V},$$

we say that the vector space $\mathcal{V}$ is finite dimensional. If no such finite set of vectors exists we say the vector space is infinite dimensional. Our focus for the remainder of the chapter is *exclusively* finite dimensional vector spaces. We will pursue the study of some infinite dimensional spaces in Chapter 3.

If a vector space $\mathcal{V}$ is finite dimensional we define its *dimension*, denoted $\dim(\mathcal{V})$, to be the smallest number n such that there exist vectors $v_1, \ldots, v_n$ satisfying

$$\text{span}\{v_1, \ldots, v_n\} = \mathcal{V}.$$

In that case we say that the set

$$\{v_1, \ldots, v_n\} \text{ is a } basis \text{ for } \mathcal{V}.$$

Notice that a basis will automatically satisfy the *linear independence* property, which means that the only solution to the equation

$$\alpha_1 v_1 + \cdots + \alpha_n v_n = 0$$

is $\alpha_1 = \cdots = \alpha_n = 0$. Otherwise, one of the elements v_i could be expressed as a linear combination of the others and $\mathcal{V}$ would be spanned by fewer than n vectors. Given this observation, it follows easily that for a given $v \in \mathcal{V}$, the scalars $(\alpha_1, \ldots, \alpha_n)$ satisfying

$$\alpha_1 v_1 + \cdots + \alpha_n v_n = v$$

are unique; they are termed the *coordinates* of v in the basis $\{v_1, \ldots, v_n\}$.

Linear independence is defined analogously for any set of vectors $\{v_1, \ldots, v_m\}$; it is equivalent to saying the vectors are a basis for their span. The maximal number of linearly independent vectors is n, the dimension of the space; in fact any linearly independent set can be extended with additional vectors to form a basis.

Examples:

From our examples so far $\mathbb{R}^n, \mathbb{C}^{m \times n}$, and $P_m^{[n]}$ are all finite dimensional vector spaces; however, $\mathcal{F}(\mathbb{R}^m, \mathbb{R}^n)$ is infinite dimensional. The real vector space $\mathbb{R}^n$ and complex vector space $\mathbb{C}^{m \times n}$ are n and mn dimensional, respectively. The dimension of $P_m^{[n]}$ is more challenging to compute and its determination is an exercise at the end of the chapter.

An important computational concept in vector space analysis is associating a general k dimensional vector space $\mathcal{V}$ with the vector space $\mathbb{F}^k$. This is done by taking a basis $\{v_1, \ldots, v_k\}$ for $\mathcal{V}$, and associating each vector v in $\mathcal{V}$ with the vector of coordinates in the given basis,

$$\begin{bmatrix} \alpha_1 \\ \vdots \\ \alpha_k \end{bmatrix} \in \mathbb{F}^k.$$

Equivalently, each vector v_i in the basis is associated with the vector

$$e_i = \begin{bmatrix} 0 \\ \vdots \\ 0 \\ 1 \\ 0 \\ \vdots \\ 0 \end{bmatrix} \in \mathbb{F}^k.$$

That is, e_i is the vector with zeros everywhere excepts its ith entry, which is equal to one. Thus we are identifying the basis $\{v_1, \ldots, v_k\}$ in $\mathcal{V}$ with the set $\{e_1, \ldots, e_k\}$ which is in fact a basis of $\mathbb{F}^k$, called the *canonical* basis.

To see how this type of identification is made, suppose we are dealing with $\mathbb{R}^{n \times m}$, which has dimension $k = nm$. Then a basis for this vector space is

$$E_{ir} = \begin{bmatrix} 0 & \cdots & & 0 \\ \vdots & \ddots & 1 & \vdots \\ 0 & \cdots & & 0 \end{bmatrix},$$

which are the matrices that are zero everywhere but their (i,r)th-entry, which is one. Then we identify each of these with the vector $e_{n(r-1)+i} \in \mathbb{R}^k$. Thus addition or scalar multiplication on $\mathbb{R}^{n \times m}$ can be translated to equivalent operations on $\mathbb{R}^k$. $\qquad\square$

1.1.4 Mappings and matrix representations

We now introduce the concept of a *linear* mapping between vector spaces. The mapping $A : \mathcal{V} \to \mathcal{W}$ is linear if

$$A(\alpha v_1 + \beta v_2) = \alpha A v_1 + \beta A v_2$$

for all v_1, v_2 in $\mathcal{V}$, and all scalars α_1 and α_2. Here $\mathcal{V}$ and $\mathcal{W}$ are vector spaces with the same associated field $\mathbb{F}$. The space $\mathcal{V}$ is called the *domain* of the mapping, and $\mathcal{W}$ its *codomain*.

Given bases $\{v_1, \ldots, v_n\}$ and $\{w_1, \ldots, w_m\}$ for $\mathcal{V}$ and $\mathcal{W}$, respectively, we associate scalars a_{ik} with the mapping A, defining them such that they satisfy

$$A v_k = a_{1k} w_1 + a_{2k} w_2 + \cdots + a_{mk} w_m,$$

for each $1 \le k \le n$. Namely, given any basis vector v_k, the coefficients a_{ik} are the coordinates of $A v_k$ in the chosen basis for $\mathcal{W}$. It turns out that these mn numbers a_{ik} completely specify the linear mapping A. To see this is true consider any vector $v \in \mathcal{V}$, and let $w = Av$. We can express both vectors in their respective bases as $v = \alpha_1 v_1 + \cdots + \alpha_n v_n$ and $w = \beta_1 w_1 + \cdots + \beta_m w_m$. Now we have

$$
\begin{aligned}
w = Av &= A(\alpha_1 v_1 + \cdots + \alpha_n v_n) \\
&= \alpha_1 A v_1 + \cdots + \alpha_n A v_n \\
&= \sum_{k=1}^{n} \sum_{i=1}^{m} \alpha_k a_{ik} w_i = \sum_{i=1}^{m} \left(\sum_{k=1}^{n} \alpha_k a_{ik} \right) w_i,
\end{aligned}
$$

and therefore by uniqueness of the coordinates we must have

$$
\beta_i = \sum_{k=1}^{n} \alpha_k a_{ik}, \quad i = 1, \ldots, m.
$$

To express this relationship in a more convenient form, we can write the set of numbers a_{ik} as the $m \times n$ matrix

$$
[A] = \begin{bmatrix} a_{11} & \cdots & a_{1n} \\ \vdots & \ddots & \vdots \\ a_{m1} & \cdots & a_{mn} \end{bmatrix}.
$$

Then via the standard matrix product we have

$$
\begin{bmatrix} \beta_1 \\ \vdots \\ \beta_m \end{bmatrix} = \begin{bmatrix} a_{11} & \cdots & a_{1n} \\ \vdots & \ddots & \vdots \\ a_{m1} & \cdots & a_{mn} \end{bmatrix} \begin{bmatrix} \alpha_1 \\ \vdots \\ \alpha_n \end{bmatrix}.
$$

In summary any linear mapping A between vector spaces can be regarded as a matrix $[A]$ mapping $\mathbb{F}^n$ to $\mathbb{F}^m$ via matrix multiplication.

Notice that the numbers a_{ik} depend intimately on the bases $\{v_1, \ldots, v_n\}$ and $\{w_1, \ldots, w_m\}$. Frequently we use only one basis for $\mathcal{V}$ and one for $\mathcal{W}$ and thus there is no need to distinguish between the map A and the basis dependent matrix $[A]$. Therefore after this section we will simply write A to denote either the map or the matrix, making which is meant context dependent.

We now give two examples to illustrate the above discussion more clearly.

Examples:

Given matrices $B \in \mathbb{C}^{k \times k}$ and $D \in \mathbb{C}^{l \times l}$ we define the map $\Pi : \mathbb{C}^{k \times l} \to \mathbb{C}^{k \times l}$ by

$$
\Pi(X) = BX - XD,
$$

where the right-hand side is in terms of matrix addition and multiplication. Clearly Π is a linear mapping since

$$
\begin{aligned}
\Pi(\alpha X_1 + \beta X_2) &= B(\alpha X_1 + \beta X_2) - (\alpha X_1 + \beta X_2)D \\
&= \alpha(BX_1 - X_1 D) + \beta(BX_2 - X_2 D) \\
&= \alpha\Pi(X_1) + \beta\Pi(X_2).
\end{aligned}
$$

If we now consider the identification between the matrix space $\mathbb{C}^{k \times l}$ and the product space $\mathbb{C}^{kl}$, then Π can be thought of as a map from $\mathbb{C}^{kl}$ to $\mathbb{C}^{kl}$, and can accordingly be represented by a complex matrix, which is $kl \times kl$. We now do an explicit 2×2 example for illustration. Suppose $k = l = 2$ and that

$$B = \begin{bmatrix} 1 & 2 \\ 3 & 4 \end{bmatrix} \text{ and } D = \begin{bmatrix} 5 & 0 \\ 0 & 0 \end{bmatrix}.$$

We would like to find a matrix representation for Π. Since the domain and codomain of Π are equal, we will use the standard basis for $\mathbb{C}^{2 \times 2}$ for each. This basis is given by the matrices E_{ir} defined earlier. We have

$$\Pi(E_{11}) = \begin{bmatrix} -4 & 0 \\ 3 & 0 \end{bmatrix} = -4E_{11} + 3E_{21};$$

$$\Pi(E_{12}) = \begin{bmatrix} 0 & 1 \\ 0 & 3 \end{bmatrix} = E_{12} + 3E_{22};$$

$$\Pi(E_{21}) = \begin{bmatrix} 2 & 0 \\ -1 & 0 \end{bmatrix} = 2E_{11} - E_{21};$$

$$\Pi(E_{22}) = \begin{bmatrix} 0 & 2 \\ 0 & 4 \end{bmatrix} = 2E_{12} + 4E_{22}.$$

Now we identify the basis $\{E_{11}, E_{12}, E_{21}, E_{22}\}$ with the standard basis for $\mathbb{C}^4$ given by $\{e_1, e_2, e_3, e_4\}$. Therefore we get that

$$[\Pi] = \begin{bmatrix} -4 & 0 & 2 & 0 \\ 0 & 1 & 0 & 2 \\ 3 & 0 & -1 & 0 \\ 0 & 3 & 0 & 4 \end{bmatrix}$$

in this basis.

Another linear operator involves the multinomial function $P_m^{[n]}$ defined earlier in this section. Given an element $a \in P_m^{[k]}$ we can define the mapping $\Omega : P_m^{[n]} \to P_m^{[n+k]}$ by function multiplication

$$\Omega(p)(x_1, x_2, \ldots, x_m) := a(x_1, x_2, \ldots, x_m)p(x_1, x_2, \ldots, x_m).$$

Again Ω can be regarded as a matrix, which maps $\mathbb{R}^{d_1} \to \mathbb{R}^{d_2}$, where d_1 and d_2 are the dimensions of $P_m^{[n]}$ and $P_m^{[n+k]}$, respectively.   □

Associated with any linear map $A : \mathcal{V} \to \mathcal{W}$ is its *image space*, which is defined by

$$\operatorname{Im} A = \{w \in \mathcal{W} : \text{ there exists } v \in \mathcal{V} \text{ satisfying } Av = w\}.$$

This set contains all the elements of $\mathcal{W}$ which are the image of some point in $\mathcal{V}$. Clearly if $\{v_1, \ldots, v_n\}$ is a basis for $\mathcal{V}$, then

$$\operatorname{Im} A = \operatorname{span}\{Av_1, \ldots, Av_n\}$$

and is thus a subspace. The map A is called *surjective* when $\operatorname{Im} A = \mathcal{W}$.

The dimension of the image space is called the *rank* of the linear mapping A, and the concept is applied as well to the associated matrix $[A]$. Namely,

$$\text{rank}[A] = \dim(\text{Im}\,A).$$

If S is a subspace of V, then the image of S under the mapping A is denoted AS. That is,

$$AS = \{w \in W : \text{ there exists } s \in S \text{ satisfying } As = w\}.$$

In particular, this means that $AV = \text{Im}\,A$.

Another important set related to A is its *kernel*, or null space, defined by

$$\text{Ker}\,A = \{v \in V : \quad Av = 0\}.$$

In words, $\text{Ker}\,A$ is the set of vectors in V which get mapped by A to the zero element in W, and is easily verified to be a subspace of V.

If we consider the equation $Av = w$, suppose v_a and v_b are both solutions; then

$$A(v_a - v_b) = 0.$$

Plainly, the difference between any two solutions is in the kernel of A. Thus given any solution v_a to the equation, all solutions are parametrized by

$$v_a + v_0,$$

where v_0 is any element in $\text{Ker}\,A$.

In particular, when $\text{Ker}\,A$ is the zero subspace, there is at most a unique solution to the equation $Av = w$. This means $Av_a = Av_b$ only when $v_a = v_b$; a mapping with this property is called *injective*.

In summary, a solution to the equation $Av = w$ will exist if and only if $w \in \text{Im}\,A$; it will be unique only when $\text{Ker}\,A$ is the zero subspace.

The dimensions of the image and kernel of A are linked by the relationship

$$\dim(V) = \dim(\text{Im}\,A) + \dim(\text{Ker}\,A),$$

proved in the exercises at the end of the chapter.

A mapping is called *bijective* when it is both injective and surjective; that is, for every $w \in W$ there exists a unique v satisfying $Av = w$. In this case there is a well-defined inverse mapping $A^{-1} : W \to V$, such that

$$A^{-1}A = I_V, \qquad AA^{-1} = I_W.$$

In the above, I denotes the identity mapping in each space, that is the map that leaves elements unchanged. For instance, $I_V : v \mapsto v$ for every $v \in V$.

From the above property on dimensions we see that if there exists a bijective linear mapping between two spaces V and W, then the spaces must have the same dimension. Also, if a mapping A is from V back to itself, namely, $A : \mathcal{V} \to \mathcal{V}$, then one of the two properties (injectivity or surjectivity) suffices to guarantee the other.

We will also use the terms *nonsingular* or *invertible* to describe bijective mappings, and apply these terms as well to their associated matrices. Notice that invertibility of the mapping A is equivalent to invertibility of $[A]$ in terms of the standard matrix product; this holds true regardless of the chosen bases.

Examples:

To illustrate these notions let us return to the mappings Π and Ω defined above. For the 2×2 numerical example given, Π maps $\mathbb{C}^{2 \times 2}$ back to itself. It is easily checked that it is invertible by showing either

$$\text{Im}\,\Pi = \mathbb{C}^{2 \times 2}, \quad \text{or equivalently } \text{Ker}\,\Pi = 0.$$

In contrast Ω is not a map on the same space, instead taking $P_m^{[n]}$ to the larger space $P_m^{[n+k]}$. And we see that the dimension of the image of Ω is at most n, and the dimension of its kernel at least k. Thus assuming $k > 0$, there are at least some elements $w \in P_m^{[n+k]}$ for which

$$\Omega v = w$$

cannot be solved. These are exactly the values of w that are not in $\text{Im}\,\Omega$.

□

1.1.5   Change of basis and invariance

We have already discussed the idea of choosing a basis $\{v_1, \ldots, v_n\}$ for the vector space $\mathcal{V}$, and then associating every vector x in $\mathcal{V}$ with its coordinates

$$x_v = \begin{bmatrix} \alpha_1 \\ \vdots \\ \alpha_n \end{bmatrix} \in \mathbb{F}^n,$$

which are the unique scalars satisfying $x = \alpha_1 v_1 + \cdots + \alpha_n v_n$. This raises the question, suppose we choose another basis $u_1, \ldots, u_n$ for $\mathcal{V}$, how can we effectively move between these basis representations? That is, given $x \in \mathcal{V}$, how are the coordinate vectors $x_v, x_u \in \mathbb{F}^n$ related?

The answer is as follows. Suppose that each basis vector u_k is expressed by

$$u_k = t_{1k} v_1 + \cdots + t_{nk} v_n,$$

in the basis $\{v_1, \ldots, v_n\}$. Then the coefficients t_{ik} define the matrix

$$T = \begin{bmatrix} t_{11} & \cdots & t_{1n} \\ \vdots & \ddots & \vdots \\ t_{n1} & \cdots & t_{nn} \end{bmatrix}.$$

Notice that such a matrix is nonsingular, since it represents the identity mapping I_V in the bases $\{v_1, \ldots, v_n\}$ and $\{u_1, \ldots, u_n\}$. Then the relationship between the two coordinate vectors is

$$T x_u = x_v.$$

Now suppose $A : \mathcal{V} \to \mathcal{V}$ and that $A_v : \mathbb{F}^n \to \mathbb{F}^n$ is the representation of A on the basis $v_1, \ldots, v_n$, and A_u is the representation of A using the basis $u_1, \ldots, u_n$. How is A_u related to A_v?

To study this, take any $x \in \mathcal{V}$ and let x_v, x_u be its coordinates in the respective bases, and z_v, z_u be the coordinates of Ax. Then we have

$$z_u = T^{-1} z_v = T^{-1} A_v x_v = T^{-1} A_v T x_u.$$

Since the above identity and

$$z_u = A_u x_u$$

both hold for every x_u, we conclude that

$$A_u = T^{-1} A_v T.$$

The above relationship is called a *similarity transformation*. This discussion can be summarized in the following commutative diagram. Let $E : \mathcal{V} \to \mathbb{F}^n$ be the map that takes elements of $\mathcal{V}$ to their representation in $\mathbb{F}^n$ with respect to the basis $\{v_1, \ldots, v_n\}$. Then

$$
\begin{array}{ccccc}
\mathcal{V} & \xrightarrow{\;\;E\;\;} & \mathbb{F}^n & \xrightarrow{\;\;T^{-1}\;\;} & \mathbb{F}^n \\
\downarrow A & & \downarrow A_v & & \downarrow A_u = T^{-1} A_v T \\
\mathcal{V} & \xrightarrow{\;\;E\;\;} & \mathbb{F}^n & \xrightarrow{\;\;T^{-1}\;\;} & \mathbb{F}^n
\end{array}
$$

Next we examine mappings when viewed with respect to a subspace. Suppose that $\mathcal{S} \subset \mathcal{V}$ is a k-dimensional subspace of $\mathcal{V}$, and that $v_1, \ldots, v_n$ is a basis for $\mathcal{V}$ with

$$\text{span}\{v_1, \ldots, v_k\} = \mathcal{S}.$$

That is the first k vectors of this basis forms a basis for $\mathcal{S}$. If $E : \mathcal{V} \to \mathbb{F}^n$ is the associated map which maps the basis vectors in $\mathcal{V}$ to the standard basis on $\mathbb{F}^n$, then

$$ES = \mathbb{F}^k \times \{0\} \subset \mathbb{F}^n.$$

Thus in $\mathbb{F}^n$ we can view S as the elements of the form

$$\begin{bmatrix} x \\ 0 \end{bmatrix} \text{ where } x \in \mathbb{F}^k.$$

From the point of view of a linear mapping $A : \mathcal{V} \to \mathcal{V}$ this partitioning of $\mathbb{F}^n$ gives a useful decomposition of the corresponding matrix $[A]$. Namely, we can regard $[A]$ as

$$[A] = \begin{bmatrix} A_1 & A_2 \\ A_3 & A_4 \end{bmatrix},$$

where $A_1 : \mathbb{F}^k \to \mathbb{F}^k, A_2 : \mathbb{F}^{n-k} \to \mathbb{F}^k, A_3 : \mathbb{F}^k \to \mathbb{F}^{n-k}$, and $A_4 : \mathbb{F}^{n-k} \to \mathbb{F}^{n-k}$. We have that

$$EAS = \text{Im} \begin{bmatrix} A_1 \\ A_3 \end{bmatrix}.$$

Finally to end this section we have the notion of *invariance* of a subspace to a mapping. We say that a subspace $S \subset \mathcal{V}$ is A-invariant if $A : \mathcal{V} \to \mathcal{V}$ and

$$AS \subset S.$$

Clearly every map has at least two invariant subspaces, the zero subspace and entire domain $\mathcal{V}$. For subspaces S of intermediate dimension, the invariance property is expressed most clearly by saying the associated matrix has the form

$$[A] = \begin{bmatrix} A_1 & A_2 \\ 0 & A_4 \end{bmatrix}.$$

Here we are assuming, as above, that our basis for $\mathcal{V}$ is obtained by extending a basis for S. Similarly if a matrix has this form the subspace $\mathbb{F}^k \times \{0\}$ is $[A]$-invariant.

We will revisit the question of finding non-trivial invariant subspaces later in the chapter, when studying eigenvectors and the Jordan decomposition.

In the next section we will pursue some of the geometric properties of sets in linear spaces.

1.2   Subsets and convexity

Up to now in this chapter, the only sets we have encountered are those that are also vector spaces. In this section we will consider more general subsets of a vector space $\mathcal{V}$, with an emphasis of introducing and examining *convexity*. Convexity is one of the main geometric ideas underlying much of global optimization theory, and in particular is the foundation for a major analytical tool used in this course. The results presented here are for work in finite dimensional vector spaces, but in many cases they have infinite dimensional versions as well. Throughout this section we have the standing assumption that $\mathcal{V}$ is a finite dimensional *real* vector space. Before we address convexity it will be useful to introduce some topological aspects of vector spaces.

1.2.1   Some basic topology

To start we seek to define the notion of a *neighborhood* of a point in the vector space $\mathcal{V}$. We do this by first defining the unit ball with respect to a basis. Suppose that $\{u_1, \ldots, u_n\}$ is a basis for the vector space $\mathcal{V}$. Then the open unit ball in this basis is defined by

$$\mathcal{B}(u_1, \ldots, u_n) = \{\alpha_1 u_1 + \cdots + \alpha_n u_n \in \mathcal{V} : \alpha_i \in \mathbb{R}, \alpha_1^2 + \cdots + \alpha_n^2 < 1\}.$$

This set contains all the points that can be expressed, in the basis, with the vector of coefficients α inside the unit sphere of $\mathbb{R}^n$. Clearly this set is basis-dependent. In particular, since $\{tu_1, \ldots, tu_n\}$ is also a basis for every $t \neq 0$, then given any nonzero element v in $\mathcal{V}$, there always exists a basis such that v is in the corresponding unit ball, and another basis such that v is not in the associated unit ball. The zero vector is the only element that belongs to every unit ball.

We now define the notion a neighborhood of a point, which intuitively means any set that totally surrounds the given point in the vector space.

Definition 1.2. *A subset $\mathcal{N}(0)$ of the vector space $\mathcal{V}$ is a neighborhood of the zero element if there exists a basis $u_1, \ldots, u_n$ for $\mathcal{V}$ such that*

$$\mathcal{B}(u_1, \ldots, u_n) \subset \mathcal{N}(0).$$

Further, a subset $\mathcal{N}(w) \subset \mathcal{V}$ is a neighborhood of the point $w \in \mathcal{V}$ if the set

$$\mathcal{N} = \{v \in \mathcal{V} : v + w \in \mathcal{N}(w)\}$$

is a neighborhood of the zero element.

This says that a set is a neighborhood of zero provided that one of its subsets is the unit ball in some basis. A neighborhood of a general point $w \in \mathcal{V}$ is any set that is equal to a neighborhood of zero that has been translated by w. Notice that while a basis was used in the definition, since its choice is allowed to be arbitrary the resulting definition of a neighborhood is in fact basis-independent.

If $\mathcal{Q}$ is a subset of $\mathcal{V}$, we say that it is *open* if for every $v \in \mathcal{Q}$, there exists a neighborhood of v which is a subset of $\mathcal{Q}$.

We denote the complement of the set Q by Q^c, which we recall is defined by

$$Q^c = \{v \in \mathcal{V} : \ v \notin Q\}.$$

A set Q is a *closed* set if its complement is open.

Next we define the closure of a set. The *closure* of Q, denoted $\bar{Q}$, is

$$\bar{Q} = \{v \in \mathcal{V} : \text{for every neighborhood } \mathcal{N}(v) \text{ of } v, \ \mathcal{N}(v) \cap Q \text{ is nonempty}\}.$$

So the closure contains all points that are arbitrarily near to Q. Clearly, $Q \subset \bar{Q}$. The closure is always a closed set, and is the smallest closed set that contains Q; these facts are left as exercises.

One of the major objectives of this section is to develop tests for determining when two subsets of $\mathcal{V}$ do not intersect. In addition to being disjoint, we would like to have a notion of when disjoint sets are not arbitrarily close to each other. We call this property strict separation of the sets. We say that two subsets Q_1, $Q_2 \subset \mathcal{V}$ are *strictly separated* if there exists a neighborhood $\mathcal{N}(0)$ of zero, such that for all $v_1 \in Q_1$ and $v_2 \in Q_2$ the condition

$$v_1 - v_2 \notin \mathcal{N}(0) \text{ holds.}$$

Clearly $\mathcal{N}(0)$ can be chosen such that $v_2 - v_1 \notin \mathcal{N}(0)$ simultaneously holds, so this definition is symmetric with respect to the sets. Therefore sets are strictly separated if the difference between any two elements chosen from each set is uniformly outside a neighborhood of the origin.

The condition of strict separation is not always easy to check, and so we present an alternative condition for a special case. First we say that a subset $Q \in \mathcal{V}$ is *bounded* if there exists a basis $u_1, \ldots, u_n$ for $\mathcal{V}$ such that

$$Q \subset \mathcal{B}(u_1, \ldots, u_n).$$

A set that is both closed and bounded is called *compact*.

We can now state the following property.

Proposition 1.3. *Suppose Q_1 and Q_2 are subsets of $\mathcal{V}$, and that Q_1 is bounded. Then Q_1 and Q_2 are strictly separated if and only if the intersection of their closures $\bar{Q}_1 \cap \bar{Q}_2$ is empty.*

The result states that provided that one set is bounded, two sets are strictly separated if their closures are disjoint. Thus if two sets are closed, with one of them compact, they are strictly separated exactly when they are disjoint.

Having introduced some topology for vector spaces, we are ready to move on to discussing convexity.

1.2.2 Convex sets

Let us begin by defining the line segment that joins two points in $\mathcal{V}$. Suppose that v_1 and v_2 are in $\mathcal{V}$, then we define the line segment $L(v_1, v_2)$ between them as the set of points

$$L(v_1, v_2) = \{v \in V : v = \mu v_1 + (1 - \mu)v_2 \text{ for some } \mu \in [0,1]\}.$$

Clearly the end points of the line segment are v_1 and v_2, which occur in the parametrization when $\mu = 1$ and $\mu = 0$, respectively.

We can now turn to the idea of convexity. Suppose that Q is a nonempty subset of the vector space V. Then Q is defined to be a *convex set* if

for any $v_1, v_2 \in Q$, the line segment $L(v_1, v_2)$ is a subset of Q.

That is Q is convex if it contains all the line segments between its points. Geometrically we have the intuitive picture shown in Figure 1.1.

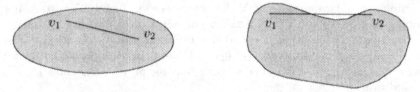

Figure 1.1. Convex and nonconvex sets

Clearly any vector space is convex, as is any subset $\{v\}$ of a vector space containing only a single element.

We can think of the expression $\mu v_1 + (1 - \mu)v_2$ for a point on the line $L(v_1, v_2)$ as a weighted average. To see this instead write equivalently

$$v = \mu_1 v_1 + \mu_2 v_2,$$

where $\mu_1, \mu_2 \in [0,1]$ and satisfy $\mu_1 + \mu_2 = 1$. Then we see that if both μ_1 and μ_2 are both equal to one-half we have our usual notion of the average. And if they take other values the weighted average "favors" one of the points. Thus the clear generalization of such an average to n points $v_1, \ldots, v_n$ is

$$v = \mu_1 v_1 + \cdots + \mu_n v_n,$$

where $\mu_1 + \cdots + \mu_n = 1$ and $\mu_1, \ldots, \mu_n \in [0,1]$. A line segment gave us geometrically a point on the line between the two endpoints. The generalization of this to an average of n points in the plane, yields a point inside the perimeter defined by the points $v_1, \ldots, v_n$. This is illustrated in Figure 1.2. Building on this intuition from $\mathbb{R}^2$, we extend the idea to an arbitrary vector space V. Given $v_1, \ldots, v_n$ we define the *convex hull* of these points by

$$\text{co}(\{v_1, \ldots, v_n\}) = \{v \in V : v = \sum_{k=1}^{n} \mu_k v_k, \text{ with } \mu_k \in [0,1], \sum_{k=1}^{n} \mu_k = 1\}.$$

In the case of Figure 1.2 this set is made of all the points inside the perimeter. In words the convex hull of the points $v_1, \ldots, v_n$ is simply the set composed of all weighted averages of these points. In particular we have

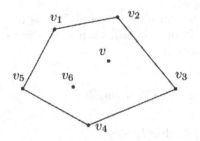

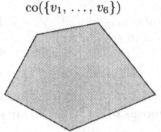

$$\mathrm{co}(\{v_1, \ldots, v_6\})$$

Figure 1.2. Convex hull of finite number of points in $\mathbb{R}^2$

that for two points $L(v_1, v_2) = \mathrm{co}(\{v_1, v_2\})$. It is a straightforward exercise to show that if $\mathcal{Q}$ is convex, then it necessarily contains any convex hull formed from a collection of its points.

So far we have only defined the convex hull in terms of a finite number of points. We now generalize this to an arbitrary set. Given a set $\mathcal{Q}$, we define its convex hull $\mathrm{co}(\mathcal{Q})$ by

$$\mathrm{co}(\mathcal{Q}) = \{v \in \mathcal{V} : \text{ there exist } n \text{ and } v_1, \ldots, v_n \in \mathcal{Q}$$
$$\text{such that } v \in \mathrm{co}(\{v_1, \ldots, v_n\})\}.$$

So the convex hull of $\mathcal{Q}$ is the collection of all possible weighted averages of points in $\mathcal{Q}$. It is straightforward to demonstrate that for any set $\mathcal{Q}$:

- the subset condition $\mathcal{Q} \subset \mathrm{co}(\mathcal{Q})$ is satisfied;

- the convex hull $\mathrm{co}(\mathcal{Q})$ is convex;

- the relationship $\mathrm{co}(\mathcal{Q}) = \mathrm{co}\,(\mathrm{co}(\mathcal{Q}))$ holds.

We also have the following results which relate convexity of a set to its convex hull.

Proposition 1.4. *A set $\mathcal{Q}$ is convex if and only if $\mathrm{co}(\mathcal{Q}) = \mathcal{Q}$ is satisfied.*

Notice that, by definition, the intersection of convex sets is always convex; therefore, given a set $\mathcal{Q}$, there exists a smallest convex set that contains $\mathcal{Q}$; it follows easily that this is precisely $\mathrm{co}(\mathcal{Q})$; in other words, if $\mathcal{Y}$ is convex and $\mathcal{Q} \subset \mathcal{Y}$, then necessarily $\mathrm{co}(\mathcal{Q})$ is a subset of $\mathcal{Y}$. Pictorially we have Figure 1.3 to visualize $\mathcal{Q}$ and its convex hull.

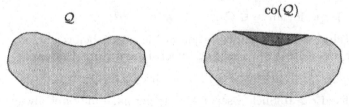

Figure 1.3. Convex hull of a set $\mathcal{Q}$

A linear mapping $F : \mathcal{V} \to \mathbb{R}$ is called a (linear) functional. If it is not identical to zero (a standing assumption from now on), then it is always surjective; namely, for fixed $a \in \mathbb{R}$ the equation

$$F(v) = a$$

always has a solution in the variable $v \in \mathcal{V}$. Also if v_1 satisfies $F(v_1) = a$, all solutions to the equation are given by

$$v = v_1 + v_0, \text{ where } v_0 \in \operatorname{Ker} F.$$

Thus we can view this set of solutions as the kernel subspace shifted away from the origin. Also, the dimension of $\operatorname{Ker} F$ is $\dim(\mathcal{V}) - 1$. A set in an n-dimensional space, obtained by shifting an $n - 1$ dimensional subspace away from the origin, is called a *hyperplane*. It is not difficult to show that every hyperplane can be generated by a linear functional F and a real number a as described above. A hyperplane in $\mathbb{R}^3$ is depicted in Figure 1.4.

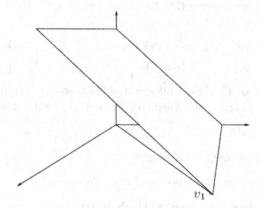

Figure 1.4. A hyperplane

An important property of a hyperplane, which is clear in the above geometric case, is that it always breaks up the space into two half-spaces: these have the form $\{v : F(v) \le a\}$ and $\{v : F(v) \ge a\}$.

This leads to the notion of *separating* two sets with a hyperplane. Given two sets $\mathcal{Q}_1$ and $\mathcal{Q}_2$ in $\mathcal{V}$, we say that the hyperplane defined by (F, a) *separates* the sets if

(a) $F(v_1) \le a$, for all $v_1 \in \mathcal{Q}_1$;

(b) $F(v_2) \ge a$, for all $v_2 \in \mathcal{Q}_2$.

Geometrically we have the illustration in Figure 1.5 below.

Further it is a *strictly separating* hyperplane if (b) is changed to

$$F(v_2) \ge a + \epsilon, \text{ for all } v_2 \in \mathcal{Q}_2,$$

for some fixed $\epsilon > 0$. Such a separating hyperplane may not always exist: in the figure if we move the sets sufficiently close together it will not be

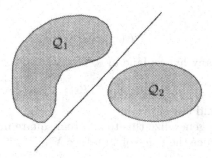

Figure 1.5. Separating hyperplane

possible to find any hyperplane which separates the sets. However, we have the following major result, which says that if two sets are convex and disjoint, then there always exists a separating hyperplane between them.

Theorem 1.5. *Suppose $\mathcal{Q}_1$ and $\mathcal{Q}_2$ are two nonempty convex subsets of the vector space $\mathcal{V}$.*

(a) *If the intersection $\mathcal{Q}_1 \cap \mathcal{Q}_2$ is empty, then there exists a hyperplane which separates $\mathcal{Q}_1$ from $\mathcal{Q}_2$.*

(b) *There exists a strictly separating hyperplane between the sets if and only if the sets are strictly separated.*

This is a very powerful theorem which we will make critical use of later in the course. While we defer the proof to the references, readers should gain confidence of its validity by drawing some geometrical pictures.

At this point we have developed the notions of convex sets and separation in an abstract setting. We next consider two explicit examples which will be key to us later.

Examples:

Here we will consider representing hyperplanes for two explicit vector spaces, the spaces $\mathbb{R}^n$ and the symmetric matrices $\mathbb{S}^n$. Let us first consider $\mathbb{R}^n$.

Given a linear functional $F : \mathbb{R}^n \to \mathbb{R}$ we see that it is completely defined by the numbers

$$F(e_1), \ldots, F(e_n).$$

where $\{e_1, \ldots, e_n\}$ is the canonical basis for $\mathbb{R}^n$. Thus given any vector $x \in \mathbb{R}^n$ we have

$$F(x) = x_1 F(e_1) + \cdots + x_n F(e_n).$$

Conversely given any $y \in \mathbb{R}^n$ a linear functional F is defined on $\mathbb{R}^n$ by $F(x) = y_1 x_1 + \ldots + y_n x_n$, which we express more concisely using matrix multiplication by

$$F(x) = y^* x.$$

Thus we see that any hyperplane in $\mathbb{R}^n$ is characterized by the equation

$$y^* x = a,$$

for some $y \in \mathbb{R}^n$ and $a \in \mathbb{R}$.

We now wish to generalize this to the real square matrices $\mathbb{R}^{n \times n}$. To do this we first introduce the *trace* of a matrix $X \in \mathbb{R}^{n \times n}$. The trace operation is defined by

$$\mathrm{Tr} X = x_{11} + \cdots + x_{nn},$$

the sum of the diagonal entries. Therefore $\mathrm{Tr} : \mathbb{R}^{n \times n} \to \mathbb{R}$. Also for X, $Z \in \mathbb{R}^{n \times n}$ and $\alpha, \beta \in \mathbb{R}$ we have

$$\mathrm{Tr}(\alpha X + \beta Z) = \sum_{k=1}^{n} \alpha x_{kk} + \sum_{k=1}^{n} \beta z_{kk} = \alpha \mathrm{Tr} X + \beta \mathrm{Tr} Z,$$

and so the trace operation defines a particular linear functional on $\mathbb{R}^{n \times n}$. Now given an element $Y \in \mathbb{R}^{n \times n}$ it is also routine to show that the mapping $F : \mathbb{R}^{n \times n} \to \mathbb{R}$ defined by

$$F(X) = \mathrm{Tr}(Y^* X) = \sum_{i,k=1}^{n} x_{ik} y_{ik} \qquad (1.1)$$

is a linear functional. The last identity is a consequence of the definitions of trace and matrix product.

The question we now ask is whether every linear functional on $\mathbb{R}^{n \times n}$ is defined in this way? The answer is yes and can be seen by expanding X in the standard basis of $\mathbb{R}^{n \times n}$, as

$$X = \sum_{i,k=1}^{n} x_{ik} E_{ik}.$$

Given a linear functional F, we have

$$F(X) = \sum_{i,k=1}^{n} x_{ik} F(E_{ik}) = \mathrm{Tr}(Y^* X),$$

where we have defined the matrix Y by $y_{ik} = F(E_{ik})$. Similarly it is straightforward to show that all linear functionals F on the symmetric matrices $\mathbb{S}^n$ are of the form in (1.1) where $Y \in \mathbb{S}^n$, and an analogous situation occurs over the space $\mathbb{H}^n$.

Later in the course we will specifically require separating hyperplanes for sets constructed from Cartesian products of the above spaces. We summarize what we will require in the following result.

Proposition 1.6. *Suppose that the vector space* $\mathbb{V}$ *is given by the Cartesian product*

$$\mathbb{V} = \mathbb{H}^{n_1} \times \cdots \times \mathbb{H}^{n_s} \times \mathbb{R} \times \cdots \times \mathbb{R}.$$

Then F *is a linear functional on* $\mathbb{V}$ *if and only if there exists* $Y = (Y_1, \ldots, Y_s, y_{s+1}, \ldots, y_{s+f}) \in \mathbb{V}$ *such that*

$$F(V) = \sum_{i=1}^{s} \mathrm{Tr}(Y_i^* V_i) + \sum_{i=s+1}^{s+f} y_i^* v_i,$$

for all $V = (V_1, \ldots, V_s, v_{s+1}, \ldots, v_{s+f}) \in \mathbb{V}$.

$\square$

As a final point in this section, we introduce the notion of *cones* in vector space.

A set $\mathcal{Q} \subset \mathcal{V}$ is called a cone if it is closed under positive scalar multiplication, that is if

$$v \in \mathcal{Q} \text{ implies } tv \in \mathcal{Q} \text{ for every } t > 0.$$

Clearly subspaces are cones, but the latter definition is broader since it includes, for example, the half-line $\mathcal{C}_v = \{\alpha v : \alpha > 0\}$ for a fixed vector v.

Of particular interest are cones which have the convexity property. In fact convex cones are precisely those which are closed under addition (i.e., $v_1, v_2 \in \mathcal{C}$ implies $v_1 + v_2 \in \mathcal{C}$). A canonical example of a convex cone is a half-space

$$F(v) \geq 0$$

that goes through the origin, or an intersection of half-spaces of this form. More illustrations are given in Figure 1.6.

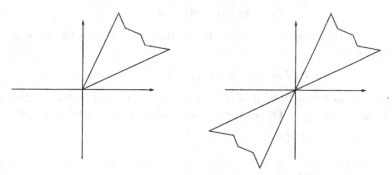

Figure 1.6. Convex and non-convex cones

Up to this point we have introduced a number of concepts related to vector spaces and have for the most part developed our results in a basis independent way, emphasizing the structure of linear spaces. This provides

a powerful way to visualize linear algebra. However, for most parts of the
course it is much more efficient to work in a particular basis, especially
since we are interested in developing results with computation in mind. In
this spirit we now turn to results about matrices.

1.3   Matrix theory

The material of this section is aimed directly at both analysis and computation. Our goals will be to review some basic facts about matrices,
and present some additional results for later reference, including two matrix decompositions which have tremendous application, the *Jordan* form
and *singular value* decomposition. Both are extremely useful for analytical purposes, and the singular value decomposition is also very important
in computations. We will also present some results about self-adjoint and
positive definite matrices.

1.3.1   Eigenvalues and Jordan form

In this section we are concerned exclusively with complex square matrices.
We begin with a definition: if $A \in \mathbb{C}^{n \times n}$, we say that $\lambda \in \mathbb{C}$ is an *eigenvalue*
of A if

$$Ax = \lambda x \tag{1.2}$$

can be satisfied for some nonzero vector x in $\mathbb{C}^n$. Such a vector x is called
an *eigenvector*. Equivalently this means that $\mathrm{Ker}(\lambda I - A) \neq 0$ or $\lambda I - A$
is singular. A matrix is singular exactly when its determinant is zero, and
therefore we have that λ is an eigenvalue if and only if

$$\det(\lambda I - A) = 0,$$

where $\det(\cdot)$ denotes determinant. Regarding λ as a variable we call the
polynomial

$$\det(\lambda I - A) = \lambda^n + a_{n-1}\lambda^{n-1} + \cdots + a_0$$

the *characteristic polynomial* of A. If A is a real matrix then the coefficients
a_k will be real as well. The characteristic polynomial can be factored as

$$\det(\lambda I - A) = (\lambda - \lambda_1) \cdots (\lambda - \lambda_n).$$

The n complex roots λ_k, which need not be distinct, are the eigenvalues
of A, and are collectively denoted by $\mathrm{eig}(A)$. Furthermore if A is a real
matrix, then any nonreal eigenvalues must appear in conjugate pairs. Also,
a matrix has the eigenvalue zero if and only if it is singular.

   Associated with every eigenvalue λ_k is the subspace

$$\mathcal{E}_k = \mathrm{Ker}(\lambda_k I - A);$$

every nonzero element in $\mathcal{E}_k$ is an eigenvector corresponding to the eigenvalue λ_k. Now suppose that a set of eigenvectors satisfies

$$\text{span}\{x_1, \ldots, x_n\} = \mathbb{C}^n.$$

Then we can define the invertible matrix $X = \begin{bmatrix} x_1 & \cdots & x_n \end{bmatrix}$, and from the matrix product we find

$$AX = \begin{bmatrix} Ax_1 & \cdots & Ax_n \end{bmatrix} = \begin{bmatrix} \lambda_1 x_1 & \cdots & \lambda_n x_n \end{bmatrix} = X\Lambda$$

where Λ is the diagonal matrix

$$\Lambda = \begin{bmatrix} \lambda_1 & & 0 \\ & \ddots & \\ 0 & & \lambda_n \end{bmatrix}.$$

Thus in this case we have a similarity transformation X such that $X^{-1}AX = \Lambda$ is diagonal, and we say that the matrix A is *diagonalizable*.

Summarizing we have the following result.

Proposition 1.7. *A matrix A is diagonalizable if and only if*

$$\mathcal{E}_1 + \mathcal{E}_2 + \cdots + \mathcal{E}_n = \mathbb{C}^n \ holds.$$

The following example shows that not all matrices can be diagonalized. Consider the 2×2 matrix

$$\begin{bmatrix} 0 & 1 \\ 0 & 0 \end{bmatrix}.$$

It has a repeated eigenvalue at zero, but only one linearly independent eigenvector. Thus it cannot be diagonalized. Matrices of this form have a special role in the decomposition we are about to introduce: define the $n \times n$ matrix N by

$$N = \begin{bmatrix} 0 & 1 & & 0 \\ & & \ddots & \\ & & & 1 \\ 0 & & & 0 \end{bmatrix},$$

where $N = 0$ if the dimension $n = 1$. Such matrices are called nilpotent because $N^n = 0$. Using these we define a matrix to be a *Jordan block* if it is of the form

$$J = \lambda I + N = \begin{bmatrix} \lambda & 1 & & 0 \\ & & \ddots & \\ & & & 1 \\ 0 & & & \lambda \end{bmatrix}.$$

Notice all scalars are 1×1 Jordan blocks. A Jordan block has one eigenvalue λ of multiplicity n. However, it has only one linearly independent eigenvector. A key feature of a Jordan block is that it has precisely n subspaces which are J-invariant. They are given by

$$\mathbb{C}^k \times \{0\},$$

for $1 \le k \le n$. When $k = 1$ this corresponds exactly to the subspace associated with its eigenvector. We can now state the Jordan decomposition theorem.

Theorem 1.8. *Suppose $A \in \mathbb{C}^{n \times n}$. Then there exists a nonsingular matrix $T \in \mathbb{C}^{n \times n}$, and an integer $1 \le p \le n$, such that*

$$T^{-1}AT = J = \begin{bmatrix} J_1 & & & 0 \\ & J_2 & & \\ & & \ddots & \\ 0 & & & J_p \end{bmatrix},$$

where the matrices J_k are Jordan blocks.

This theorem states that a matrix can be transformed to one that is block-diagonal, where each of the diagonal matrices is a Jordan block. Clearly if a matrix is diagonalizable each Jordan block J_k will simply be a scalar equal to an eigenvalue of A. In general each block J_k has a single eigenvalue of A in all its diagonal entries; however, a given eigenvalue of A may occur in several blocks.

The relevance of the Jordan decomposition is that it provides a canonical form to characterize matrix similarity; namely, two matrices are similar if and only if they share the same Jordan form. Another related feature is that the Jordan form exhibits the structure of *invariant subspaces* of a given matrix. This is best seen by writing the above equation as

$$AT = TJ.$$

Now suppose we denote by T_1 the submatrix of T formed by its first n_1 columns, where n_1 is the dimension of the block J_1. Then the first n_1 columns of the preceding equation give

$$AT_1 = T_1 J_1,$$

which implies that $\mathcal{S}_1 = \mathrm{Im} T_1$ is invariant under A. Furthermore, we can use this formula to study the linear mapping on $\mathcal{S}_1$ obtained by restriction of A. In fact we find that in the basis defined by the columns of T_1, this linear mapping has the associated matrix J_1; in particular, the only eigenvalue of A restricted to $\mathcal{S}_1$ is λ_1.

The preceding idea can be extended by selecting T_1 to contain the columns corresponding to more than one Jordan block. The resulting invariant subspace will be such that the restriction of A to it has only the eigenvalues of the chosen blocks. Even more generally, we can pick any invariant subspace of J and generate from it invariant subspaces of A. Indeed there are exactly n_k invariant subspaces of A associated with the $n_k \times n_k$

Jordan block J_k, and all invariant subspaces of A can be constructed from this collection.

We will not explicitly require a constructive method for transforming a matrix to Jordan form, and will use this result solely for analysis.

1.3.2 Self-adjoint, unitary, and positive definite matrices

We have already introduced the adjoint A^* of a complex matrix A; in this section we study in more detail the structure given to the space of matrices by this operation. A first observation, which will be used extensively below, is that

$$(AB)^* = B^* A^*$$

for matrices A and B of compatible dimensions; this follows directly by definition.

Another basic concept closely related to the adjoint is the Euclidean length of a vector $x \in \mathbb{C}^n$, defined by

$$|x| = \sqrt{x^* x}$$

This extends the usual definition of magnitude of a complex number, so our notation will not cause any ambiguity. In particular,

$$|x|^2 = x^* x = \sum_{i=1}^{n} |x_i|^2.$$

Clearly $|x|$ is never negative, and is zero only when the vector $x = 0$. Later in the course we will discuss generalizations of this concept in more general vector spaces.

We have already encountered the notion of a Hermitian matrix, characterized by the self-adjoint property $Q^* = Q$. Recall the notation $\mathbb{H}^n$ for the real vector space of complex Hermitian matrices. We now collect some properties and introduce some new definitions, for later use. Everything we will state will apply as well to the set $\mathbb{S}^n$ of real, symmetric matrices.

Our first result about self-adjoint matrices is that their eigenvalues are always *real*. Suppose $Ax = \lambda x$ for nonzero x. Then we have

$$\lambda x^* x = x^* A x = (Ax)^* x = \lambda^* x^* x.$$

Since $x^* x > 0$ we conclude that $\lambda = \lambda^*$.

We say that two vectors $x, y \in \mathbb{C}^n$ are *orthogonal* if

$$y^* x = 0.$$

Given a set of vectors $\{v_1, \ldots, v_k\}$ in $\mathbb{C}^n$ we say the vectors are *orthonormal* if

$$v_i^* v_r = \begin{cases} 1, & \text{if } i = r; \\ 0, & \text{if } i \neq r. \end{cases}$$

The vectors are orthonormal if each has unit length and is orthogonal to all the others. It is easy to show that orthonormal vectors are linearly independent, so such a set can have at most n members. If $k < n$, then it is always possible to find a vector v_{k+1} such that $\{v_1, \ldots, v_{k+1}\}$ is an orthonormal set. To see this, form the $k \times n$ matrix

$$V_k^* = \begin{bmatrix} v_1^* \\ \vdots \\ v_k^* \end{bmatrix}.$$

The kernel of V_k^* has the nonzero dimension $n - k$, and therefore any element of the kernel is orthogonal to the vectors $\{v_1, \ldots, v_k\}$. We conclude that any element of unit length in $\mathrm{Ker}\, V_k^*$ is a suitable candidate for v_{k+1}. Applying this procedure repeatedly we can generate an orthonormal basis $\{v_1, \ldots, v_n\}$ for $\mathbb{C}^n$.

A square matrix $U \in \mathbb{C}^{n \times n}$ is called *unitary* if it satisfies

$$U^*U = I.$$

From this definition we see that the columns of any unitary matrix forms an orthonormal basis for $\mathbb{C}^n$. Further, since U is square it must be that $U^* = U^{-1}$ and therefore $UU^* = I$. So the columns of U^* also form an orthonormal basis. A key property of unitary matrices is that if $y = Ux$, for some $x \in \mathbb{C}^n$, then the length of y is equal to that of x:

$$|y| = \sqrt{y^*y} = \sqrt{(Ux)^*(Ux)} = \sqrt{x^*U^*Ux} = |x|.$$

Unitary matrices are the only matrices that leave the length of every vector unchanged. We are now ready to state the *spectral theorem* for Hermitian matrices.

Theorem 1.9. *Suppose H is a matrix in $\mathbb{H}^n$. Then there exist a unitary matrix U and a real diagonal matrix Λ such that*

$$H = U\Lambda U^*.$$

Notice that since $U^* = U^{-1}$ for a unitary U, the above expression is a similarity transformation. Therefore the theorem says that a self-adjoint matrix can be diagonalized by a unitary similarity transformation. Thus the columns of U are all eigenvectors of H. Since the proof of this result assembles a number of concepts from this chapter we provide it below.

Proof. We will use an induction argument. Clearly the result is true if H is simply a scalar, and it is therefore sufficient to show that if the result holds for matrices in $\mathbb{H}^{n-1}$ then it holds for $H \in \mathbb{H}^n$. We proceed with the assumption that the decomposition result holds for $(n-1) \times (n-1)$ Hermitian matrices.

The matrix H has at least one eigenvalue λ_1, and λ_1 is real since H is Hermitian. Let x_1 be an eigenvector associated with this eigenvalue, and without loss of generality we assume it to have length one. Define X to be any unitary matrix with x_1 as its first column, namely,

$$X = [x_1 \cdots x_n].$$

Now consider the product X^*HX. Its first column is given by $X^*Hx_1 = \lambda_1 X^* x_1 = \lambda_1 e_1$, where e_1 is the first element of the canonical basis. Its first row is described by $x_1^* HX$, which is equal to $\lambda_1 x_1^* X = \lambda_1 e_1^*$, since $x_1^* H = \lambda_1 x_1^*$ because H is self-adjoint. Thus we have

$$X^*HX = \begin{bmatrix} \lambda_1 & 0 \\ 0 & H_2 \end{bmatrix},$$

where H_2 a Hermitian matrix in $\mathbb{H}^{n-1}$. By the inductive hypothesis there exists a unitary matrix X_2 in $\mathbb{C}^{(n-1)\times(n-1)}$ such that $H_2 = X_2 \Lambda_2 X_2^*$, where Λ_2 is both diagonal and real. We conclude that

$$H = \left(X \begin{bmatrix} I & 0 \\ 0 & X_2 \end{bmatrix} \right) \begin{bmatrix} \lambda_1 & 0 \\ 0 & \Lambda_2 \end{bmatrix} \left(\begin{bmatrix} I & 0 \\ 0 & X_2^* \end{bmatrix} X^* \right).$$

The right-hand side gives the desired decomposition. ∎

We remark, additionally, that the eigenvalues of H can be arranged in decreasing order in the diagonal of Λ. This follows directly from the above induction argument: just take λ_1 to be the largest eigenvalue.

We now focus on the case where these eigenvalues have a definite sign. Given $Q \in \mathbb{H}^n$, we say it is *positive definite*, denoted $Q > 0$, if

$$x^*Qx > 0,$$

for all nonzero $x \in \mathbb{C}^n$. Similarly Q is positive *semidefinite*, denoted $Q \geq 0$, if the inequality is nonstrict; and negative definite and negative semidefinite are similarly defined. If a matrix is not positive or negative semidefinite, then it is *indefinite*.

The following properties of positive matrices follow directly from the definition, and are left as exercises:

- If $Q > 0$ and $A \in \mathbb{C}^{n\times m}$, then $A^*QA \geq 0$. If $\text{Ker}(A) = \{0\}$, then $A^*QA > 0$.

- If $Q_1 > 0$, $Q_2 > 0$, then $\mu_1 Q_1 + \mu_2 Q_2 > 0$ whenever $\mu_1 > 0$, $\mu_2 \geq 0$. In particular, the set of positive definite matrices is a *convex cone* in $\mathbb{H}^n$, as defined in the previous section.

At this point we may well ask, how can we check whether a matrix is positive definite? The following answer is derived from Theorem 1.9:

If $Q \in \mathbb{H}^n$, then $Q > 0$ if and only if the eigenvalues of Q are all positive.

Notice in particular that a positive definite matrix is always invertible, and its inverse is also positive definite. Also a matrix is positive semidefinite exactly when none of its eigenvalues are negative; in that case the number of strictly positive eigenvalues is equal to the rank of the matrix.

An additional useful property for positive matrices is the existence of a square root. Let $Q = U\Lambda U^* \geq 0$, in other words the diagonal elements of Λ are non-negative. Then we can define $\Lambda^{\frac{1}{2}}$ to be the matrix with diagonal elements $\lambda_k^{\frac{1}{2}}$, and

$$Q^{\frac{1}{2}} := U\Lambda^{\frac{1}{2}}U^*.$$

Then $Q^{\frac{1}{2}} \geq 0$ (also $Q^{\frac{1}{2}} > 0$ when $Q > 0$) and it is easily verified that $Q^{\frac{1}{2}}Q^{\frac{1}{2}} = Q$.

Having defined a notion of positivity, our next aim is to generalize the idea of ordering to matrices: namely, what does it mean for a matrix to be larger than another matrix? We write

$$Q > S$$

for matrices $Q, S \in \mathbb{H}^n$ to denote that $Q - S > 0$. We refer to such expressions generally as matrix inequalities. Note that for matrices it may be that neither $Q \leq S$ nor $Q \geq S$ holds; that is, not all matrices are comparable.

We conclude our discussion by establishing a very useful result, known as the Schur complement formula.

Theorem 1.10. *Suppose that Q, M, and R are matrices and that M and Q are self-adjoint. Then the following are equivalent:*

(a) The matrix inequalities $Q > 0$ and

$$M - RQ^{-1}R^* > 0 \text{ both hold.}$$

(b) The matrix inequality

$$\begin{bmatrix} M & R \\ R^* & Q \end{bmatrix} > 0 \text{ is satisfied.}$$

Proof. The two inequalities listed in (a) are equivalent to the single block inequality.

$$\begin{bmatrix} M - RQ^{-1}R^* & 0 \\ 0 & Q \end{bmatrix} > 0 \, .$$

Now left- and right-multiply this inequality by the nonsingular matrix

$$\begin{bmatrix} I & RQ^{-1} \\ 0 & I \end{bmatrix}$$

and its adjoint, respectively, to get

$$\begin{bmatrix} M & R \\ R^* & Q \end{bmatrix} = \begin{bmatrix} I & RQ^{-1} \\ 0 & I \end{bmatrix} \cdot \begin{bmatrix} M - RQ^{-1}R^* & 0 \\ 0 & Q \end{bmatrix} \begin{bmatrix} I & 0 \\ Q^{-1}R^* & I \end{bmatrix} > 0.$$

Therefore inequality (b) holds if and only if (a) holds.   ∎

We remark that an identical result holds in the negative definite case, replacing all ">" by "<".

Having assembled some facts about self-adjoint matrices, we move on to our final matrix theory topic.

1.3.3   Singular value decomposition

Here we introduce the singular value decomposition of a rectangular matrix, which will have many applications in our analysis, and is of very significant computational value. The term *singular value decomposition*, or SVD, refers to the product $U\Sigma V^*$ in the statement of the theorem below.

Theorem 1.11. *Suppose $A \in \mathbb{C}^{m \times n}$ and that $p = \min\{m, n\}$. Then there exist unitary matrices $U \in \mathbb{C}^{m \times m}$ and $V \in \mathbb{C}^{n \times n}$ such that*

$$A = U\Sigma V^*,$$

where $\Sigma \in \mathbb{R}^{m \times n}$ and its scalar entries satisfy

(a) the condition $\sigma_{ir} = 0$, for $i \neq r$;

(b) the ordering $\sigma_{11} \geq \sigma_{22} \geq \cdots \geq \sigma_{pp} \geq 0$.

Proof. Since the result holds for A if and only if it holds for A^*, we assume without loss of generality that $n \geq m$. To start let r be the rank of A^*A, which is Hermitian and therefore by Theorem 1.9 we have

$$A^*A = V \begin{bmatrix} \bar{\Sigma}^2 & 0 \\ 0 & 0 \end{bmatrix} V^*, \text{ where } \bar{\Sigma} = \begin{bmatrix} \sigma_1 & & 0 \\ & \ddots & \\ 0 & & \sigma_r \end{bmatrix} > 0 \text{ and } V \text{ is unitary.}$$

We also assume that the nonstrict ordering $\sigma_1 \geq \cdots \geq \sigma_r$ holds. Now define

$$J = \begin{bmatrix} \bar{\Sigma} & 0 \\ 0 & I \end{bmatrix}$$

and we have

$$J^{-1}V^*A^*AVJ^{-1} = (AVJ^{-1})^*(AVJ^{-1}) = \begin{bmatrix} I_r & 0 \\ 0 & 0 \end{bmatrix},$$

where I_r denotes the $r \times r$ identity matrix. From the right-hand side we see that the first r columns of AVJ^{-1} form an orthonormal set, and the remaining columns must be zero. Thus

$$AVJ^{-1} = \begin{bmatrix} U_1 & 0 \end{bmatrix},$$

where $U_1 \in \mathbb{C}^{m \times r}$. This leads to

$$A = \begin{bmatrix} U_1 & 0 \end{bmatrix} \begin{bmatrix} \bar{\Sigma} & 0 \\ 0 & I \end{bmatrix} V^* = \begin{bmatrix} U_1 & U_2 \end{bmatrix} \begin{bmatrix} \bar{\Sigma} & 0 \\ 0 & 0 \end{bmatrix} V^*,$$

where the right-hand side is valid for any $U_2 \in \mathbb{C}^{m \times (m-r)}$. So choose U_2 such that $[U_1 \quad U_2]$ is unitary.   ∎

When $n = m$ the matrix Σ in the SVD is diagonal. When these dimensions are not equal Σ has the form of either

$$\begin{bmatrix} \sigma_{11} & & & 0 \\ & \ddots & & \\ 0 & & \sigma_{mm} & 0 \end{bmatrix} \text{ when } n > m, \text{ or } \begin{bmatrix} \sigma_{11} & & 0 \\ & \ddots & \\ & & \sigma_{nn} \\ 0 & & 0 \end{bmatrix} \text{ when } n < m.$$

The first p non-negative scalars σ_{kk} are called the *singular values* of the matrix A, and are denoted by the ordered set $\sigma_1, \ldots \sigma_p$, where $\sigma_k = \sigma_{kk}$. As we already saw in the proof, the decomposition of the theorem immediately gives us that

$$A^*A = V(\Sigma^*\Sigma)V^* \text{ and } AA^* = U(\Sigma\Sigma^*)U^*,$$

which are singular value decompositions of A^*A and AA^*. But since $V^* = V^{-1}$ and $U^* = U^{-1}$ it follows that these are also the diagonalizations of the matrices. Thus

$$\sigma_1^2 \geq \sigma_2^2 \geq \cdots \geq \sigma_p^2 \geq 0$$

are exactly the p largest eigenvalues of A^*A and AA^*; the remaining eigenvalues of either matrix are all necessarily equal to zero. This observation provides a straightforward method to obtain the singular value decomposition of any matrix A, by diagonalizing the Hermitian matrices A^*A and AA^*.

The SVD of a matrix has many useful properties. We use $\bar{\sigma}(A)$ to denote the largest singular value σ_1, which from the SVD has the following property.

$$\bar{\sigma}(A) = \max\{|Av| : v \in \mathbb{C}^n \text{ and } |v| = 1\}.$$

Namely, it gives the maximum magnification of length a vector v can experience when acted upon by A.

Finally, partition $U = [u_1 \quad \cdots \quad u_m]$ and $V = [v_1 \quad \cdots \quad v_n]$ and suppose that A has r nonzero singular values. Then

$$\text{Im}\, A = \text{Im} \begin{bmatrix} u_1 & \cdots u_r \end{bmatrix} \text{ and } \text{Ker}\, A = \text{Im} \begin{bmatrix} v_{r+1} & \cdots & v_n \end{bmatrix}.$$

That is, the SVD provides an orthonormal basis for both the image and kernel of A. Furthermore notice that the rank of A is equal to r, precisely the number of nonzero singular values.

1.4   Linear matrix inequalities

This section is devoted to introducing the central concept of a *linear matrix inequality* which we use throughout the course. The reasons for its importance are the following:

- From an analytical point of view, many important problems can be reduced to this form. Such reductions will be presented as we go along during the rest of the course.

- The resulting computational problem can be efficiently and completely solved by recently established numerical algorithms. While this second issue is beyond the scope of this course, we will devote some space in this section to explain some of the properties behind this computational tractability.

A *linear matrix inequality*, abbreviated LMI, in the variable X is an inequality of the form

$$F(X) < Q,$$

where

- the variable X takes values in a real vector space $\mathcal{X}$;

- the mapping $F : \mathcal{X} \to \mathbb{H}^n$ is linear;

- the matrix Q is in the set of Hermitian matrices $\mathbb{H}^n$.

The above is a strict inequality and $F(X) \leq Q$ is a nonstrict linear matrix inequality. Thus to determine whether an inequality is an LMI, we simply see whether the above conditions are satisfied. Let us consider some explicit examples.

Examples:

To start we note that every LMI can be written in a vector form. Suppose that $V_1, \ldots, V_m$ is a basis for the vector space $\mathcal{X}$. Then for any X in $\mathcal{X}$ we have that there exist scalars $x_1, \ldots, x_m$ such that

$$X = x_1 V_1 + \cdots + x_m V_m.$$

If we substitute into the LMI $F(X) < Q$, and use the linearity of F, we have

$$x_1 F(V_1) + \cdots + x_m F(V_m) < Q.$$

Thus the variables are the scalars x_k, and the $F(V_k)$ are fixed Hermitian matrices. In other words, every LMI can be converted to the form

$$x_1 F_1 + \cdots + x_m F_m < Q,$$

where $x \in \mathbb{R}^m$ is the variable. While this coordinate form could also be taken as a definition of an LMI, it is not how we typically encounter LMIs in our course, and is often cumbersome for analysis.

Consider the inequality

$$A^*XA - X < Q, \tag{1.3}$$

where $A \in \mathbb{R}^{n \times n}, Q \in \mathbb{H}^n$ and the variable X is in $\mathbb{S}^n$. If we define

$$F(X) = A^*XA - X,$$

then clearly this is a linear mapping $\mathbb{S}^n \to \mathbb{S}^n$. Therefore we see that (1.3) is an LMI.

Now look at the matrix inequality

$$A^*XA + BY + Y^*B^* + T < 0,$$

where $A \in \mathbb{C}^{n \times n}, B \in \mathbb{C}^{n \times m}, T \in \mathbb{H}^n$, and the variables X and Y are in $\mathbb{S}^n$ and $\mathbb{R}^{m \times n}$, respectively. This too is an LMI in the variables X and Y. To see this explicitly let

$$Z := (X, Y) \in \mathbb{S}^n \times \mathbb{R}^{m \times n}$$

and define $F(Z) = A^*XA + BY + Y^*B^*$. Then $F : \mathbb{S}^n \times \mathbb{R}^{m \times n} \to \mathbb{H}^n$ is a linear map and the LMI can be written compactly as

$$F(Z) < -T.$$

With these examples and definition in hand, we will easily be able to recognize an LMI. □

Here we have formulated LMIs in terms of the Hermitian matrices, which is the most general situation for our later analysis. In some problems LMIs are written over the space of symmetric matrices $\mathbb{S}^n$, and this is the usual form employed for computation. This change is inconsequential in regard to the following discussion, and furthermore in the exercises we will see that the Hermitian form can always be converted to the symmetric form.

An important property of LMIs is that a list of them can always be converted to *one* LMI. To see this suppose we have the two LMIs

$$F_0(X_0) < Q_0 \quad \text{and} \quad F_1(X_1) < Q_1$$

in the variables X_0 and X_1 in vector spaces $\mathcal{X}_0$ and $\mathcal{X}_1$, respectively. Then this is equivalent to the *single* LMI

$$F(X) < Q,$$

where

$$X = (X_0, X_1) \in \mathcal{X}_0 \times \mathcal{X}_1 \text{ and } Q = \begin{bmatrix} Q_0 & 0 \\ 0 & Q_1 \end{bmatrix}$$

and we define the function $F : \mathcal{X}_0 \times \mathcal{X}_1 \to \mathbb{H}^{n_0+n_1}$ by

$$F(X) = \begin{bmatrix} F_0(X_0) & 0 \\ 0 & F_1(X_1) \end{bmatrix}.$$

Notice that if X_0 and X_1 are equal, or have common components, then X should be defined in a corresponding appropriate manner.

An important point, which will appear repeatedly in this course, is that conditions that do not look like LMIs at first glance can sometimes be converted to them. The Schur complement given in Theorem 1.10 will be very useful in this regard. We illustrate this by a simple example.

Example:

Let $A \in \mathbb{R}^{n \times n}$, $b, c \in \mathbb{R}^n$, and $d \in \mathbb{R}$. The inequality

$$(Ax + b)^*(Ax + b) - c^*x - d < 0$$

is *not* an LMI, since the expression is quadratic in the variable $x \in \mathbb{R}^n$. However, the Schur complement formula implies it is equivalent to

$$\begin{bmatrix} c^*x + d & (Ax + b)^* \\ Ax + b & I \end{bmatrix} > 0.$$

Since the left hand side matrix now depends affinely on x, the latter is an LMI and can easily be expressed in the form $F(x) < Q$.   □

As our course progresses we will find that many control problems can be formulated in terms of finding solutions to LMIs. More generally we will have optimization problems with LMI constraints on the solution space. This class of optimization is known as *semidefinite programming*. The simplest semidefinite program is a decision problem, and is of the form

Does there exist $X \in \mathcal{X}$, satisfying $F(X) < Q$?

This is known as a *feasibility problem* and asks only whether there is any element which satisfies the LMI. Computational aspects will be discussed later, but we can already note an important property of the solution set.

Proposition 1.12. *The set* $\mathcal{C} := \{X \in \mathcal{X} \text{ such that } F(X) < Q\}$ *is convex in* $\mathcal{X}$.

Proof. Suppose X_1, $X_2 \in \mathcal{C}$, which means they satisfy $F(X_1) < Q$ and $F(X_2) < Q$. Consider any point X_3 in $L(X_1, X_2)$, namely, $X_3 = \mu X_1 + (1 - \mu)X_2$, for some value $\mu \in [0, 1]$. Using linearity of the function F we have

$$F(X_3) = \mu F(X_1) + (1 - \mu)F(X_2) < \mu Q + (1 - \mu)Q = Q.$$

The inequality follows from the fact that positive definite matrices are a convex cone. Therefore $X_3 \in \mathcal{C}$.   ■

We remark that the above proposition does not say anything about the LMI being feasible; indeed $\mathcal{C}$ could be empty.

We now turn to a more general semidefinite optimization problem:

$$\text{minimize: } c(X);$$
$$\text{subject to: } F(X) \leq Q \text{ and } X \in \mathcal{X},$$

where $c(X)$ is a linear functional on $\mathcal{X}$. It is referred to as the *linear objective* problem. In this formulation we are being informal about the meaning of "minimize"; in rigor we want to compute the infimum, which may or may not be achieved at a given X. Alternatively, the constraint could be specified as the strict version $F(X) < Q$, in which case we are always dealing with an infimum.

Example:

Let $\mathcal{X} = \mathbb{R}^2$; find the infimum of x_1 subject to

$$\begin{bmatrix} -x_1 & 0 \\ 0 & -x_2 \end{bmatrix} \leq \begin{bmatrix} 0 & 1 \\ 1 & 0 \end{bmatrix}.$$

It is easy to show that the answer is zero, but this value is not achieved by any matrix satisfying the constraint; notice this happens even in the case of non-strict LMI constraints.   □

Clearly the linear objective problem makes sense only if the LMI is feasible, which seems to imply that such problems are of higher difficulty than feasibility. In fact, both are of very similar nature:

- In the exercises you will show how the linear objective problem can be tackled by a family of feasibility questions.

- Conversely, the feasibility problem can be recast as the problem

$$\text{find: } J := \inf \ t$$
$$\text{subject to: } F(X) - tI \leq Q.$$

The latter is a linear objective problem in the variables $t \in \mathbb{R}$ and $X \in \mathcal{X}$, and the corresponding LMI constraint is automatically (strictly) feasible. It is an easy exercise to see that $J < 0$ if and only if the LMI $F(X) < Q$ is feasible.

Given these relationships, we focus for the rest of the section on the linear objective problem, with the assumption that the strict LMI $F(X) < Q$ is feasible. A pictorial view of the problem is given in Figure 1.7, for the case

$$X = \begin{bmatrix} x_1 \\ x_2 \end{bmatrix} \in \mathbb{R}^2.$$

The convex set depicted in the figure represents the feasibility set $\overline{\mathcal{C}} = \{X : F(X) \leq Q\}$ for the linear objective problem; while we have drawn a bounded set, we remark that this is not necessarily the case.

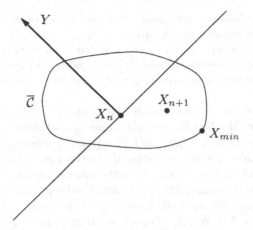

Figure 1.7. Illustration of semidefinite programming

Since $c(X)$ is a linear functional on $\mathbb{R}^2$, it has the form

$$c(X) = Y^*X = y_1x_1 + y_2x_2$$

for a fixed vector $Y \in \mathbb{R}^2$; therefore the point X_{min} that solves the problem is the element of $\overline{C}$ with the most negative projection in the direction of Y, as depicted in Figure 1.7. Also the picture suggests that there are no other local minima for the function in the set; namely, for every other point there is a so-called descent direction. This property, fundamental to convex optimization problems, is now stated precisely.

Proposition 1.13. *Suppose X_0 is a local minimum of the linear objective problem; that is, $c(X_0) \leq c(X)$ for every $X \in \mathcal{N}(X_0) \cap \overline{C}$. Then X_0 is the global minimum of the problem over $\overline{C}$.*

Proof. Let X_1 be another point in $\overline{C}$. Since $\overline{C}$ is convex, it contains the line segment

$$L(X_1, X_0) = \{\mu X_1 + (1 - \mu)X_0, \mu \in [0, 1]\}.$$

Also by definition the neighborhood $\mathcal{N}(X_0)$ will contain the points $\mu X_1 + (1 - \mu)X_0$ for sufficiently small μ, say $\mu \in [0, \epsilon)$. Now the function $f(\mu) = c(\mu X_1 + (1 - \mu)X_0)$ is affine in $\mu \in [0, 1]$, and $f(0) \leq f(\mu)$ for $\mu \in [0, \epsilon)$ by hypothesis. Then $f(\mu)$ is non-decreasing and $f(0) \leq f(1)$, or equivalently $c(X_0) \leq c(X_1)$, which concludes our proof.   ∎

The above property generalizes (see the exercises) to any convex optimization problem, and plays a strong role in ensuring that these problems can be solved globally, and not just locally, by numerical algorithms.

At this point the reader may be wondering about the nature of semidefinite programming algorithms; we will purposely stay away from any detailed discussion of these methods, which would take us far afield, and defer to the authoritative references provided at the end of the chapter. It is indeed an interesting feature of this area that theory and computation decouple nicely so that detailed knowledge of the latter is not necessary to appreciate the former. We will, however provide a few remarks aimed mainly at reinforcing the idea that these problems are fundamentally tractable.

A first observation is that clearly the minimum, if it exists, must lie on the boundary of the feasible set; thus one could think one restricting the search to the boundary. This is in fact a popular alternative in the case of *linear* programming, which corresponds to the case where the feasible set is a *polytope*, the intersection of a finite number of half-spaces. However, in the semidefinite programming problem the boundary is in general quite complicated, and thus methods involving *interior* points are favored.

Going back to Figure 1.7, suppose we have a point X_n in our feasible set; an immediate consequence is that we need only keep the set

$$\{X \in \bar{\mathcal{C}} : c(X) \leq c(X_n)\}$$

for our remaining search for the global minimum. This amounts to intersecting $\bar{\mathcal{C}}$ with a half-space; thus we can progressively shrink the feasibility region to zero, provided we are able to generate successively a "good" new feasible point X_{n+1}. Many optimization algorithms are based on this principle; one of the simplest is the so-called ellipsoid algorithm that alternates between cutting and overbounding the resulting set by an ellipsoid; X_{n+1} would then be the center of such an ellipsoid; for details see the references.

More efficient methods for semidefinite programming are based on *barrier functions* to impose the feasibility constraint. The idea is to minimize the function

$$c(X) + \alpha\phi(X) \tag{1.4}$$

where $\alpha > 0$, and the barrier function $\phi(X)$ is convex and approaches infinity on the boundary of the feasible set. For a definition of a convex function, and some basic properties see the exercises at the end of the chapter. Also you will see there that the function

$$\phi(X) = -\log(\det[Q - F(X)])$$

serves as a barrier function for the set $\mathcal{C}$.

Provided we start from a feasible point, the minimization in (1.4) can be globally solved by unconstrained optimization methods (e.g., Newton's algorithm). By successively reducing the weight of the barrier function, an iteration is produced which can be shown to converge to the global minimum. The computational complexity of these algorithms has moderate (polynomial) growth with problem size, the latter being characterized by the dimensionality of the variable space $\mathcal{X}$ and of the constraint set $\mathbb{H}^n$. For extensive details on this active area of optimization, as well as many other alternative algorithms, the reader is encouraged to consult the references.

We have now completed our preliminary preparation for the course, and are ready to investigate some control theory problems.

Exercises

1.1. Find a basis for $\mathbb{C}^{m \times n}$ as a real vector space. What is the dimension of this real vector space?

1.2. The spaces $\mathbb{H}^n$ and $\mathbb{S}^n$ are both real vector spaces; find bases for each. How are they related?

1.3. Determine the dimension of the set of homogeneous multinomials $P_3^{[4]}$. Show the general formula for the dimension of $P_m^{[n]}$ is $\frac{(m+n-1)!}{n!(m-1)!}$; namely, "$(m + n - 1)$ choose n." *Hint:* try using induction.

1.4. We consider the mapping Ω defined in §1.1.4. Let $a \in P_3^{[1]}$ be $a(x_1, x_2, x_3) = x_2$, and consider $\Omega : P_3^{[1]} \to P_3^{[2]}$, which is defined by

$$(\Omega p)(x_1, x_2, x_3) = a(x_1, x_2, x_3)p(x_1, x_2, x_3).$$

Choose bases for $P_3^{[1]}$ and $P_3^{[2]}$, and represent Ω as the corresponding matrix $[\Omega]$.

1.5. Suppose $A : \mathcal{V} \to \mathcal{W}$. Let $\{Av_1, \ldots, Av_r\}$ be a basis for $\mathrm{Im}A$ and $\{u_1, \ldots, u_k\}$ be a basis for $\mathrm{Ker}A$. Show that $\{v_1, \ldots, v_r, u_1, \ldots, u_k\}$ is a basis for $\mathcal{V}$ and deduce that $\dim(\mathcal{V}) = \dim(\mathrm{Ker}A) + \dim(\mathrm{Im}A)$.

1.6. Given a mapping $A : \mathcal{V} \to \mathcal{V}$ show that both $\mathrm{Ker}A$ and $\mathrm{Im}A$ are A-invariant.

1.7. Suppose $A : \mathbb{C}^n \to \mathbb{C}^n$. Show that if the subspace S is A-invariant, then S contains an eigenvector of A.

1.8. Use the Jordan form to show that any matrix is similar to its transpose.

1.9. By direct calculation, show that given $A \in \mathbb{C}^{m \times n}$ and $B \in \mathbb{C}^{n \times m}$, the identity $\mathrm{Tr}AB = \mathrm{Tr}BA$ holds. Use this to prove that the trace of any square matrix is the sum of its eigenvalues.

1.10. Determine whether the set of matrices in $\mathbb{R}^{n \times n}$ which have all their eigenvalues in the open left half-plane $\mathbb{C}^-$ is convex.

1.11. Provide an example of two closed sets which are disjoint and convex, but are *not* strictly separated.

1.12. Prove that every linear functional F on $\mathbb{S}^n$ can be expressed as $F(X) = \mathrm{Tr}(YX)$ for some fixed $Y \in \mathbb{S}^n$.

1.13. Suppose A is a Hermitian matrix, and that λ_0 and λ_1 are two eigenvalues with corresponding eigenvectors x_0 and x_1. Prove that if $\lambda_0 \neq \lambda_1$, then x_0 and x_1 are orthogonal.

1.14. Show that if $P \geq 0$ and $Q \geq 0$, then $\text{Tr}(PQ) \geq 0$.

1.15. Suppose $A \in \mathbb{C}^{m \times n}$ has the singular value decomposition $U\Sigma V^*$.

   (a) Show that if $x \in \text{span}\{v_{k+1}, \ldots, v_n\}$ then $|Ax| \leq \sigma_{k+1}|x|$.
   (b) If A is $n \times n$, let $\sigma_1 \geq \sigma_2 \geq \cdots \sigma_n$ be the diagonal elements of Σ. Denote $\underline{\sigma}(A) = \sigma_n$. Show that A is invertible if and only if $\underline{\sigma}(A) > 0$, and in that case

$$\bar{\sigma}(A^{-1}) = \frac{1}{\underline{\sigma}(A)}.$$

   (c) If $A \in \mathbb{H}^n$, then $-\bar{\sigma}(A)\,I \leq A \leq \bar{\sigma}(A)\,I$.

1.16. Given an $n \times n$ Hermitian matrix A, denote its ordered eigenvalues by $\lambda_1(A) \geq \lambda_2(A) \geq \cdots \geq \lambda_n(A)$. Show that these satisfy

$$\lambda_k(A) = \max_{\dim(\mathcal{S})=k} \min_{0 \neq y \in \mathcal{S}} \frac{y^* A y}{y^* y}.$$

This is called the Courant-Fischer formula.

1.17. Suppose that $\gamma > 0$ and that $X \in \mathbb{R}^{n \times m}$.

   (a) Show that $\bar{\sigma}(X) \leq \gamma$ if and only if $X^* X \leq \gamma^2 I$.
   (b) Convert the constraint $\bar{\sigma}(X) \leq \gamma$ to an equivalent LMI condition.

1.18. The *spectral radius* of a matrix $M \in \mathbb{C}^{n \times n}$ is defined as

$$\rho(M) := \max\{|\lambda| \text{ such that } \lambda \text{ is an eigenvalue of } M\}.$$

   (a) Show that $\rho(M) \leq \bar{\sigma}(M)$ and find both numbers for

$$M = \begin{bmatrix} 0 & 1 & & 0 \\ & \ddots & \ddots & \\ & & & 1 \\ 0 & & & 0 \end{bmatrix}.$$

   (b) Show that

$$\rho(M) \leq \inf_{D \text{ invertible}} \bar{\sigma}(DMD^{-1}).$$

   (c) Prove that there is equality in (b). *Hint:* use the Jordan form.
   (d) Deduce that $\rho(M) < 1$ if and only if the set of LMIs

$$X > 0, \qquad M^* X M - X < 0$$

is feasible over $X \in \mathbb{H}^n$.

1.19. Consider a *real* LMI given by

$$\bar{F}(X) < \bar{Q},$$

where the linear map $\bar{F} : \mathcal{X} \to \mathbb{S}^{\bar{n}}$, $\bar{Q} \in \mathbb{S}^{\bar{n}}$, and $\mathcal{X}$ is a real vector space. Show that any LMI of the standard form given in §1.4 with respect to the Hermitian matrices $\mathbb{H}^n$ can be converted to a real LMI with $\bar{n} = 2n$. *Hint:* first find a condition for a matrix $A \in \mathbb{H}^n$ to be positive definite, in the form of an LMI on the real and imaginary parts of A.

1.20. In §1.4 two types of LMI problems were introduced, the feasibility problem and the linear objective problem. Show how the feasibility problem can be used with iteration to solve the latter.

1.21. Another type of LMI optimization problem is the so-called *generalized eigenvalue* minimization problem, which is

$$\text{minimize: } \gamma;$$
$$\text{subject to: } F_0(X) + \gamma F_1(X) < Q_0 + \gamma Q_1$$
$$F_1(X) < Q_1,$$
$$\text{and } X \in \mathcal{X}.$$

where F_1 and F_2 are linear mapping from $\mathcal{X}$ to $\mathbb{H}^n$, and $Q_0, Q_1 \in \mathbb{H}^n$. Show that the linear objective problem can be reformulated in this format. Further show that this problem can be solved by iteration using the feasibility problem.

1.22. Let $\mathcal{C}$ be a convex set in a real vector space $\mathcal{X}$. A function $\phi : \mathcal{C} \to \mathbb{R}$ is said to be convex if it satisfies

$$\phi(\mu x_1 + (1 - \mu)x_2) \le \mu\phi(x_1) + (1 - \mu)\phi(x_2)$$

for every x_1, x_2 in $\mathcal{C}$ and every $\mu \in [0, 1]$. The minimization of such a function is called a convex optimization problem. As an important example, the function $\phi(x) = -\log(x)$ is convex in $(0, \infty)$. Clearly, any linear function is convex.

(a) Prove that for a convex function, every local minimum is a global minimum.
(b) Show a function ϕ is convex if and only if for any x_1, x_2 in $\mathcal{C}$, the function $f(\mu) := \phi(\mu x_1 + (1 - \mu)x_2)$ is convex in $\mu \in [0, 1]$.
(c) Prove that $\phi(X) = -\log(\det(X))$ is convex in the set of positive matrices $\{X > 0\}$. *Hint:* use the identity

$$\mu X_1 + (1 - \mu)X_2 = X_2^{\frac{1}{2}}\left(I + \mu X_2^{-\frac{1}{2}}(X_1 - X_2)X_2^{-\frac{1}{2}}\right)X_2^{\frac{1}{2}}$$

and express $\det(\mu X_1 + (1 - \mu)X_2)$ in terms of the eigenvalues of the Hermitian matrix $H := X_2^{-\frac{1}{2}}(X_1 - X_2)X_2^{-\frac{1}{2}}$.

(d) Deduce that if $F : \mathcal{X} \to \mathbb{H}^n$ is linear, $-\log(\det[Q - F(X)])$ is a convex barrier function for the set $\mathcal{C} = \{X \in \mathcal{X} : F(X) < Q\}$.

Notes and references

Given its ubiquitous presence in analytical subjects, introductory linear algebra is the subject of many excellent books; one choice is [156]. For an advanced treatment from a geometric perspective the reader is referred to [68].

For many more details on convexity see the standard reference [134].

Two excellent sources for matrix theory are [75] and the companion work [76]. For information and algorithms for computing with matrices see [66].

LMIs have a long history as an analytical tool, but it is only with recent advances in semidefinite programming that they have acquired a more central role in computation. These advances have been largely motivated by the pioneering work [108] on interior point methods for general convex optimization problems. Semidefinite programming is still undergoing rapid development; a recent survey is [163]. Also see [21] for an introduction to LMI optimization with many examples from control theory, and [61] for recent advances on theory, computation and applications of these methods.

Reference [98] is a more general work on optimization primarily aimed at infinite dimensional vector spaces.

2
State Space System Theory

We will now begin our study of system theory. This chapter is devoted to examining one of the building blocks used in the foundation of this course, the continuous time, state space system. Our goal is to cover the fundamentals of state space systems, and we will consider and answer questions about their basic structure, controlling and observing them, and representations of them.

The following two equations define a state space system.

$$\dot{x}(t) = Ax(t) + Bu(t), \quad \text{with} \quad x(0) = x_0 \tag{2.1}$$
$$y(t) = Cx(t) + Du(t),$$

where $u(t), x(t)$, and $y(t)$ are vector valued functions; and A, B, C, and D are real matrices. We recall that the derivative $\dot{x}(t)$ is simply the vector formed from the derivatives of each scalar entry of $x(t)$. The first of the above equations is called the state equation and the other the output equation. The variable $t \geq 0$ is *time* and the function $u(t)$ is referred to as the system *input*. The functions $x(t)$ and $y(t)$ are called the *state* and *output* of the system, respectively, and depend on the input.

For later reference we define the dimensions of the vectors by

$$u(t) \in \mathbb{R}^m, \ x(t) \in \mathbb{R}^n \text{ and } y(t) \in \mathbb{R}^p \ .$$

Thus A is an $n \times n$ matrix; the matrix B is $n \times m$; matrices C and D are $p \times n$ and $p \times m$, respectively. Notice that the system given above is a first-order linear differential equation with an initial condition, and therefore has a unique solution. The state space formulation above is very general because many systems of higher-order linear differential equations can be reformulated in this way. This generality motivates our study of state space systems. Before considering this system as a whole we will examine simpler versions of it to build our understanding successively.

2.1   The autonomous system

In this short section we will develop some intuition about the state equation by considering one of its simplifications. This will lead us to introducing and studying the generalization of the exponential function to matrices, a function we will find repeatedly useful in the chapter.

Let us focus on the autonomous system

$$\dot{x}(t) = Ax(t), \quad \text{with} \quad x(0) = x_0 . \tag{2.2}$$

This system is a special case of (2.1) in which there are no inputs and no outputs. The state function $x(t)$, for $t \geq 0$, is completely specified by the initial condition $x(0)$. The immediate question that comes to mind is, can we explicitly state the solution to this differential equation?

Consider first the case where $x(t)$ is a scalar, namely, $n = 1$. Then A is equal to some scalar a. Thus we have that

$$x(t) = e^{at}x(0)$$

is the unique solution for $x(t)$. There are a number of equivalent ways to define the meaning of the scalar function e^{at}, and one of these is using its power or Taylor series expansion.

We follow this lead for the multivariable case when $n > 1$ and make the definition: the *matrix exponential* e^M of a square matrix M is defined to be the matrix sum of the power series

$$e^M = I + M + \tfrac{1}{2!}M^2 + \tfrac{1}{3!}M^3 + \tfrac{1}{4!}M^4 + \cdots .$$

It is not difficult to show that e^M is well-defined; namely, the above series always converges. Notice that if M is a scalar then this definition agrees with the familiar power series expansion of the scalar exponential function. Our main focus will be the time dependent function

$$e^{At} = I + At + \tfrac{1}{2!}A^2t^2 + \tfrac{1}{3!}A^3t^3 + \cdots$$

which is defined for every t. The following are some basic properties of the matrix exponential.

(a) $e^0 = I$, where 0 denotes the zero element in $\mathbb{R}^{n \times n}$;

(b) $e^{M^*} = (e^M)^*$;

(c) $\frac{d}{dt}e^{At} = Ae^{At} = e^{At}A$;

(d) If the square matrices M and N commute, that is, $MN = NM$, then the relationship $e^{M+N} = e^M e^N$ holds.

The first three properties follow immediately from the definition; for the last one see the Exercise 2.3 at the end of the chapter. We will now illustrate the matrix types which result from this definition by looking at two special cases; by using the Jordan form, these two examples can be used to specify the exponential of any matrix.

Examples:

Our first example pertains to the case where A is a diagonalizable matrix; in this case, a similarity transformation T exists such that

$$T^{-1}AT = \Lambda = \begin{bmatrix} \lambda_1 & & 0 \\ & \ddots & \\ 0 & & \lambda_n \end{bmatrix} \tag{2.3}$$

holds, where λ_k are the eigenvalues of A. This is a convenient transformation because, for $k \geq 0$, we have $A^k = T\Lambda^k T^{-1}$. Therefore

$$e^{At} = T\left(I + \Lambda t + \tfrac{1}{2!}\Lambda^2 t^2 + \cdots\right)T^{-1} = T \begin{bmatrix} e^{\lambda_1 t} & & 0 \\ & \ddots & \\ 0 & & e^{\lambda_n t} \end{bmatrix} T^{-1}.$$

Thus we see that the matrix exponential consists of linear combinations of exponentials whose exponents are the eigenvalues of the matrix.

The second case we consider is where A is exactly a Jordan block, and so has the form

$$A = \begin{bmatrix} \lambda & 1 & & 0 \\ & \ddots & \ddots & \\ & & & 1 \\ 0 & & & \lambda \end{bmatrix} = \lambda I + N, \tag{2.4}$$

where N is the matrix with ones above the diagonal and zeros in its remaining entries. An interesting and useful feature of N is that its powers are easily computed. In particular its square is

$$N^2 = \begin{bmatrix} 0 & 0 & 1 & & 0 \\ & \ddots & \ddots & \ddots & \\ & & 0 & 0 & 1 \\ & & & 0 & 0 \\ 0 & & & & 0 \end{bmatrix}$$

and the successive powers of N have analogous structure, with the diagonal of ones shifting upwards, until eventually we find $N^n = 0$. That is N is *nilpotent* of order n.

This feature of N can be exploited to compute the exponential e^{At}. Since the matrices λI and N commute we can invoke property (d) above to find that

$$e^{At} = e^{\lambda I t} e^{Nt} = e^{\lambda t} \left(I + Nt + \tfrac{1}{2!} N^2 t^2 + \cdots \tfrac{1}{(n-1)!} N^{n-1} t^{n-1} \right).$$

This gives us the final result

$$e^{At} = e^{\lambda t} \begin{bmatrix} 1 & t & \frac{t^2}{2} & \cdots & \frac{t^{n-1}}{(n-1)!} \\ & \ddots & \ddots & \ddots & \vdots \\ & & 1 & t & \frac{t^2}{2} \\ 0 & & & 1 & t \\ 0 & & & & 1 \end{bmatrix}.$$

So we have shown that the matrix exponential of a single Jordan block consists of functions of the form $\left(\frac{t^k}{k!} \right) e^{\lambda t}$. □

Having gained some familiarity with the matrix exponential we now return to the differential equation (2.2). Using the above properties of the matrix exponential, we can see that it follows immediately that the unique solution to this autonomous equation is given by

$$x(t) = e^{At} x(0), \text{ for } t \geq 0.$$

One of the main properties of interest for such a system is that of stability. We say that the system described by (2.2) is *internally stable* if, for every initial condition $x(0) \in \mathbb{R}^n$, the limit

$$x(t) \overset{t \to \infty}{\longrightarrow} 0 \text{ holds.}$$

This limit simply says that each scalar entry of $x(t)$ tends to zero with time. In the exercises you will show that the system is internally stable if and only if each eigenvalue λ of A satisfies

$$\text{Re}\lambda < 0;$$

namely, they are all in the left half of the complex plane. A matrix that has this property is called a *Hurwitz matrix*. Thus we have the following summarizing result.

Proposition 2.1. *The autonomous system in (2.2) is internally stable if and only if A is Hurwitz.*

At this point we have covered all the key properties of the autonomous state equation, and have seen that these follow directly from those of the matrix exponential function. We are now ready to consider systems with inputs — controlled systems.

2.2 Controllability

This section is the most important of the chapter as it contains machinery and concepts that are crucial to the rest of the course. The problem we intend to pose and answer is a fundamental question about the control of state space systems. We will study the equation

$$\dot{x}(t) = Ax(t) + Bu(t), \text{ with } x(0) = 0, \tag{2.5}$$

which is the state equation from (2.1). Notice that for now we set the initial condition of the state to zero. Thus it is easy to verify that the solution to this equation is

$$x(t) = \int_0^t e^{A(t-\tau)} Bu(\tau) d\tau, \text{ for } t \geq 0.$$

The integrand in the equation is a vector, so each scalar entry of $x(t)$ is just equal to the integral of of the corresponding scalar entry of $e^{A(t-\tau)} Bu(\tau)$. Our first objective will be to determine what states can be reached by manipulating the input $u(t)$. In other words how much control do we have over the values of $x(t)$ through choice of the function u? We will find that this question can be answered in very precise terms.

2.2.1 Reachability

We begin our study by asking, given a *fixed* time t, what are the possible values of the state vector $x(t)$? Or asked another way: given a vector in $\mathbb{R}^n$ is it possible to steer $x(t)$ to this value by choosing an appropriate input function $u(t)$? We will answer this question completely, and will find that it has a surprisingly simple answer. To do this we require three related concepts.

Set of reachable states

For a *fixed* time $t > 0$, let $\mathcal{R}_t$ denote the states that are reachable at time t by some input function u. Namely, $\mathcal{R}_t$ is the set

$$\mathcal{R}_t := \{\xi \in \mathbb{R}^n : \text{ there exists } u \text{ such that } x(t) = \xi\}.$$

This definition is made respecting the state equation given in (2.5), where the initial condition is zero. It turns out that $\mathcal{R}_t$ is a *subspace* of $\mathbb{R}^n$. This is a simple consequence of the linearity of (2.5), which we leave as an exercise.

Controllable subspace

Next we define a subspace associated with any state equation. Given the state equation $\dot{x}(t) = Ax(t) + Bu(t)$ we call the matrix

$$\begin{bmatrix} B & AB & A^2B & \cdots & A^{n-1}B \end{bmatrix}$$

the associated *controllability matrix*. Equivalently we say it is the controllability matrix of the matrix pair (A, B). Recall that A is an $n \times n$ matrix, and notice that the dimension n plays an important role in the above definition.

Associated with the controllability matrix is the *controllable subspace*, denoted $\mathcal{C}_{AB}$, which is defined to be the image of the controllability matrix.

$$\mathcal{C}_{AB} := \mathrm{Im} \begin{bmatrix} B & AB & \cdots & A^{n-1}B \end{bmatrix} .$$

Thus we see that $\mathcal{C}_{AB}$, like $\mathcal{R}_t$ above is a subspace of $\mathbb{R}^n$. When dimension of $\mathcal{C}_{AB}$ is n, or equivalently the controllability matrix has the full rank of n, we say that (A, B) is a *controllable* pair. In the same vein we refer to a state space system as being *controllable* when the associated matrix pair (A, B) is controllable. We shall soon see the motivation for this terminology.

Controllability gramian

Here we define yet another object associated with the state equation, a matrix which depends on time. For each $t > 0$, the time dependent *controllability gramian* $n \times n$ matrix

$$W_t := \int_0^t e^{A\tau} BB^* e^{A^*\tau} d\tau .$$

Having defined the set of reachable states, the controllability subspace, and the controllability gramian, we can now state the main result of this section. It will take a number of steps to prove.

Theorem 2.2. *For each time $t > 0$ the set equality*

$$\mathcal{R}_t = \mathcal{C}_{AB} = \mathrm{Im}\, W_t \; holds.$$

This theorem says that the set of reachable states is always equal to the controllability subspace, and is also equal to the image of the controllability gramian. Since the controllability subspace $\mathcal{C}_{AB}$ is independent of time so is the set of reachable states: if a state can be reached at a particular time, then it can be reached at any $t > 0$, no matter how small. According to the theorem, if (A, B) is controllable, then $\mathcal{R}_t$ is equal to the entire state space $\mathbb{R}^n$; that is, all the states are reachable by appropriate choice of the input function u.

Let us now move on to proving Theorem 2.2. We will accomplish this by proving three lemmas, showing sequentially that

- $\mathcal{R}_t$ is a subset of $\mathcal{C}_{AB}$;

- $\mathcal{C}_{AB}$ is a subset of $\mathrm{Im} W_t$;

- $\mathrm{Im} W_t$ is a subset of $\mathcal{R}_t$.

These facts will be proved in lemmas 2.6, 2.8 and 2.9, respectively.

Before our first result we require a fact about matrices known as the Cayley–Hamilton theorem. Denote the characteristic polynomial of matrix A by $\text{char}_A(s)$. That is,

$$\text{char}_A(s) := \det(sI - A) =: s^n + a_{n-1}s^{n-1} + \cdots a_0,$$

where $\det(\cdot)$ denotes the determinant of the argument matrix. Recall that the eigenvalues of A are the same as the roots of $\text{char}_A(s)$. We now state the Cayley–Hamilton theorem.

Theorem 2.3. *Given a square matrix A the following matrix equation is satisfied*

$$A^n + a_{n-1}A^{n-1} + a_{n-2}A^{n-2} + \cdots + a_0 I = 0,$$

where a_k denote the scalar coefficients of the characteristic polynomial of A.

That is the Cayley–Hamilton theorem says a matrix satisfies its own characteristic equation. In shorthand notation we write

$$\text{char}_A(A) = 0.$$

We will not prove this result here but instead illustrate the idea behind the proof, using the example matrices considered above.

Examples:

First consider the case of a diagonalizable matrix A as in (2.3). Then it follows that

$$\text{char}_A(A) = T\text{char}_A(\Lambda)T^{-1} = T \begin{bmatrix} \text{char}_A(\lambda_1) & & 0 \\ & \ddots & \\ 0 & & \text{char}_A(\lambda_n) \end{bmatrix} T^{-1}.$$

Now by definition each of the eigenvalues is a root of $\text{char}_A(s)$, and so we see in this case $\text{char}_A(A) = 0$.

Next we turn to the case of a Jordan block (2.4). Clearly in this case A has n identical eigenvalues, and has characteristic polynomial

$$\text{char}_A(s) = (s - \lambda)^n .$$

Now it is easy to see that $\text{char}_A(A) = 0$ since

$$\text{char}_A(A) = (A - \lambda I)^n = N^n = 0,$$

where N is the nilpotent matrix defined in the earlier examples.

The general case of the Cayley–Hamilton theorem can be proved using the ideas from these examples and the Jordan decomposition, and you are asked to do this in the exercises.   □

The significance of the Cayley–Hamilton theorem for our purposes is that it says the matrix A^n is a linear combination of the matrix set

$\{A^{n-1}, A^{n-2}, \ldots, A, I\}$. Namely,

$$A^n \in \text{span}\{A^{n-1}, A^{n-2}, \ldots, A, I\}.$$

More generally we have the next proposition.

Proposition 2.4. *Suppose* $k \geq n$. *Then there exist scalar constants* $\alpha_0, \ldots, \alpha_{n-1}$ *satisfying*

$$A^k = \alpha_0 I + \alpha_1 A + \cdots + \alpha_{n-1} A^{n-1}$$

We now return for a moment to the matrix exponential and have the following result.

Lemma 2.5. *There exist scalar functions* $\phi_0(t), \ldots, \phi_{n-1}(t)$ *such that*

$$e^{At} = \phi_0(t)I + \phi_1(t)A + \cdots + \phi_{n-1}(t)A^{n-1} ,$$

for every $t \geq 0$.

The result says that the time dependent matrix exponential e^{At} can be written as a finite sum, where the time dependence is isolated in the *scalar* functions $\phi_k(t)$. The result is easily proved by observing that, for each $t \geq 0$, the matrix exponential has the expansion

$$e^{At} = I + At + \frac{(At)^2}{2!} + \frac{(At)^3}{3!} + \cdots .$$

The result then follows by expanding A^k, for $k \geq n$, using Proposition 2.4.

We are ready for the first step in the proof of our main result, which is to prove that the reachable states are a subset of the image of the controllability matrix.

Lemma 2.6. *The set of reachable states* $\mathcal{R}_t$ *is a subset of the controllability subspace* $\mathcal{C}_{AB}$.

Proof. Fix $t > 0$ and choose any reachable state $\xi \in \mathcal{R}_t$. It is sufficient to show that $\xi \in \mathcal{C}_{AB}$. Since $\xi \in \mathcal{R}_t$ there exists an input function u such that

$$\xi = \int_0^t e^{A(t-\tau)} Bu(\tau) d\tau .$$

Now substitute the expansion for the matrix exponential from Lemma 2.5 to get

$$\xi = \int_0^t \phi_0(t-\tau) Bu(\tau) d\tau + \cdots + A^{n-1} \int_0^t \phi_{n-1}(t-\tau) Bu(\tau) d\tau .$$

Writing this as a product we get

$$\xi = \begin{bmatrix} B & AB & \cdots & A^{n-1}B \end{bmatrix} \begin{bmatrix} \int_0^t \phi_0(t-\tau)u(\tau)d\tau \\ \vdots \\ \int_0^t \phi_{n-1}(t-\tau)u(\tau)d\tau \end{bmatrix} .$$

The rightmost factor is a vector in $\mathbb{R}^{nm}$, and so we see that ξ is in the image of the controllability matrix. ∎

Next in our plan is to show that $\mathcal{C}_{AB}$ is contained in the image of W_t. Again we require some preliminaries. Given a subspace S we define its *orthogonal complement* $S^\perp$ by

$$S^\perp = \{x \in \mathbb{R}^n : x^*y = 0, \text{ for all } y \in S\}.$$

It is not hard to show that the dimension of $S^\perp$ is n minus the dimension of S. We have the following elementary result which can be proved as an exercise.

Proposition 2.7. *Given any matrix W the subspace equality*

$$(\mathrm{Im}W)^\perp = \mathrm{Ker}\, W^* \quad holds.$$

We now prove the next step in the demonstration of Theorem 2.2.

Lemma 2.8. *The controllability subspace satisfies*

$$\mathcal{C}_{AB} \subset \mathrm{Im}W_t .$$

Proof. From Exercise 2.6 the above inclusion is equivalent to

$$\mathcal{C}_{AB}^\perp \supset (\mathrm{Im}W_t)^\perp.$$

We focus on showing this equivalent condition. From Proposition 2.7 we have that

$$(\mathrm{Im}W_t)^\perp = \mathrm{Ker}\, W_t^* = \mathrm{Ker}\, W_t ,$$

where the latter identity is true since the controllability gramian is symmetric. So we need to show that if $\xi \in \mathrm{Ker}\, W_t$ then $\xi \in \mathcal{C}_{AB}^\perp$.

Let $\xi \in \mathrm{Ker}\, W_t$ and thus we have

$$\xi^* W_t \xi = 0 .$$

Applying this to the definition of W_t we have

$$0 = \xi^* \left(\int_0^t e^{A\tau} BB^* e^{A^*\tau} d\tau \right) \xi = \int_0^t (\xi^* e^{A\tau} B)(B^* e^{A^*\tau} \xi) d\tau .$$

Let $y(\tau) = B^* e^{A^*\tau} \xi$ and the last equation says

$$\int_0^t y^*(\tau) y(\tau) d\tau = 0 .$$

Since the integrand above is non-negative, it follows that

$$y^*(\tau) = \xi^* e^{A\tau} B = 0$$

for each $0 \leq \tau \leq t$. Hence we have that the right-sided derivatives at zero satisfy

$$\left. \frac{d^k y^*}{d\tau^k} \right|_{\tau=0} = 0$$

for all $k \geq 0$. Now

$$\left. \frac{d^k y^*}{d\tau^k} \right|_{\tau=0} = \xi^* A^k B \, ,$$

and so it follows that

$$\xi^* \begin{bmatrix} B & AB & \cdots & A^{n-1}B \end{bmatrix} = 0,$$

where we have written n vector equations as a single matrix equation. The latter equation says that the vector ξ is orthogonal to all vectors in the image of the controllability matrix. This means ξ is in $\mathcal{C}_{AB}^\perp$ as required. $\blacksquare$

To finish our proof of Theorem 2.2 we complete the planned chain of containments proving that any element in the image of the gramian is necessarily a reachable state.

Lemma 2.9. *The image* $\mathrm{Im}W_t \subset \mathcal{R}_t$, *where* $\mathcal{R}_t$ *is the set of reachable states.*

Proof. Select any time $t > 0$ and $\xi \in \mathrm{Im}W_t$. Then by definition there exists υ in $\mathbb{R}^m$ so that

$$\xi = W_t \upsilon \, .$$

Now define

$$u(\tau) = B^* e^{A^*(t-\tau)} \upsilon, \quad \text{for } 0 \leq \tau \leq t \, .$$

Then the solution to $\dot{x} = Ax + Bu$, $x(0) = 0$ at time t is

$$x(t) = \int_0^t e^{A(t-\tau)} Bu(\tau)d\tau = \int_0^t e^{A(t-\tau)} BB^* e^{A^*(t-\tau)} \upsilon d\tau \qquad (2.6)$$

$$= \int_0^t e^{A\tau} BB^* e^{A^*\tau} d\tau \, \upsilon = W_t \upsilon = \xi \, . \qquad (2.7)$$

By definition this means that ξ is indeed in the set of reachable states. $\blacksquare$

Thus we have successfully proved the theorem of this section, which says that $\mathcal{R}_t = \mathcal{C}_{AB} = \mathrm{Im}W_t$. Since $\mathcal{C}_{AB}$ is the image of the controllability matrix this gives us a simple way to compute the reachable states of the system. Also notice that in the proof of Lemma 2.9 we constructed an explicit input u for reaching a given state in the set of reachable states. Summarizing, we have shown precisely which states are reachable and how to reach them.

2.2.2   Properties of controllability

Recall that we said the pair (A, B) is controllable if $\mathcal{C}_{AB}$ has maximum dimension n; in other words, all states are reachable. This is an important situation and we now investigate this special case further. Let us now return to the full state space system of (2.1), which is

$$\dot{x}(t) = Ax(t) + Bu(t), \quad x(0) = x_0$$
$$y(t) = Cx(t) + Du(t).$$

This set of equations specifies a relationship between the input u and the output y. It is natural to ask whether this is the unique set of equations of this form which relate u to y. The answer is, on the contrary, there are many state space systems which provide the same relationship. We elaborate on this point later, but for the moment we will exhibit a family of state space systems obtained from the given one by a change of coordinates in the state. By defining

$$\tilde{x}(t) = Tx(t),$$

where $T \in \mathbb{R}^{n \times n}$ is a nonsingular matrix, we have

$$\dot{\tilde{x}}(t) = T\dot{x}(t) = T(Ax(t) + Bu(t))$$
$$= TAT^{-1}\tilde{x}(t) + TBu(t) ,$$

which has the initial condition $\tilde{x}(0) = Tx(0)$. Similarly we have that

$$y(t) = Cx(t) + Du(t) = CT^{-1}\tilde{x}(t) + Du(t) .$$

So we see that a coordinate transformation defines a new state and thus a new *realization* for the same input–output relationship. We will use the term *state transformation* to refer to both the matrix T and to the mapping

$$(A, B, C, D) \mapsto (TAT^{-1}, TB, CT^{-1}, D)$$

it induces from one realization to another. Notice that the A matrix is transformed by *similarity*, whereas D is unaffected by a state transformation.

We have the first property of controllability now stated, which says that systems remain controllable under state transformations.

Proposition 2.10. *The pair (A, B) is controllable, if and only if the matrix pair (TAT^{-1}, TB) is controllable.*

Proof. The controllability matrix of (TAT^{-1}, TB) is

$$\begin{bmatrix} TB & (TAT^{-1})TB & \cdots & (TAT^{-1})^{n-1}TB \end{bmatrix} = T \begin{bmatrix} B & AB & \cdots & A^{n-1}B \end{bmatrix}$$

Since T is a state transformation the rank of this matrix is clearly equal to that of

$$\begin{bmatrix} B & AB & \cdots & A^{n-1}B \end{bmatrix} ,$$

which of course is the controllability matrix associated with the matrix pair (A, B). Thus either of these controllability matrices has rank n exactly when the other does. ∎

In the language of subspaces the above result says

$$\mathcal{C}_{\tilde{A}\tilde{B}} = T\mathcal{C}_{AB},$$

where $\tilde{A} = TAT^{-1}$ and $\tilde{B} = TB$. This should be readily clear from the proof and the definition of the controllable subspace.

Recall the definition of a subspace being *invariant* to a matrix: suppose $\mathcal{W}$ is a subspace of $\mathbb{R}^n$ and A is an $n \times n$ matrix, then we say $\mathcal{W}$ is A-invariant if

$$A\mathcal{W} \subset \mathcal{W}.$$

Also recall that if $\mathcal{W}$ is r dimensional and A-invariant, then there exists a state transformation T such that

$$TAT^{-1} = \begin{bmatrix} \tilde{A}_{11} & \tilde{A}_{12} \\ 0 & \tilde{A}_{21} \end{bmatrix}$$

where $\tilde{A}_{11} \in \mathbb{R}^{r \times r}$ and

$$T\mathcal{W} = \mathrm{Im}\begin{bmatrix} I_r \\ 0 \end{bmatrix}.$$

We use this definition and the latter fact in the next two very important results.

Proposition 2.11. *Suppose $A \in \mathbb{R}^{n \times n}$ and $B \in \mathbb{R}^{n \times m}$. Then the controllable subspace $\mathcal{C}_{AB}$ is A-invariant.*

Proof. We need to show that $A\mathcal{C}_{AB} \subset \mathcal{C}_{AB}$. Using the definition of $\mathcal{C}_{AB}$ we have

$$AC_{AB} = A\,\mathrm{Im}\begin{bmatrix} B & AB & \cdots & A^{n-1}B \end{bmatrix}$$
$$= \mathrm{Im}\begin{bmatrix} AB & A^2B & \cdots & A^nB \end{bmatrix}.$$

Now by the Cayley–Hamilton theorem we see that

$$\mathrm{Im}\begin{bmatrix} AB & A^2B & \cdots & A^nB \end{bmatrix} \subset \mathrm{Im}\begin{bmatrix} B & AB & \cdots & A^{n-1}B \end{bmatrix}$$

and therefore the result follows.   ∎

An immediate consequence of this proposition and the preceding discussion is the following.

Theorem 2.12. *Given a matrix pair (A, B) with the $\dim(\mathcal{C}_{AB}) = r$, there exists a state transformation T such that*

(a) The transformed pair has the form

$$TAT^{-1} = \begin{bmatrix} \tilde{A}_{11} & \tilde{A}_{12} \\ 0 & \tilde{A}_{22} \end{bmatrix} \quad and \quad TB = \begin{bmatrix} \tilde{B}_1 \\ 0 \end{bmatrix},$$

where $\tilde{A}_{11} \in \mathbb{R}^{r \times r}$ and $\tilde{B}_1 \in \mathbb{R}^{r \times m}$.

(b) The transformed controllable subspace is

$$TC_{AB} = \text{Im} \begin{bmatrix} I_r \\ 0 \end{bmatrix} \subset \mathbb{R}^n.$$

(c) The pair $(\tilde{A}_{11}, \tilde{B}_1)$ is controllable.

The first two parts of the theorem are essentially complete since we already know that $\mathcal{C}_{AB}$ is invariant to A, and that $\text{Im} B \subset \mathcal{C}_{AB}$ holds by definition. Part (c) follows by evaluating the rank of the controllability matrix for the transformed system. We leave this as an exercise. When a matrix pair is in the form given in (a) of the theorem, with (c) satisfied we say it is in *controllability form*.

We now summarize the procedure for transforming a pair (A, B) into this form.

General procedure for controllability form

(a) Find a basis $\{v_1, \ldots, v_r\}$ in $\mathbb{R}^n$ for $\mathcal{C}_{AB}$.

(b) Augment this basis to get $\{v_1, \ldots, v_n\}$ a complete basis for $\mathbb{R}^n$.

(c) Define $T^{-1} = [v_1, \ldots, v_n]$ and set

$$\tilde{A} = TAT^{-1} \quad \text{and} \quad \tilde{B} = TB.$$

(d) Partition these matrices as

$$\tilde{A} = \begin{bmatrix} \tilde{A}_{11} & \tilde{A}_{12} \\ 0 & \tilde{A}_{22} \end{bmatrix} \quad \text{and} \quad \tilde{B} = \begin{bmatrix} \tilde{B}_1 \\ 0 \end{bmatrix}.$$

It is important to understand exactly why the above procedure works, as it collects a number of key ideas. We now move on to examine the implications of the controllability form.

2.2.3 Stabilizability

When a system is put in controllability form, as the name suggests, the controllable and uncontrollable parts of the system are isolated. We can see this explicitly. Consider the state equation

$$\dot{x}(t) = Ax(t) + Bu(t), \quad x(0) = x_0 .$$

Define a new state $\tilde{x}(t) = Tx(t)$, where T is a state transformation that takes the system to controllability form. Thus the new state equations are

$$\dot{\tilde{x}}(t) = \begin{bmatrix} \dot{\tilde{x}}_1(t) \\ \dot{\tilde{x}}_2(t) \end{bmatrix} = \begin{bmatrix} \tilde{A}_{11} & \tilde{A}_{12} \\ 0 & \tilde{A}_{22} \end{bmatrix} \begin{bmatrix} \tilde{x}_1(t) \\ \tilde{x}_2(t) \end{bmatrix} + \begin{bmatrix} \tilde{B}_1 \\ 0 \end{bmatrix} u(t),$$

with the initial condition given by

$$\tilde{x}(0) = \begin{bmatrix} \tilde{x}_1(0) \\ \tilde{x}_2(0) \end{bmatrix} = Tx(0).$$

Notice that $\tilde{x}_2(t)$ evolves according to the equation

$$\dot{\tilde{x}}_2(t) = \tilde{A}_{22}\tilde{x}_2(t) \ ,$$

which only depends on the initial condition $\tilde{x}_2(0)$, and is entirely independent of both $\tilde{x}_1(0)$ and the input u. Therefore $\tilde{x}_2$ is said to be the *uncontrollable* part of the system. Now suppose that the matrix $\tilde{A}_{22}$ is not Hurwitz, then there exists an initial condition $\tilde{x}_2(0)$ such that

$$\tilde{x}_2(t) \ \ \text{does not tend to zero as } t \to \infty \ .$$

Namely, there exists an initial condition $x(0)$ so that no matter what u is, the state $x(t)$ does not tend to zero as the time t tends to infinity. Conversely, it turns out that if $\tilde{A}_{22}$ above is Hurwitz, then for every initial condition $x(0)$, there exists an input u such that $x(t)$ tends asymptotically to zero. This latter fact will become clear to us later, and motivates the following definition.

Definition 2.13. *Suppose (A, B) is a matrix pair. If $\tilde{A}_{22}$ in a controllability form is Hurwitz, then we say the pair is stabilizable.*

Thus a matrix pair is said to be stabilizable if for every initial condition it is possible to find an input that asymptotically steers the state $x(t)$ to the origin; namely, this is a weak form of stability for a controlled system. If the matrix pair is not stabilizable there exists an initial condition so that irrespective of $u(t)$, the state function $x(t)$ does not tend to the origin.

So far the only way we have to determine stabilizability of a pair (A, B) is to convert the system to controllability form. This is both inconvenient and unnecessary, and our goal is now to develop a more sophisticated test. To do this we first develop an alternate method for checking controllability, the Popov–Belevitch–Hautus (PBH) test. The test is stated below.

Theorem 2.14. *The pair (A, B) is controllable if and only if, for each $\lambda \in \mathbb{C}$, the rank condition*

$$\operatorname{rank}\begin{bmatrix} A - \lambda I & B \end{bmatrix} = n \quad \text{holds} \ .$$

Proof. First notice that this condition need only be satisfied at the eigenvalues of A, since the rank of $A - \lambda I$ is n otherwise.

We first prove "only if," by a contrapositive argument. Suppose λ is an eigenvalue of A and

$$\operatorname{rank}\begin{bmatrix} A - \lambda I & B \end{bmatrix} < n \ .$$

Then there exists a nonzero vector x such that

$$x^* \begin{bmatrix} A - \lambda I & B \end{bmatrix} = 0 \ .$$

Thus we have that both

$$x^* A = \lambda x^* \quad \text{and} \quad x^* B = 0 \quad \text{hold}.$$

From the former it follows that

$$x^* A^2 = \lambda^2 x^*$$

or more generally $x^* A^k = \lambda^k x^*$ for each $k \geq 1$. Hence

$$x^* \begin{bmatrix} B & AB & \cdots & A^{n-1}B \end{bmatrix} = 0 ,$$

and we see (A, B) is not controllable.

Now we prove "if," again using the contrapositive. Suppose (A, B) is not controllable, then by Theorem 2.12 there exists a state transformation such that

$$\tilde{A} = TAT^{-1} = \begin{bmatrix} \tilde{A}_{11} & \tilde{A}_{21} \\ 0 & \tilde{A}_{22} \end{bmatrix} \text{ and } \begin{bmatrix} \tilde{B}_1 \\ 0 \end{bmatrix} ,$$

where the dimension of the square matrix $\tilde{A}_{22}$ is nonzero. Let λ be an eigenvalue of $\tilde{A}_{22}$ and we see that

$$\text{rank} \begin{bmatrix} \tilde{A} - \lambda I & \tilde{B} \end{bmatrix} < n .$$

Therefore $\begin{bmatrix} A - \lambda I & B \end{bmatrix}$ has rank less than n since

$$\begin{bmatrix} A - \lambda I & B \end{bmatrix} = T^{-1} \begin{bmatrix} \tilde{A} - \lambda I & \tilde{B} \end{bmatrix} \begin{bmatrix} T & 0 \\ 0 & I \end{bmatrix} .$$

This completes the contrapositive argument.

∎

In view of this theorem we say that an eigenvalue λ of matrix A is *controllable* if

$$\text{rank} \begin{bmatrix} A - \lambda I & B \end{bmatrix} = n ,$$

and so every eigenvalue of A is controllable if and only if (A, B) is a controllable pair. Thinking back on matrix $\tilde{A}$ in the above proof, we see that $\tilde{A}_{11}$ corresponds to the controllable states, whereas all the eigenvalues of $\tilde{A}_{22}$ are uncontrollable. Note that the eigenvalues of $\tilde{A}_{11}$ and $\tilde{A}_{22}$ need not be distinct. From this discussion and the proof of Theorem 2.14 the next result follows readily.

Corollary 2.15. *The matrix pair (A, B) is stabilizable if and only if the condition*

$$rank \begin{bmatrix} A - \lambda I & B \end{bmatrix} = n \quad \text{holds for all } \lambda \in \bar{\mathbb{C}}^+ .$$

This corollary states that one need only check the above rank condition at the unstable eigenvalues of A, that is those in the closed right half-plane $\bar{\mathbb{C}}^+$. This is a much simpler task than determining stabilizability from the definition.

2.2.4   Controllability from a single input

This section examines the special case of a pair (A, B) where $B \in \mathbb{R}^{n \times 1}$. That is our state space system has only a single scalar input. This investigation will furnish us with a new system realization which has important applications.

A matrix of the form

$$\begin{bmatrix} 0 & 1 & & & 0 \\ & \ddots & \ddots & \\ 0 & & 0 & 1 \\ -a_0 & -a_1 & & & -a_{n-1} \end{bmatrix}$$

is called a *companion matrix*, and has the useful property that its characteristic polynomial is

$$s^n + a_{n-1}s^{n-1} + \cdots + a_0 .$$

The latter fact is easily verified. We now prove an important theorem which relates single-input systems to companion matrices.

Theorem 2.16. *Suppose* (A, B) *is a controllable pair, and B is a column vector. Then there exists a state transformation T such that*

$$TAT^{-1} = \begin{bmatrix} 0 & 1 & & & 0 \\ & \ddots & \ddots & \\ 0 & & 0 & 1 \\ -a_0 & -a_1 & & & -a_{n-1} \end{bmatrix} \quad and \quad TB = \begin{bmatrix} 0 \\ \vdots \\ 0 \\ 1 \end{bmatrix} .$$

The theorem states that if a single-input system is controllable, then it can be transformed to the special realization above, where A is in companion form and B is the nth standard basis vector of $\mathbb{R}^n$. This is called the *controllable canonical realization*. To prove this result we will again use the Cayley–Hamilton theorem.

Proof. Referring to the characteristic polynomial of A and invoking the Cayley–Hamilton theorem we get

$$A^n + a_{n-1}A^{n-1} + \cdots + a_0 I = 0 .$$

Post-multiplying this by B gives

$$A^n B = -a_0 B - \cdots - a_{n-1}A^{n-1}B .$$

From this equation, keeping in mind that B is a column vector, it is easy to verify that

$$A \begin{bmatrix} B & \cdots & A^{n-1}B \end{bmatrix} = \underbrace{\begin{bmatrix} B & AB & \cdots & A^{n-1}B \end{bmatrix}}_{P} \underbrace{\begin{bmatrix} 0 & & 0 & -a_0 \\ 1 & \ddots & & -a_1 \\ & \ddots & 0 & \vdots \\ 0 & & 1 & -a_{n-1} \end{bmatrix}}_{M} .$$

Note that P is both square and nonsingular. From the last equation we have

$$P^{-1}AP = M .$$

Also since

$$B = \begin{bmatrix} B & AB & \cdots & A^{n-1}B \end{bmatrix} \begin{bmatrix} 1 \\ 0 \\ \vdots \\ 0 \end{bmatrix} ,$$

we see that

$$P^{-1}B = \begin{bmatrix} 1 \\ 0 \\ \vdots \\ 0 \end{bmatrix} .$$

Let us hold these forms in abeyance for the moment, and introduce two new matrices.

$$\tilde{A} := \begin{bmatrix} 0 & 1 & & 0 \\ & \ddots & \ddots & \\ 0 & & 0 & 1 \\ -a_0 & -a_1 & & -a_{n-1} \end{bmatrix} \quad \text{and} \quad \tilde{B} = \begin{bmatrix} 0 \\ \vdots \\ 0 \\ 1 \end{bmatrix} .$$

These are the matrices to which we want to transform. Define

$$\tilde{P} := \begin{bmatrix} \tilde{B} & \tilde{A}\tilde{B} & \cdots & \tilde{A}^{n-1}\tilde{B} \end{bmatrix} .$$

Now $(\tilde{A}, \tilde{B})$ is controllable; to see this simply write out the controllability matrix. Therefore since $\tilde{A}$ has the same characteristic polynomial as A, we can use the same argument already followed to show that

$$\tilde{P}^{-1}\tilde{A}\tilde{P} = M \quad \text{and} \quad \tilde{P}^{-1}\tilde{B} = \begin{bmatrix} 1 \\ 0 \\ \vdots \\ 0 \end{bmatrix} .$$

Returning to our expressions earlier we have that

$$P^{-1}AP = \tilde{P}^{-1}\tilde{A}\tilde{P} \quad \text{and} \quad P^{-1}B = \tilde{P}^{-1}\tilde{B}.$$

Set $T = \tilde{P}P^{-1}$ and we arrive at the desired result of

$$TAT^{-1} = \tilde{A} \quad \text{and} \quad TB = \tilde{B} .$$

■

Realize that the proof above provides us with an explicit procedure for putting a matrix pair in the controllable canonical form. The transformation given here will have a direct role in the next section, where we move on to one of the basic types of closed-loop control.

2.3   Eigenvalue assignment

In this section we consider a state space problem associated with a special type of closed-loop control called state feedback. Take the state equation

$$\dot{x}(t) = Ax(t) + Bu(t), \quad x(0) = x_0$$

and suppose we use the feedback law $u(t) = Fx(t)$ where F is a fixed matrix in $\mathbb{R}^{m \times n}$. Then the state equation becomes

$$\dot{x}(t) = (A + BF)x(t), \quad x(0) = x_0 .$$

This is an autonomous system whose behavior depends only on $A + BF$. Thus to control the system we might wish to stabilize it, meaning that $A + BF$ is Hurwitz. More generally we could try to specify the eigenvalues of $A + BF$ exactly by selecting F, thus achieving certain dynamical characteristics. The question we answer in this section is, when are the eigenvalues of $A + BF$ freely assignable through choice of F? We answer this by first treating the single-input case, and then tackling the more challenging multi-input case.

2.3.1   Single-input case

Here we consider the case where $B \in \mathbb{R}^{n \times 1}$. Suppose (A, B) is controllable; then by Theorem 2.16 we know there exists a state transformation such that

$$\tilde{A} = \begin{bmatrix} 0 & I \\ -a_0 & [-a_1 \ \cdots \ -a_{n-1}] \end{bmatrix} \quad \text{and} \quad \tilde{B} = \begin{bmatrix} 0 \\ \vdots \\ 0 \\ 1 \end{bmatrix} .$$

Let $\mathcal{E} = \{\lambda_0, \ldots, \lambda_{n-1}\}$ be the desired set of eigenvalues for $A + BF$. Since the matrix $A + BF$ is real we necessarily restrict $\mathcal{E}$ to be a set which is symmetric with respect to the real line; explicitly, if λ is in $\mathcal{E}$, then so is the complex conjugate of λ. Next let $\tilde{F} = \begin{bmatrix} \tilde{F}_0 & \cdots & \tilde{F}_{n-1} \end{bmatrix}$ where each $\tilde{F}_k$ is a real scalar to be chosen. Then

$$\tilde{A} + \tilde{B}\tilde{F} = \begin{bmatrix} 0 & & I & \\ -a_0 + \tilde{F}_0 & [-a_1 + \tilde{F}_1 & \cdots & -a_{n-1} + \tilde{F}_{n-1}] \end{bmatrix}.$$

Now this matrix is in companion form and so its characteristic polynomial is

$$s^n + (a_{n-1} - \tilde{F}_{n-1})s^{n-1} + \cdots + (a_0 - \tilde{F}_0).$$

Clearly by appropriate choice of the scalars $\tilde{F}_k$ we can arrange for the characteristic polynomial to be equal to

$$(s - \lambda_0)(s - \lambda_1) \cdots (s - \lambda_{n-1}),$$

which is a polynomial with *real* coefficient since $\mathcal{E}$ is symmetric. Note that the choice of these scalars $\tilde{F}_k$ is unique. Thus we have assigned the eigenvalues of $\tilde{A} + \tilde{B}\tilde{F}$ to be the set $\mathcal{E}$. Recalling that $\tilde{A}$ is related to A by a state transformation we have

$$T(\tilde{A} + \tilde{B}\tilde{F})T^{-1} = A + B(\tilde{F}T^{-1}) =: A + BF,$$

where F is defined to by $\tilde{F}T^{-1}$. Therefore the eigenvalues of $A + BF$ have been successfully assigned. We have the following general result.

Theorem 2.17. *Suppose $B \in \mathbb{R}^{n \times 1}$. The eigenvalues of $A + BF$ can be arbitrarily assigned if and only if (A, B) is controllable.*

That controllability ensures the assignability of eigenvalues can be seen from the preceding discussion. We leave the "only if" part of the proof as an exercise. We are now ready to deal with the harder proof of the multi-input case.

2.3.2 Multi-input case

In this section we treat the case of eigenvalue assignment when $B \in \mathbb{R}^{n \times m}$ and $m \geq 1$; so far we have only looked at this problem with $m = 1$. As we showed for the single-input case, it turns out that (A, B) controllable means we can assign $A + BF$ freely. Also if (A, B) is not controllable, then some eigenvalues of $A + BF$ remain fixed no matter how we choose F.

The first step to proving the general multi-input result is the following key lemma.

Lemma 2.18. *Suppose (A, B) is controllable. If $B_1 \in \mathbb{R}^{n \times 1}$ is any nonzero column of B; then there exists an F_1 such that the matrix pair $(A + BF_1, B_1)$ is controllable.*

The proof of this lemma is rather involved, and so we show its application in the following major theorem first. This theorem is called the eigenvalue or pole placement theorem.

Theorem 2.19. *The eigenvalues of $A + BF$ are freely assignable, by choosing $F \in \mathbb{R}^{m \times n}$, if and only if the pair (A, B) is controllable.*

Proof. To show "if" start by choosing any nonzero column of B. Now by Lemma 2.18 there exists a matrix F_1 such that $(A+BF_1, B_1)$ is controllable.

Having chosen such an F_1 we can invoke Theorem 2.17 to ensure the existence of a matrix F_2 such that

$$(A + BF_1) + B_1 F_2$$

has any desired eigenvalue assignment. Finally using the definition of B_1, choose a matrix F such that $A + BF$ equals $(A + BF_1) + B_1 F_2$ above.

As with the single input case we leave the "only if" part of the proof as an exercise.

■

We remark here that unlike the single-input case, in the multi-input case the state feedback matrix F may not be unique. Having established that Lemma 2.18 is indeed worth proving, we are ready to proceed to its proof.

Proof of Lemma 2.18. The proof revolves around the artificial discrete time system

$$x_{k+1} = Ax_k + Bu_k, \text{ with the initial condition } x_0 = B_1, \qquad (2.8)$$

which we define for the sole purpose of simplifying our notation in this proof. Suppose we can prove that there exists an input sequence $\{u_0, \ldots, u_{n-2}\}$ such that the resulting state trajectory

$$\{x_0, \ldots, x_{n-1}\} \quad \text{spans} \quad \mathbb{R}^n .$$

Then let the matrix F_1 be a solution to

$$F_1 \begin{bmatrix} x_0 & x_1 & \cdots & x_{n-1} \end{bmatrix} = \begin{bmatrix} u_0 & u_1 & \cdots & u_{n-1} \end{bmatrix} , \qquad (2.9)$$

where u_{n-1} is any arbitrary vector. It is then easy to see that $(A+BF_1, B_1)$ is controllable: observe from (2.8) and (2.9) that

$$x_{k+1} = (A + BF_1)x_k \quad x_0 = B_1 ,$$

and therefore

$$\begin{bmatrix} x_0 & \cdots & x_{n-1} \end{bmatrix} = \begin{bmatrix} B_1 & (A + BF_1)B_1 & \cdots & (A + BF_1)^{n-1}B_1 \end{bmatrix} .$$

Since the left-hand side has full rank so must the right-hand side, which establishes the claim.

Thus it is sufficient to show there exists input sequence $\{u_0, \ldots, u_{n-2}\}$ such that $\{x_0, \ldots, x_{n-1}\}$ are linearly independent. We use induction. Suppose that $k < n$ and $\{u_0, \ldots, u_{k-1}\}$ results in the linearly independent sequence

$$\{x_0, \ldots, x_k\}.$$

Define $\mathcal{X} = \mathrm{Im}\begin{bmatrix} x_0 & \cdots & x_k \end{bmatrix} \subset \mathbb{R}^n$, and therefore we need to show that there exists a new input vector u_k such that x_{k+1} is not in the subspace $\mathcal{X}$. That is there exists u_k such that

$$Ax_k + Bu_k \notin \mathcal{X} . \tag{2.10}$$

We will prove this by contradiction.

Suppose on the contrary that for any choice of the kth input $u_k \in \mathbb{R}^n$ we have

$$Ax_k + Bu_k \in \mathcal{X} .$$

Then in particular Ax_k is in $\mathcal{X}$ for $u_k = 0$, and since $\mathcal{X}$ is a subspace we have that

$$Bu_k \in \mathcal{X}$$

for all $u_k \in \mathbb{R}^m$. This latter condition means $\mathrm{Im}\, B \subset \mathcal{X}$.

We now show that $\mathcal{X}$ must be A-invariant, or equivalently

$$Ax_i \in \mathcal{X} \quad \text{for} \quad 0 \leq i \leq k .$$

This holds for $i = k$ from our assumption above. For $i < k$ we note that

$$Ax_i = x_{i+1} - Bu_i .$$

Observe that this means the right-hand side is in $\mathcal{X}$, since $x_{i+1} \in \mathcal{X}$ and $Bu_i \in \mathcal{X}$ by assumption. Therefore $\mathcal{X}$ is A-invariant. Now this means $\mathrm{Im}\, AB \subset A\mathcal{X} \subset \mathcal{X}$ and more generally that

$$\mathrm{Im}\, A^i B \subset \mathcal{X} \quad \text{for} \quad 0 \leq i \leq n - 1 .$$

This implies

$$\mathrm{Im}\begin{bmatrix} B & AB & \cdots & A^{n-1}B \end{bmatrix} \subset \mathcal{X} .$$

Since $\mathcal{X}$ is of dimension less than n we have a contradiction since (A, B) is by hypothesis controllable. Therefore (2.10) must be true. ∎

This proof tells us how to explicitly construct F; however, much better numerical methods exist. A complete discussion of numerically reliable techniques would take us too far afield.

As a final question, we inquire what happens when (A, B) is not controllable. In the exercises you will see that the uncontrollable eigenvalues (i.e. the eigenvalues of $\tilde{A}_{22}$ in the controllability form) always remain fixed, no matter which F is used for feedback; however, the remaining eigenvalues can indeed be freely assigned. As an important special case, we note that if (A, B) is stabilizable, then $A + BF$ can be made Hurwitz by appropriate choice of F. This follows up on the comments made regarding the definition of stabilizability: indeed we see in the exercises that the state of a stabilizable system can be driven to zero by an appropriate choice of input.

This concludes our study of controllability.

2.4   Observability

In the last section we studied a special case of the system equations presented in (2.1), which had an input but no output. We will now in contrast consider a system with an output but no input; the system is

$$\dot{x}(t) = Ax(t), \text{ with } x(0) = x_0$$
$$y(t) = Cx(t).$$

This system has no input, and its solution depends entirely on the initial condition $x(0) = x_0$. The solution of this equation is clearly

$$y(t) = Ce^{At}x_0, \text{ for } t \geq 0.$$

We regard the function $y(t)$ as an output and will now focus on the question of whether we can determine the value of x_0 by observing the variable $y(t)$ over a time interval.

2.4.1   The unobservable subspace

In this section we take the perspective of an onlooker who is only able to measure or observe the output function $y(t)$ over a finite interval $[0, T]$, but wants to find the value of x_0. We can regard the expression for y as a map Ψ from the state space $\mathbb{R}^n$ to the vector space $\mathcal{F}(\mathbb{R}, \mathbb{R}^p)$ of functions that are $\mathbb{R}^p$-valued; that is $\Psi : \mathbb{R}^n \rightarrow \mathcal{F}(\mathbb{R}, \mathbb{R}^p)$ and is defined by

$$x_0 \overset{\Psi}{\mapsto} Ce^{At}x_0 .$$

Here we will only consider the case where t is in the interval $[0, T]$. So we have that

$$y = \Psi x_0$$

and our task is, if possible, to determine x_0 given the function y. Now this is possible exactly when there is only one solution to the above equation. When y is a given solution we can be sure that it is the only solution, if and only if the kernel of Ψ is zero. Namely,

$$\text{Ker}\, \Psi = 0 .$$

Thus if this condition is met we will be able to determine x_0, and if it is violated then the initial conditions cannot be unambiguously discerned by observing y. For this reason we say the matrix pair (C, A) is *observable* if $\text{Ker}\, \Psi = 0$. The following result gives us an explicit test for checking observability.

Theorem 2.20. *The kernel of* Ψ *is given by*

$$\operatorname{Ker}\Psi = \operatorname{Ker}C \cap \cdots \cap \operatorname{Ker}CA^{n-1} = \operatorname{Ker}\begin{bmatrix} C \\ CA \\ \vdots \\ CA^{n-1} \end{bmatrix}.$$

The subspace $\operatorname{Ker}\Psi$ is usually denoted $\mathcal{N}_{CA}$ and is called the *unobservable subspace* of the pair (C, A) for reasons we will soon see. Thus a system is observable when $\mathcal{N}_{CA} = 0$. The matrix on the right in the theorem is named the *observability matrix*; from the theorem, observability is equivalent to this matrix having full column rank.

Proof. We first show that $\operatorname{Ker}\Psi \subset \operatorname{Ker}C \cap \cdots \cap CA^{n-1}$. Let $x_0 \in \operatorname{Ker}\Psi$ and therefore by definition we know that

$$Ce^{At}x_0 = 0 \quad \text{for } t \ge 0.$$

It is straightforward to verify that for each $k \ge 0$ the following holds:

$$\frac{d^k}{dt^k}Ce^{At}x_0\bigg|_{t=0} = CA^k x_0.$$

But by the equation that precedes the above we see that the left-hand side must be zero, and so x_0 must be in the kernel of any matrix CA^k where k is non-negative.

To complete our proof we must show that $\operatorname{Ker}C \cap \cdots \cap \operatorname{Ker}CA^{n-1} \subset \operatorname{Ker}\Psi$. By Lemma 2.5 we know there exist scalar functions $\phi_k(t)$ such that

$$e^{At} = \phi_0(t)I + \cdots + \phi_{n-1}(t)A^{n-1}$$

for $t \ge 0$. From this it is clear that if x_0 is in $\operatorname{Ker}C \cap \cdots \cap \operatorname{Ker}CA^{n-1}$, then $Ce^{At}x_0 = 0$. ∎

Although observability and controllability have been motivated in very different ways they are intimately related algebraically.

Proposition 2.21. *The following are equivalent:*

(a) (C, A) is observable;

(b) (A^*, C^*) is controllable.

Proof. Condition (b) holds when the controllability matrix

$$P := \begin{bmatrix} C^* & A^*C^* & \cdots & (A^*)^{n-1}C^* \end{bmatrix}$$

has the full rank of n. This is true if and only if the transpose matrix P^* has rank n. Clearly, P^* is exactly the observability matrix associated with (C, A). ∎

Having established this algebraic link between the concepts of controllability and observability, we can now easily find analogous versions of all our results on controllability for observability. These analogs are listed in the following proposition, and their proofs are left as an exercise and review.

Proposition 2.22.

(a) *The pair* (C, A) *is observable if and only if*

$$rank \begin{bmatrix} A - \lambda I \\ C \end{bmatrix} = n, \quad for\ all\ \lambda \in \mathbb{C} \ ;$$

(b) *the eigenvalues of* $A + LC$ *are freely assignable by choosing* L, *if and only if,* (C, A) *is observable;*

(c) *the unobservable subspace* $\mathcal{N}_{CA}$ *is* A-*invariant;*

(d) *there exists a state transformation* T *such that*

$$TAT^{-1} = \begin{bmatrix} \tilde{A}_{11} & 0 \\ \tilde{A}_{21} & \tilde{A}_{22} \end{bmatrix} \ ; \quad CT^{-1} = \begin{bmatrix} \tilde{C}_1 & 0 \end{bmatrix}$$

where $(\tilde{C}_1, \tilde{A}_{11})$ *is observable;*

(e) *the matrix* $\tilde{A}_{22}$ *in part (d) is Hurwitz, if and only if, the condition*

$$rank \begin{bmatrix} A - \lambda I \\ C \end{bmatrix} = n \quad for\ all\ \lambda \in \bar{\mathbb{C}}^+\ holds \ .$$

All these properties can be obtained from our work on controllability. Notice that in part (d) the following holds.

$$T\mathcal{N}_{CA} = \text{Im} \begin{bmatrix} 0 \\ I \end{bmatrix} \ .$$

We will call the decomposition of part (d) an *observability form*, because it explicitly isolates the invariant subspace $\mathcal{N}_{CA}$. Writing out the state equations in this form gives us particular insight:

$$\dot{\tilde{x}}_1(t) = \tilde{A}_{11}\tilde{x}_1(t),$$
$$\dot{\tilde{x}}_2(t) = \tilde{A}_{21}\tilde{x}_1(t) + \tilde{A}_{22}\tilde{x}_2(t),$$
$$y(t) = \tilde{C}_1\tilde{x}_1(t), \quad \text{with the initial condition} \quad \begin{bmatrix} \tilde{x}_1(0) \\ \tilde{x}_2(0) \end{bmatrix} = Tx.$$

From here it is clear that $\tilde{x}_1$ only depends on the initial condition $\tilde{x}_1(0)$, and is completely unaffected by $\tilde{x}_2$; therefore y is entirely independent of $\tilde{x}_2$. For this reason we say

- the vector $\tilde{x}_1$ contains the observable state;
- the vector $\tilde{x}_2$ contains the unobservable state.

Now suppose $\tilde{x}_1(0) = 0$. Then $\tilde{x}_1(t)$ is zero for all time,

$$\dot{\tilde{x}}_2(t) = \tilde{A}_{22}\tilde{x}_2(t),$$

and $y = 0$. If $\tilde{A}_{22}$ is not Hurwitz we have for some initial state $\tilde{x}_2(0)$, that y is zero for all time yet $\tilde{x}_2(t)$ does not tend to zero. In contrast if $\tilde{A}_{22}$ is Hurwitz we know that $\tilde{x}_2(t)$ will tend to zero.

Because this situation is important, it is worthwhile to define a term: when $\tilde{A}_{22}$ is Hurwitz we say that (C, A) is a *detectable* matrix pair. Part (e) of the proposition provides a way to check detectability via a PBH-type test.

Here we have introduced the notion of observability, motivated by the fact that we wish to determine $x(0)$ given the function y. We now turn to issues related to actually constructing a method for determining the state given the system output.

2.4.2 Observers

Frequently in control problems it is of interest to obtain an estimate of the state $x(t)$ based on past values of the output y. From our investigation in the last section we learned under what conditions $x(0)$ can be determined exactly by measuring y. If they are met, there are various approaches to accomplish this; we will investigate an important one in the exercises. Right now we focus on determining asymptotic estimates for $x(t)$.

We will pursue this goal using our full state space system

$$\dot{x}(t) = Ax(t) + Bu(t), \quad x(0) = x_0,$$
$$y(t) = Cx(t) + Du(t).$$

Our objective is to find an asymptotic approximation of $x(t)$ given the input u and the output y, but without knowledge of the initial condition $x(0)$.

There are a number of ways to address this problem, and the one we pursue here is using a dynamical system called an *observer*. The system equations for an observer are

$$\dot{w}(t) = Mw(t) + Ny(t) + Pu(t), \quad w(0) = w_0$$
$$\hat{x}(t) = Qw(t) + Ry(t) + Su(t). \tag{2.11}$$

So the inputs to the observer are u and y, and the output is $\hat{x}$. The key property that the observer must have is that the error

$$\hat{x}(t) - x(t) \xrightarrow{t \to \infty} 0,$$

for all initial conditions $x(0)$ and $w(0)$, and all system inputs u.

Theorem 2.23. *An observer exists if and only if (C, A) is detectable. Furthermore, in that case one such observer is given by*

$$\dot{\hat{x}}(t) = (A + LC)\hat{x}(t) - Ly(t) + (B + LD)u(t), \qquad (2.12)$$

where the matrix L is chosen such that $A + LC$ is Hurwitz.

Notice in the observer equation (2.12) that we have $\hat{x} = w$, so we have removed w for simplicity. An observer with this structure is called a full order Luenberger observer.

Proof. We first show the necessity of the detectability condition, by contrapositive. We will show that if (C, A) is not detectable, there always exist initial conditions $x(0)$ and $w(0)$, and an input $u(t)$, where the observer error does not decay to zero. In particular, we will select $u = 0$, $w(0) = 0$, and $x(0)$ to excite only the unobservable states, as is now explained.

Without loss of generality we will assume the system is already in observability form. For the general case, a state transformation must be added to the following argument; this is left as an exercise. With this assumption, and $u = 0$, we have the equations

$$\dot{x}_1(t) = A_{11}x_1(t)$$
$$\dot{x}_2(t) = A_{21}x_1(t) + A_{22}x_2(t)$$
$$y(t) = C_1x_1(t).$$

Notice that if (C, A) is not detectable, A_{22} is not a Hurwitz matrix. Therefore $x_2(0)$ can be chosen so that the solution to

$$\dot{x}_2(t) = A_{22}x_2(t)$$

does not tend to zero as $t \to \infty$. Now choosing $x_1(0) = 0$, it is clear that $x_1(t)$, and also $y(t)$, will be identically zero for all time, while $x_2(t)$ evolves according to the autonomous equation and does not tend to zero.

Now turning to a generic observer equation of the form (2.11), we see that if $w(0) = 0$, then $w(t)$ and therefore $\hat{x}(t)$ are identically zero. This contradicts the fact that the error $\hat{x}(t) - x(t)$ tends to zero, so we have shown necessity.

The proof of sufficiency is constructive. We know that if (C, A) is detectable, L can be chosen to make $A + LC$ a Hurwitz matrix. With this choice, we construct the observer in (2.12). Substituting the expression for y into (2.12) we have

$$\dot{\hat{x}}(t) = (A + LC)\hat{x}(t) - L(Cx(t) + Du(t)) + (B + LD)u(t)$$
$$= (A + LC)\hat{x}(t) - LCx(t) + Bu(t).$$

Now if we subtract the state equation from this, we obtain

$$\dot{\hat{x}}(t) - \dot{x}(t) = (A + LC)\hat{x}(t) - LCx(t) + Bu(t) - Ax(t) - Bu(t)$$
$$= (A + LC)(\hat{x}(t) - x(t)).$$

This means that the error $e = \hat{x} - x$ satisfies the autonomous differential equation

$$\dot{e}(t) = (A + LC)e(t).$$

Since $A + LC$ is a Hurwitz matrix, $e(t)$ tends to zero as required.   ∎

2.4.3 Observer-based Controllers

We will now briefly exhibit a first example of feedback control, combining the ideas of the previous sections. Feedback consists of connecting a *controller* of the form

$$\dot{x}_K(t) = A_K x_K(t) + B_K y(t)$$
$$u(t) = C_K x_K(t) + D_K y(t)$$

to the open-loop system (2.1), so that the combined system achieves some desired properties. A fundamental requirement is the stability of the resulting autonomous system. We have already seen that in the case where the state is available for feedback (i.e., when $y = x$), then a static control law of the form

$$u(t) = Fx(t)$$

can be used to achieve desired closed-loop eigenvalues, provided that (A, B) is controllable. Or if it is at least stabilizable, the closed loop can be made internally stable.

The question arises as to whether we can achieve similar properties for the general case, where the output contains only partial information on the state. The answer, not surprisingly, will combine controllability properties with observability of the state from the given output. We state the following result.

Proposition 2.24. *Given open-loop system (2.1), the controller*

$$\dot{\hat{x}}(t) = (A + LC + BF + LDF)\hat{x}(t) - Ly(t)$$
$$u(t) = F\hat{x}(t)$$

is such that the closed-loop system has eigenvalues exactly at the eigenvalues of the matrices $A + BF$ and $A + LC$.

Notice that given the choice of u, the first equation can be rewritten as

$$\dot{\hat{x}}(t) = (A + LC)\hat{x}(t) - Ly(t) + (B + LD)u(t)$$

and has exactly the structure of a Luenberger observer (2.12) for the state x. Given this observation, the expression for u can be interpreted as being analogous to the state feedback law considered above, except that the estimate $\hat{x}$ is used in place of the unavailable state.

Proof. Combining the open-loop system with the controller, and eliminating the variables u, y, leads to the combined equations

$$\begin{bmatrix} \dot{x}(t) \\ \dot{\hat{x}}(t) \end{bmatrix} = \begin{bmatrix} A & BF \\ -LC & A + LC + BF \end{bmatrix} \begin{bmatrix} x(t) \\ \hat{x}(t) \end{bmatrix}.$$

Now the state transformation

$$T = \begin{bmatrix} I & -I \\ 0 & I \end{bmatrix}$$

(which amounts to replacing x by $x - \hat{x}$ as a state variable) leads to

$$T \begin{bmatrix} A & BF \\ -LC & A + LC + BF \end{bmatrix} T^{-1} = \begin{bmatrix} A + LC & 0 \\ -LC & A + BF \end{bmatrix},$$

which exhibits the eigenvalue structure.   ∎

As a consequence of this result, we conclude that controllability and observability of (A, B, C, D) suffices to place the closed-loop eigenvalues at any desired locations, whereas stabilizability and detectability suffice to obtain an internally stabilizing feedback law. In addition, notice that the state-feedback and observer components appear in a decoupled fashion, so they can be designed independently; this property is commonly called the *separation* principle.

2.5   Minimal realizations

In the rest of the chapter we look at the state space system

$$\dot{x}(t) = Ax(t) + Bu(t), \quad x(0) = x_0$$
$$y(t) = Cx(t) + Du(t)$$

from the point of view of the relationship it establishes between inputs and outputs. We focus on the case where the system is initially at rest, with $x(0) = 0$, and see that in this case the system establishes a *map* between inputs and outputs, given by the formula

$$y(t) = \int_0^t Ce^{A(t-\tau)}Bu(\tau)d\tau + Du(t).$$

In this section take the standpoint that this input–output map is the primary object of study, and the state space system is merely a *realization* of this map, that is, a way to define such a map in terms of first-order differential equations. From this point of view, we say two realizations are *equivalent* when

$$\int_0^t Ce^{A(t-\tau)}Bu(\tau)d\tau + Du(t) = \int_0^t C_1 e^{A_1(t-\tau)}B_1 u(\tau)d\tau + D_1 u(\tau)$$

holds for all input functions u and all times $t \geq 0$. In particular we will be interested in finding, among all equivalent realizations, the one with the minimal *dynamic order*, which we define as the dimension n of the square matrix A.

We first characterize equivalence in the following result.

Lemma 2.25. *Two system realizations* (A, B, C, D) *and* (A_1, B_1, C_1, D_1) *are equivalent if and only if* $D = D_1$ *and*

$$Ce^{At}B = C_1 e^{A_1 t}B_1 \quad \text{holds for all } t \geq 0.$$

Proof. It is clear from the input–output formula that the said conditions are sufficient for equivalence. We now show necessity.

First, it is clear that $D = D_1$ must hold since in particular $Du(0) = D_1 u(0)$ for all $u(0)$. With this in place we can now rewrite the equivalence condition as

$$\int_0^t \{Ce^{A(t-\tau)}B - C_1 e^{A_1(t-\tau)}B_1\}u(\tau)d\tau = 0, \qquad (2.13)$$

for all functions u. It thus remains to show that the factor multiplying $u(t)$ in (2.13) must be identically zero. For simplicity we show this in the case where $Ce^{At}B$ and $C_1 e^{A_1 t}B_1$ are scalar functions; the multivariable case follows using a very similar argument.

Suppose, by contrapositive, that for some $t_0 \geq 0$ the functions are not equal. Then define

$$u(t) = Ce^{A(t_0+1-t)}B - C_1 e^{A_1(t_0+1-t)}B_1$$

and notice that $u(1) \neq 0$. Thus applying this input to the condition in (2.13), with $t = t_0 + 1$, we see that

$$\int_0^{t_0+1} \{Ce^{A(t_0+1-\tau)}B - C_1 e^{A_1(t_0+1-\tau)}B_1\}u(\tau)d\tau = \int_0^{t_0+1} |u(\tau)|^2 d\tau > 0,$$

where the last inequality follows from the fact that u is a continuous function and $u(1) \neq 0$. Thus we have a contradiction and necessity follows. ∎

This lemma reduces the equivalence of realizations to checking that two functions of time are equal. We now provide an alternative characterization, in terms of the matrices $CA^k B$, which are called the *Markov parameters* of the system realization (A, B, C, D).

Lemma 2.26. *The two realizations* (A, B, C, D) *and* (A_1, B_1, C_1, D_1) *are equivalent, if and only if* $D = D_1$ *and*

$$CA^k B = C_1 A_1^k B_1, \quad \text{for all } k \geq 0. \qquad (2.14)$$

Proof. By Lemma 2.25 we must show the equivalence of (2.14) with

$$Ce^{At}B = C_1 e^{A_1 t}B_1,$$

for all $t \geq 0$. The sufficiency of (2.14) follows clearly by definition of the matrix exponential, since

$$Ce^{At}B = CB + CABt + CA^2B\frac{t^2}{2} + \cdots$$

For the converse, assume the time functions are equal and notice that the kth (right) derivative of $Ce^{At}B$ is given by

$$\frac{d^k}{dt^k}Ce^{At}B = CA^k e^{At}B,$$

with a similar expression for the derivatives of $C_1 e^{A_1 t}B_1$. These derivatives all must be equal, so we have

$$CA^k B = \frac{d^k}{dt^k}Ce^{At}B\bigg|_{t=0} = \frac{d^k}{dt^k}C_1 e^{A_1 t}B_1\bigg|_{t=0} = C_1 A_1^k B_1 \quad \text{as required.}$$

∎

Now we turn to the question of minimality: given a realization (A, B, C, D) of order n, does there exist a realization (A_1, B_1, C_1, D) of order $n_1 < n$ that is equivalent? The next theorem tells us that this question is related to observability and controllability.

Proposition 2.27. *Suppose (A, B, C, D) is a system realization. If either (C, A) is not observable or (A, B) is not controllable, then there exists a lower-order realization (A_1, B_1, C_1, D) for the system.*

Proof. There are two cases to consider: (A, B) not controllable and (C, A) unobservable. We will assume that (A, B) is not controllable since the unobservable case is nearly identical in proof. With this assumption we convert to controllability form. That is choose a state transformation T such that

$$TAT^{-1} = \begin{bmatrix} \tilde{A}_{11} & \tilde{A}_{12} \\ 0 & \tilde{A}_{22} \end{bmatrix} \text{ and } TB = \begin{bmatrix} \tilde{B}_1 \\ 0 \end{bmatrix},$$

where the dimensions of $\tilde{A}_{22}$ are nonzero. Let

$$CT^{-1} = \begin{bmatrix} \tilde{C}_1 & \tilde{C}_2 \end{bmatrix}$$

and we see that

$$Ce^{At}B = \begin{bmatrix} \tilde{C}_1 & \tilde{C}_2 \end{bmatrix} \begin{bmatrix} e^{\tilde{A}_{11}t} & ? \\ 0 & ? \end{bmatrix} \begin{bmatrix} \tilde{B}_1 \\ 0 \end{bmatrix}$$

$$= \tilde{C}_1 e^{\tilde{A}_{11}t} \tilde{B}_1 .$$

Thus setting $A_1 = \tilde{A}_{11}, B_1 = \tilde{B}_1$ and $C_1 = \tilde{C}_1$ and invoking Lemma 2.25 we have found a lower-order realization.   ∎

The result above says that if a realization is not both controllable and observable, then there exists a lower-order equivalent realization. Immediately we wonder whether controllability and observability of a realization guarantees that it is of the lowest order; such a lowest-order realization is called a *minimal realization*. The answer is affirmative, as we show in the following result.

Theorem 2.28. *If* (C, A) *is observable and* (A, B) *is controllable, then* (A, B, C, D) *is a minimal realization.*

Proof. We will show that if (A_1, B_1, C_1, D_1) is a realization which is equivalent to (A, B, C, D), then the dimension of A_1 is no smaller than that of A.

From Lemma 2.26 we know that

$$CA^k B = C_1 A_1^k B_1 \quad \text{for } k \geq 0 .$$

Therefore in particular we have that

$$\begin{bmatrix} C \\ CA \\ \vdots \\ CA^{n-1} \end{bmatrix} \begin{bmatrix} B & AB & \cdots & A^{n-1}B \end{bmatrix}$$

is equal to

$$\begin{bmatrix} C_1 \\ C_1 A_1 \\ \vdots \\ C_1 A_1^{n-1} \end{bmatrix} \begin{bmatrix} B_1 & A_1 B_1 & \cdots & A_1^{n-1} B_1 \end{bmatrix} .$$

Now both the matrices in the first expression have the full rank of n; thus their product has rank n. This means that the second expression has rank n, and simply by the dimensions we see that A_1 must be at least $n \times n$ in size. ∎

We have completed our study of minimal realizations for input–output systems with only time domain considerations. We now turn to an alternative characterization of systems which is of great importance and will be used throughout the course.

2.6   Transfer functions and state space

An alternative method for the study of systems from an input–output perspective, is to employ *transforms* of the relevant signals. In particular, the Laplace transform of a time function $f(t)$ is defined as

$$\hat{f}(s) := \int_0^\infty f(t) e^{-st} dt,$$

and s varies in a region of the complex plane $\mathbb{C}$ where the above integral converges. Thus we define a *linear* transformation between functions of

t and functions of s. We will not pursue here the question of describing function spaces in which this transform is well-defined; an important case will be discussed in detail in the next chapter. For the moment assume that all time functions of interest increase no faster than exponentially. More precisely, given a function f assume that there exist real constants α and β satisfying

$$|f(t)| \leq \beta e^{\alpha t} \quad \text{for all } t \geq 0.$$

In this case the above transform is well defined for $\text{Re}(s) > \alpha$.

A basic property is that if the time derivative $\dot{f}(t)$ has a well-defined Laplace transform; the latter is given by

$$s\hat{f}(s) - f(0).$$

This property can be used to derive a Laplace transform version of our basic state space system (2.1) of the form

$$s\hat{x}(s) - x(0) = A\hat{x}(s) + B\hat{u}(s),$$
$$\hat{y}(s) = C\hat{x}(s) + D\hat{u}(s).$$

In the special case of $x(0) = 0$, the input–output relationship

$$\hat{y}(s) = \hat{G}(s)\hat{u}(s) \tag{2.15}$$

follows, where

$$\hat{G}(s) = C(sI - A)^{-1}B + D$$

is called the *transfer function* of the system. This function is well defined whenever the matrix $(sI - A)$ has an inverse, and in particular the identity (2.15) will hold in some right-half plane $\text{Re}(s) > \alpha$, where this inverse, as well as the transforms $\hat{y}(s)$ and $\hat{u}(s)$, are well defined.

To more conveniently write the transfer function associated with a state space realization we introduce the notation

$$\left[\begin{array}{c|c} A & B \\ \hline C & D \end{array} \right](s) := C(sI - A)^{-1}B + D. \tag{2.16}$$

In the above we have made explicit the dependence of the transfer function on the variable s; from here onwards, however, we will adopt the more customary notation and suppress this variable from the expression on the left-hand side.

We can also relate the relationship (2.15) to the time domain description

$$y(t) = \int_0^t Ce^{A(t-\tau)}Bu(\tau)d\tau + Du(t) \tag{2.17}$$

considered in the previous section. To do this, first notice that the transform of the term $Du(t)$ is $D\hat{u}(s)$. Next we can show that $(sI-A)^{-1}$ is the Laplace transform of the matrix e^{At}. For this purpose consider

$$\text{Re}(s) > \max \, \text{Re}(\, \text{eig}(A)\,),$$

and write

$$\int_0^\infty e^{At} e^{-st} dt = \int_0^\infty e^{-(sI-A)t} dt = -(sI-A)^{-1} \int_0^\infty \frac{d}{dt} \left[e^{-(sI-A)t} \right] dt$$

$$= (sI-A)^{-1} \left[I - \lim_{t\to+\infty} e^{-(sI-A)t} \right] = (sI-A)^{-1}$$

where the last identity follows from the fact that since the eigenvalues of $-(sI - A)$ have all negative real parts, all the entries in the matrix will have exponentially decaying terms. We refer to the study of the formula for the matrix exponential, leaving details as an exercise.

It follows immediately from linearity that $C(sI-A)^{-1}B$ is the transform of $Ce^{At}B$. Then the first term in (2.17) can be interpreted as a *convolution* integral between $Ce^{At}B$ and $u(t)$, whose transform will be the product

$$\left[\begin{array}{c|c} A & B \\ \hline C & 0 \end{array} \right] \cdot \hat{u}(s)$$

completing the relationship (2.15). We will not expand here on the issue of convolutions and their relationship with transforms, we return to this topic in the next chapter.

The transfer function provides yet another version of the equivalence between two state space realizations.

Lemma 2.29. *The two realizations (A, B, C, D) and (A_1, B_1, C_1, D_1) are equivalent, if and only if*

$$\hat{G}(s) = \left[\begin{array}{c|c} A & B \\ \hline C & D \end{array} \right] = \left[\begin{array}{c|c} A_1 & B_1 \\ \hline C_1 & D_1 \end{array} \right] = \hat{G}_1(s) , \qquad (2.18)$$

for all s where the transfer functions are well defined.

Proof. One can argue that this must be true based on the input–output relationship (2.15). Alternatively, using Lemma 2.25 and the transform of $Ce^{At}B$ computed above, it suffices to show that $\hat{G}(s) = \hat{G}_1(s)$ if and only if

$$D = D_1, \quad C(sI-A)^{-1}B = C_1(sI-A_1)^{-1}B_1.$$

Here the "if" direction is obvious, and "only if" follows by noticing that

$$D = \lim_{s\to\infty} \hat{G}(s) = \lim_{s\to\infty} \hat{G}_1(s) = D_1,$$

which in turn uses the fact that

$$\lim_{s\to\infty} (sI-A)^{-1} = \lim_{s\to\infty} \frac{1}{s}(I - A/s)^{-1} = 0 ,$$

and similarly with A_1. ■

2.6.1   Rational matrices and state space realizations

We now discuss the structure of the transfer function $\hat{G}(s)$, constituted by *real rational functions* of the variable s.

A rational function is the quotient of two real polynomials, of the general form

$$\hat{g}(s) = \frac{b_m s^{m-1} + \cdots + b_1 s + b_0}{s^n + a_{n-1} s^{n-1} + \cdots + a_0}$$

for appropriate scalars a_k and b_k, where m and n are the degrees of numerator and denominator polynomials. The function is called *real rational* if the a_k, b_k are real numbers. This function is defined everywhere in the complex plane, except at the roots of the denominator polynomial. Recall that a function has a *pole* in the complex plane if it tends to infinity as that point is approached; for the rational function $\hat{g}(s)$ poles can only occur at the roots of its denominator.

We say that a rational function is *proper* if $n \geq m$, and *strictly proper* if $n > m$. Notice these are equivalent to the statements

$$\lim_{s \to \infty} \hat{g}(s) \quad \text{exists in } \mathbb{C} \quad \text{and} \quad \lim_{s \to \infty} \hat{g}(s) = 0, \text{ respectively.}$$

Our interest here is primarily in rational, proper functions. We extend our notion of such scalar functions to the matrix case: suppose

$$\hat{G}(s) = \begin{bmatrix} \hat{g}_{11}(s) & \cdots & \hat{g}_{m1}(s) \\ \vdots & \ddots & \vdots \\ \hat{g}_{1p}(s) & \cdots & \hat{g}_{mp}(s) \end{bmatrix}$$

where each $\hat{g}_{ij}(s)$ is a scalar function. Then we define $\hat{G}(s)$ to be real rational and proper if each scalar function $\hat{g}_{ij}(s)$ has these properties. Define the set

$$RP = \{\text{set of matrix functions that are both real rational, and proper}\}.$$
$$(2.19)$$

We say that $\hat{G}(s) \in RP$ has a pole at a point if *any* of its scalar entries does.

Now returning to our transfer function, we see that it is indeed rational since $(sI - A)^{-1}$ can be written using Cramer's rule as

$$(sI - A)^{-1} = \frac{1}{\det(sI - A)} \text{adj}(sI - A),$$

where adj $(\cdot)$ denotes the classical adjoint of a matrix and is formed of cofactors of elements of the matrix. Therefore adj$(sI - A)$ is a matrix of polynomial entries, each of which has order less than n. The determinant $\det(Is - A)$ is also a polynomial; indeed, the characteristic polynomial of A, and has order n. Therefore $(sI - A)^{-1}$ is a strictly proper, real rational matrix, and it follows that

$$\left[\begin{array}{c|c} A & B \\ \hline C & D \end{array}\right] \text{ is an element of the set } RP.$$

In words, every state space transfer function defines a real rational matrix function that is proper.

We now turn to the converse question: given a real rational transfer function, is it always the transfer function of some state space system? The next proposition establishes this converse in the strictly proper scalar case.

Lemma 2.30. *Suppose that*

$$\hat{g}_1(s) = \frac{c_{n-1}s^{n-1} + \cdots + c_0}{s^n + a_{n-1}s^{n-1} + \cdots + a_0},$$

then there exists a state space realization $(A, B, C, 0)$ *such that*

$$\hat{g}_1(s) = \left[\begin{array}{c|c} A & B \\ \hline C & 0 \end{array}\right], \quad \text{where } A \text{ is } n \times n.$$

Proof. Let

$$C = [c_0 \cdots c_{n-1}], \quad A = \begin{bmatrix} 0 & 1 & & 0 \\ & \ddots & \ddots & \\ 0 & & 0 & 1 \\ -a_0 & -a_1 & & -a_{n-1} \end{bmatrix}, \quad \text{and } B = \begin{bmatrix} 0 \\ \vdots \\ 0 \\ 1 \end{bmatrix}.$$

Then it is routine to verify that this realization satisfies the claim. ■

From this lemma we can prove the more general scalar result.

Proposition 2.31. *If* $\hat{g}(s)$ *is a proper rational function given by*

$$\hat{g}(s) = \frac{c_n s^n + c_{n-1}s^{n-1} + \cdots + c_0}{s^n + a_{n-1}s^{n-1} + \cdots + a_0},$$

then it has a state space realization (A, B, C, D) *where* A *is* $n \times n$.

Proof. First since $\hat{g}(s)$ is proper it is possible to write it as

$$\hat{g}(s) = \hat{g}_1(s) + D,$$

where $\hat{g}_1$ is strictly proper and D is a real constant. Now invoke Lemma 2.30. ■

Having treated the scalar case, we now turn to the matrix case, starting with the following fact. Consider two transfer functions

$$\hat{G}_1(s) = \left[\begin{array}{c|c} A_1 & B_1 \\ \hline C_1 & D_1 \end{array}\right] \text{ and } \hat{G}_2(s) = \left[\begin{array}{c|c} A_2 & B_2 \\ \hline C_2 & D_2 \end{array}\right].$$

If we stack them to form a larger transfer function the following can be verified algebraically.

$$\begin{bmatrix} \hat{G}_1(s) \\ \hat{G}_2(s) \end{bmatrix} = \begin{bmatrix} C_1(sI - A_1)^{-1}B_1 + D_1 \\ C_2(sI - A_2)^{-1}B_2 + D_2 \end{bmatrix}$$

$$= \left[\begin{array}{cc|c} A_1 & 0 & B_1 \\ 0 & A_2 & B_2 \\ \hline C_1 & 0 & D_1 \\ 0 & C_2 & D_2 \end{array} \right]$$

That is, a block column with two transfer function entries is exactly a transfer function. Similarly for appropriately dimensioned transfer functions we have

$$\begin{bmatrix} \hat{G}_1(s) & \hat{G}_2(s) \end{bmatrix} = \left[\begin{array}{cc|cc} A_1 & 0 & B_1 & 0 \\ 0 & A_2 & 0 & B_2 \\ \hline C_1 & C_2 & D_1 & D_2 \end{array} \right].$$

Consequently if we are given a matrix function $\hat{G}(s)$ in the set RP, we can begin by finding realizations

$$\hat{g}_{ij}(s) = C_{ij}(sI - A_{ij})^{-1}B_{ij} + D_{ij} .$$

for each entry using Proposition 2.31, and then use the previous rules successively to form a state space realization for $\hat{G}(s)$. We thus have the following result, which summarizes the previous discussion.

Proposition 2.32. *The set RP of proper real rational functions is equal to the set of state space transfer functions.*

The above construction provides a concrete algorithm for constructing a state space realization for a rational proper function. The advantage of using explicit state space realizations is that operations such as multiplication, addition, or inversion of functions in RP is very conveniently done; see Exercise 2.18 where we derive the formulas for the standard manipulations. However, the above procedure may not be economical from the point of view of the dimension of the state space system. These issues are discussed next.

2.6.2 Minimality

In the above algorithm we assign separate states to each entry of the matrix, which may lead to an unnecessarily high state dimension. This fact is now illustrated.

Example:

Consider the transfer function

$$\hat{G}(s) = \begin{bmatrix} \frac{1}{s+1} & \frac{2}{s+1} \end{bmatrix}.$$

The previous construction would lead to

$$A = \begin{bmatrix} -1 & 0 \\ 0 & -1 \end{bmatrix}, \quad B = \begin{bmatrix} 1 & 0 \\ 0 & 1 \end{bmatrix}, \quad C = \begin{bmatrix} 1 & 2 \end{bmatrix},$$

but clearly a lower-order realization is given by

$$A = -1, \quad B = 1, \quad C = \begin{bmatrix} 1 & 2 \end{bmatrix}.$$

$\square$

This raises once more the question of *minimality* of a state space realization, in this case as a representation of a rational transfer function. As before a realization of a rational function $\hat{G}(s)$ is minimal if it has the lowest possible dimension n for the A-matrix.

How does lack of minimality reflect itself in the realization? In principle the fact that (A, B, C, D) has more states than are necessary to describe $\hat{G}(s)$ would imply that *pole-zero cancellations* will occur if we calculate

$$C \left\{ \frac{1}{\text{char}_A(s)} \text{adj}\,(sI - A) \right\} B + D \,,$$

since this expression must be the same as the one obtained with a lower-order characteristic polynomial in the denominator.

In the case of a scalar $\hat{G}(s)$, it can be shown that minimality is equivalent to there being no pole-zero cancellations occurring when the state space transfer function is computed. For the matrix case, however, the issue is more complicated.

Examples:

For example, it is not immediately obvious that

$$\hat{G}_1(s) = \begin{bmatrix} \frac{1}{s} & 0 \\ \frac{1}{s^2} & \frac{1}{s} \end{bmatrix}$$

can be realized with $n = 2$, whereas

$$\hat{G}_2(s) = \begin{bmatrix} 0 & \frac{1}{s} \\ \frac{1}{s^2} & \frac{1}{s} \end{bmatrix}$$

requires that n be at least three, when in both cases the algorithm described above would lead to $n = 4$.

$\square$

Therefore, a necessary complement to the above constructive procedure is a test for the minimality of a given state space realization of a transfer function, and a procedure to remove unnecessary states. Fortunately we have already provided the tools to address this problem.

Theorem 2.33. *A realization (A, B, C, D) of a transfer function $\hat{G}(s)$ is minimal if and only if (A, B) is controllable and (C, A) is observable.*

Proof. We have already established that equality of transfer functions is the same as equivalence of the corresponding input–output maps. Therefore the proof reduces to Proposition 2.27 and Theorem 2.28.   ∎

In addition to being able to test for minimality, our work in the earlier sections provides a method to reduce the order of a non-minimal realization, namely, eliminating uncontrollable or unobservable states. This can be easily done by transforming the realization to the controllability form or to the observability form (these steps can be done successively). As an exercise, the reader is invited to construct minimal realizations for the above transfer functions $\hat{G}_1(s)$ and $\hat{G}_2(s)$. The order of a minimal realization of a transfer function $\hat{G}(s)$ will be called in the sequel the *order* of $\hat{G}(s)$.

This brings us to the end of our work on basic state space system theory. We have provided algebraic methods to describe a general class of linear systems, both in the time domain and in the transfer function method. As a final remark we note that the results of this chapter extend, with very minor changes to the case of *complex valued* signals; indeed by linearity this analysis is equivalent to separately studying the real and imaginary parts. Although this extension has no physical motivation, we will see that it is mathematically convenient for the analysis to follow.

In the next chapter we will move beyond the algebraic and begin to establish *quantitative* descriptions of signals and systems, from an input–output perspective.

Exercises

2.1. Consider the following set of differential equations:

$$u_1 + c\,u_2 = \ddot{q}_1 + b\,\dot{q}_2 + q_1,$$
$$u_2 = \ddot{q}_2 + d\,\ddot{q}_1 + e\,\dot{q}_2 + f\,q_1,$$

where the dependent functions are q_1 and q_2. Convert this to a 2×2 state space system.

2.2. Suppose A is Hermitian and is Hurwitz. Show that if $x(t) = e^{At}x(0)$, then $|x(t)| < |x(0)|$, for each $t > 0$. Here $|\cdot|$ denotes the Euclidean length of the vector.

2.3. Prove that $e^{(N+M)t} = e^{Nt}e^{Mt}$ if N and M commute. Use the following steps:

  (a) Show that e^{Nt} and M commute;
  (b) by taking derivatives directly show that the time derivative of $Q(t) = e^{Nt}e^{Mt}$ satisfies $\dot{Q}(t) = (N+M)e^{Nt}e^{Mt}$;
  (c) since $Q(t)$ satisfies $\dot{Q}(t) = (N+M)Q(t)$, with $Q(0) = I$, show that $Q(t) = e^{(N+M)t}$.

Notice in particular this demonstrates $e^{N+M} = e^N e^M$.

2.4. Using Jordan decomposition prove the Cayley-Hamilton Theorem and deduce Proposition 2.4.

2.5. Prove Proposition 2.7.

2.6. In the proof of Lemma 2.8 we use that fact that given two subspaces $\mathcal{V}_1$ and $\mathcal{V}_2$, then $\mathcal{V}_1 \subset \mathcal{V}_2$ if and only if $\mathcal{V}_1^\perp \supset \mathcal{V}_2^\perp$. Prove this fact. *Hint:* first show that for any subspace $\mathcal{V} = (\mathcal{V}^\perp)^\perp$.

2.7. Using a change of basis transform the pair

$$A = \begin{bmatrix} 0 & 1 & 0 & 1 \\ 0 & 0 & 1 & 0 \\ 0 & 2 & -1 & 0 \\ 0 & -1 & 1 & 1 \end{bmatrix} \qquad B = \begin{bmatrix} 1 & 1 \\ 1 & 0 \\ 1 & 0 \\ 0 & 0 \end{bmatrix}$$

to the form $\tilde{A} = \begin{bmatrix} \tilde{A}_{11} & \tilde{A}_{12} \\ 0 & \tilde{A}_{22} \end{bmatrix}$ and $\tilde{B} = \begin{bmatrix} \tilde{B}_1 \\ 0 \end{bmatrix}$ where $(\tilde{A}_{11}, \tilde{B}_1)$ is controllable. That is, controllability form.

2.8. Fill in the details of the proof of Theorem 2.12 on controllability form.

2.9. (a) Consider the discrete time state equation

$$x_{k+1} = Ax_k + Bu_k, \text{ with } x_0 = 0.$$

A state $\xi \in \mathbb{R}^n$ is said to be reachable if there exists a sequence $u_0, \dots, u_{N-1}$ such that $x_N = \xi$.

   (i) Show that ξ reachable if and only if it is in the image of the matrix $\begin{bmatrix} B & AB & \cdots & A^{n-1}B \end{bmatrix}$.

   (ii) If ξ is reachable the length N of the sequence u_k can be chosen to be at most n.

   (b) Given the uncontrolled discrete time state space system

$$x_{k+1} = Ax_k, \text{ with initial condition } x_0$$
$$y_k = Cx_k,$$

we say it is observable if each x_0 gives rise to a *unique* output sequence y_k.

   (i) Show that the above system is observable if and only if (A^*, C^*) is controllable.

   (ii) Prove that if the system is observable, the initial condition x_0 can be determined from the finite sequence $y_0, \dots, y_{n-1}$.

2.10. Provide the details of a proof for Corollary 2.15.

2.11. Consider an uncontrollable pair (A, B).

(i) Using the controllability form, discuss the possibilities for assignment of eigenvalues of the matrix $A + BF$. Deduce that controllability is necessary for arbitrary eigenvalue assignment.

(i) If (A, B) is stabilizable, show that there exists F that makes $A + BF$ Hurwitz. From here show, given an initial condition $x(0)$, how to construct an input $u(t)$ that drives the state of $\dot{x} = Ax + Bu$ asymptotically to zero.

2.12. Using our proofs on controllability and stabilizability as a guide, prove all parts of Proposition 2.22.

2.13. Given a pair (C, A), devise a test to determine whether it is possible to find a matrix L so that all the eigenvalues of $A + LC$ are equal to -1; you should not have to construct L explicitly.

2.14. In this question we derive the famous *Kalman decomposition*, which is a generalization of the controllability and observability forms we have studied in this chapter. Suppose we are given a matrix triple (C, A, B).

(a) Prove that the subspace intersection $\mathcal{C}_{AB} \cap \mathcal{N}_{CA}$ is A-invariant.

(b) Show that there exists a state transformation T so that

$$
T^{-1}AT = \begin{bmatrix} A_1 & 0 & A_6 & 0 \\ A_2 & A_3 & A_4 & A_5 \\ 0 & 0 & A_7 & 0 \\ 0 & 0 & A_8 & A_9 \end{bmatrix}, \quad T^{-1}B = \begin{bmatrix} B_1 \\ B_2 \\ 0 \\ 0 \end{bmatrix},
$$

and $CT = \begin{bmatrix} C_1 & 0 & C_2 & 0 \end{bmatrix}$. Furthermore the following properties are satisfied:

- the pair (C_1, A_1) is observable;
- the pair (A_1, B_1) is controllable;
- the pair $\left(\begin{bmatrix} A_1 & 0 \\ A_2 & A_3 \end{bmatrix}, \begin{bmatrix} B_1 \\ B_2 \end{bmatrix} \right)$ is controllable;
- the pair $\left(\begin{bmatrix} C_1, C_2 \end{bmatrix}, \begin{bmatrix} A_1 & A_6 \\ 0 & A_7 \end{bmatrix} \right)$ is observable.

To do this use the A-invariance properties of the controllable subspace $\mathcal{C}_{AB}$ and the unobservable subspace $\mathcal{N}_{CA}$, and use T to decompose the state space into the four independent subspaces that are controllable and observable; controllable and unobservable; observable and uncontrollable; unobservable and uncontrollable.

2.15. Given the system matrices

$$
A = \begin{bmatrix} 2 & 2 & 1 \\ 1 & 1 & 1 \\ 4 & 1 & 0 \end{bmatrix}; \quad B = \begin{bmatrix} 2 & 2 \\ 2 & 1 \\ 3 & 2 \end{bmatrix}
$$

$$
C = \begin{bmatrix} 2 & 1 & 0 \end{bmatrix},
$$

determine whether it is possible to construct an *observer* for the system

$$\dot{x} = Ax + Bu \quad x(0) = x_0$$
$$y = Cx.$$

2.16. Suppose (C, A) is observable. Show that it is possible to determine *exactly* the initial condition $x(0)$, by appropriately choosing n samples of the output $y(t) = Ce^{At}x(0)$, from any given positive interval $[0, T]$.

2.17.   (i) Given the system

$$\dot{x}(t) = A\,x(t) + B\,u(t), \quad x(0) = x_0,$$

provide an explicit solution for the state $x(1)$ in terms of $u(t)$.

(ii) Suppose in the system of (i) that

$$A = \begin{bmatrix} -1 & 2 & 0 \\ -4 & 5 & 0 \\ 0 & 0 & 4 \end{bmatrix} \quad B = \begin{bmatrix} 1 \\ 1 \\ 0 \end{bmatrix} \quad x(0) = \begin{bmatrix} e^{-3} \\ 2e^{-3} \\ e^{-4} \end{bmatrix}.$$

Does there exist a function $u(t)$ so that

$$x(1) = \begin{bmatrix} 11 \\ 12 \\ 1 \end{bmatrix} ?$$

Give full justification for your answer, but note that you do *not* necessarily have to find $u(t)$ explicitly.

2.18. Given two transfer functions

$$\hat{G}_1(s) = \left[\begin{array}{c|c} A_1 & B_1 \\ \hline C_1 & D_1 \end{array} \right] \quad \text{and} \quad \hat{G}_2(s) = \left[\begin{array}{c|c} A_2 & B_2 \\ \hline C_2 & D_2 \end{array} \right],$$

verify the following formulas.

(a) $\hat{G}_1(s)\,\hat{G}_2(s) = \left[\begin{array}{cc|c} A_1 & B_1 C_2 & B_1 D_2 \\ 0 & A_2 & B_2 \\ \hline C_1 & D_1 C_2 & D_1 D_2 \end{array} \right].$

(b) $\hat{G}_1^{-1} = \left[\begin{array}{c|c} A_1 - B_1 D_1^{-1} C_1 & B_1 D_1^{-1} \\ \hline -D_1^{-1} C_1 & D_1^{-1} \end{array} \right]$, assuming that the inverse $\hat{G}_1^{-1}$ is in the set RP. Furthermore, show that $\hat{G}_1^{-1}$ exists and is in RP exactly when D_1 is a nonsingular matrix.

(c) $\hat{G}_1(s) + \hat{G}_2(s) = \left[\begin{array}{cc|c} A_1 & 0 & B_1 \\ 0 & A_2 & B_2 \\ \hline C_1 & C_2 & D_1 + D_2 \end{array} \right].$

2.19. (i) Prove that if (A, B, C, D) is a minimal realization for a transfer function $\hat{G}(s)$, then every eigenvalue of A must be a pole of $\hat{G}(s)$. *Hint:* Start with the case where A has only one (possibly repeated) eigenvalue λ; show that if λ is not a pole of $\hat{G}(s)$, then $\hat{G}(s) = D$. For the general case, first transform the system to Jordan form.

  (ii) Now consider *any* realization (A, B, C, D) of $\hat{G}(s)$. Show that every eigenvalue of A is either a pole of $\hat{G}(s)$ or it is uncontrollable or unobservable.

2.20. Construct minimal realizations for

$$\hat{G}_1(s) = \begin{bmatrix} \frac{1}{s} & 0 \\ \frac{1}{s^2} & \frac{1}{s} \end{bmatrix}, \quad \hat{G}_2(s) = \begin{bmatrix} 0 & \frac{1}{s} \\ \frac{1}{s^2} & \frac{1}{s} \end{bmatrix}.$$

2.21. Recall the double pendulum system of the introductory chapter. Construct a linearized model of the pendulum around an equilibrium point. Discuss whether there are any equilibrium points around which the system linearization loses controllability. Is it stabilizable around these points?

Notes and references

The material covered in this chapter is standard, and for this reason we primarily highlight some of the major references for the concepts and ideas. The presentation here is largely based on [57]. The fundamental ideas of controllability and observability introduced in this chapter are due to Kalman [82]. The PBH test, as its name suggests, was developed independently by the three researchers in [13, 72, 126]. The well-known result on controllable canonical form presented in conjunction with single input eigenvalue assignment originates from [125]. The first proof of the multivariable eigenvalue assignment result appears in [173]. The controllability and observability forms of this chapter are special cases of the so-called Kalman decomposition, which was introduced in [62, 83]. This also led to the notion of system minimality. The key idea involved in constructing the Luenberger observer presented here was first reported in [97].

For more details and historical references on time domain state space theory than provided above the reader is referred to [23, 25, 151]. For an in-depth look at state space systems from a geometric perspective, see [174]. In the final section of this chapter we introduced the basics of rational function theory and connections with state space realizations; a more complete treatment in a systems context can be found in [80]. For coverage of infinite dimensional state space systems, for instance, see [27] and the references therein. Reference [122] provides an introduction to state space systems using a behavioral approach.

3
Linear Analysis

One of the prevailing viewpoints for the study of systems and signals is that in which a dynamical system is viewed as a mapping between input and output functions. This concept underlies most of the basic treatments of signal processing, communications, and control. Although a functional analytic perspective is implicit in this viewpoint, the associated machinery is not commonly applied to the study of dynamical systems. In this course we will see that incorporating more tools from analysis (e.g., function spaces, operators) into this conceptual picture leads to methods of key importance for the study of systems. In particular, operator norms provide a natural way to quantify the "size" of a system, a fundamental requirement for a quantitative theory of system uncertainty and model approximation.

In this chapter we introduce some of the basic concepts from analysis that are required for the development of robust control theory. This involves assembling some definitions, providing examples of the central objects, and presenting their important properties. This is the most mathematically abstract chapter of the course, and is intended to provide us with a solid and rigorous framework. At the same time, we will not attempt a completely self-contained mathematical treatment, which would constitute a course in itself. To build intuition, the reader is encouraged to supply examples and proofs in a number of places. For detailed proofs of the most technical results presented consult the references listed at the end of the chapter.

During the chapter we will encounter the terminology "for almost every" and "essential supremum" (denoted by "ess sup") which refer to mathematical concepts that are not part of the core of our course. These clauses are used to make our statements precise, but can be replaced by "for every" and "supremum," respectively, without compromising understanding of the material. If these former terms are unfamiliar, precise definitions and a short introduction to them are given in Appendix A.

3.1   Normed and inner product spaces

The most important mathematical concept in this course is that of a *normed space*, which we will use continually to measure both signals and systems. We will start by defining exactly what we mean by a norm, and then move on to some examples.

Definition 3.1. *A norm* $\| \cdot \|_{\mathcal{V}}$ *on a vector space* $\mathcal{V}$ *is a function mapping* $\mathcal{V} \to [0, \infty)$ *which, for each* $v \in \mathcal{V}$, *satisfies*

(a) $\|v\|_{\mathcal{V}} = 0$ *if and only if* $v = 0$;

(b) $|\alpha| \cdot \|v\|_{\mathcal{V}} = \|\alpha v\|_{\mathcal{V}}$, *for all scalars* α;

(c) $\|u + v\|_{\mathcal{V}} \leq \|u\|_{\mathcal{V}} + \|v\|_{\mathcal{V}}$, *for all* $u \in \mathcal{V}$.

Defining a norm on a vector space is done to make the notion of the "size" of an element precise; that is, the size of v is $\|v\|_{\mathcal{V}}$. With this in mind condition (a) says that the zero element is the only member of $\mathcal{V}$ which has size zero. The second norm requirement (b) means that the size of αv must scale linearly according to the sizes of α and v. The third condition is called the *triangle inequality*, inspired by the case of Euclidean space: if u and v are vectors that specify two sides of a triangle, then (c) states that the third side $u + v$ must have length no larger than the sum of the lengths of the other two.

A vector space together with a norm is called a *normed space* and is denoted $(\mathcal{V}, \| \cdot \|_{\mathcal{V}})$. Frequently when the norm and space is clear we simply write $\| \cdot \|$, suppressing the subscript on the norm symbol. Understanding normed spaces will provide us with a powerful point of view, and we now begin to explore this new concept.

We say a sequence $v_k \in \mathcal{V}$ *converges* in a normed space $\mathcal{V}$ if there exists $v \in \mathcal{V}$ such that

$$\|v - v_k\| \overset{k \to \infty}{\longrightarrow} 0.$$

We use $\lim_{k \to \infty} v_k = v$ to denote the above relationship; thus this preserves our usual notation for limits with numbers, although our interpretation is slightly more involved. Further, a subspace $\mathcal{S} \subset \mathcal{V}$ is defined to be *closed*, if every sequence $v_k \in \mathcal{S}$ that converges, converges to an element in $\mathcal{S}$.

Examples:

In Chapter 1 we have already encountered a norm: the Euclidean length on the finite dimensional space $\mathbb{C}^n$. A variety of other norms can be put on this space; a standard family is formed by the so-called p-norms. Given a $p \geq 1$ we define the p-norm

$$|v|_p := (|v_1|^p + \cdots + |v_n|^p)^{\frac{1}{p}}, \quad \text{where } v \in \mathbb{C}^n.$$

Notice that we use the notation $|\cdot|_p$ instead of $\|\cdot\|_p$; the reason for this convention is to avoid confusion with function norms, defined below. When the sub-index is suppressed, we will assume by default that we are referring to $p = 2$, that is, the Euclidean norm $|v| = |v|_2$.

This family of norms is extended to the case where $p = \infty$ by defining the ∞-norm to be

$$|v|_\infty := \max_{1 \leq k \leq n} |v_k|.$$

We can also define norms for matrices; in particular the Frobenius norm is defined as

$$|M|_F := (\operatorname{Tr} M^* M)^{\frac{1}{2}} \quad \text{for } M \in \mathbb{C}^{m \times n},$$

and the maximum singular value from Chapter 1

$$\bar{\sigma}(M) := (\text{maximum eigenvalue of } M^* M)^{\frac{1}{2}}$$

is also a norm. Once again we use the notations $|\cdot|_F$ and $\bar{\sigma}(\cdot)$ instead of the norm sign $\|\cdot\|$.

Our primary interest is norms defined on spaces of functions. For $p \geq 1$, let $L_p^n(-\infty, \infty)$ denote the vector space of functions mapping $\mathbb{R}$ to $\mathbb{C}^n$ that satisfy

$$\int_{-\infty}^{\infty} |u(t)|_p^p \, dt < \infty,$$

where $|\cdot|_p$ denotes the vector norm on $\mathbb{C}^n$ defined above. Then

$$\|u\|_p := \left(\int_{-\infty}^{\infty} |u(t)|_p^p \, dt \right)^{\frac{1}{p}}$$

is a norm on this space, and is the one tacitly associated with it; see Appendix A for some additional technical detail on this.

The superscript n is called the *spatial dimension* of $L_p^n(-\infty, \infty)$. In most cases we will not need to keep track of this dimension and so we simply write $L_p(-\infty, \infty)$.

We complete the family of L_p spaces with the normed space $L_\infty(-\infty, \infty)$ consisting of functions such that

$$\|u\|_\infty := \operatorname*{ess\ sup}_{t \in \mathbb{R}} |u(t)|_\infty$$

is finite. Once again $|u|_\infty$ is the previously defined vector norm.

We also define the space $L_p[0, \infty)$ to be the set of functions in $L_p(-\infty, \infty)$ with support in $[0, \infty)$, more precisely,

$$L_p[0, \infty) = \{u(t) \in L_p(-\infty, \infty) : u(t) = 0 \text{ for } t < 0\}.$$

This convention is nonstandard; more commonly $L_p[0, \infty)$ is the space of functions with restricted domain. However, it is clear that given a function on $[0, \infty)$, one can trivially identify it with a function on the above set by defining it to be zero for $t < 0$.

By adopting the above rule, all our functions will be defined for all t and we are allowed to write $L_p[0, \infty) \subset L_p(-\infty, \infty)$; in words $L_p[0, \infty)$ is a normed subspace of $L_p(-\infty, \infty)$. Analogously we define the subspace $L_p(-\infty, 0]$ of functions that are zero for $t > 0$.   □

Another very important notion that we will frequently encounter is an *inner product*, and it is closely related to the idea of a norm.

Definition 3.2. *An inner product* $\langle \cdot, \cdot \rangle_\mathcal{V}$ *on a vector space* $\mathcal{V}$ *is a function mapping* $\mathcal{V} \times \mathcal{V} \to \mathbb{F}$ *so that*

(a) *the inner product* $\langle v, v \rangle_\mathcal{V}$ *is non-negative, for all* $v \in \mathcal{V}$;

(b) $\langle v, v \rangle_\mathcal{V} = 0$ *if and only if* $v = 0$;

(c) *if* $v \in \mathcal{V}$, *then* $\langle v, \alpha_1 u_1 + \alpha_2 u_2 \rangle = \alpha_1 \langle v, u_1 \rangle + \alpha_2 \langle v, u_2 \rangle$, *for all* $u_i \in \mathcal{V}$ *and scalars* α_i; *i.e., the mapping* $u \mapsto \langle v, u \rangle$ *is linear on* $\mathcal{V}$;

(d) $\langle u, v \rangle_\mathcal{V}$ *is the complex conjugate of* $\langle v, u \rangle_\mathcal{V}$, *for all* $v, u \in \mathcal{V}$.

Geometrically the inner product captures the idea of angle between two elements of $\mathcal{V}$ in the same way that the so-called "dot product" does in Euclidean space.

A vector space $\mathcal{V}$ together with an inner product $\langle \cdot, \cdot \rangle_\mathcal{V}$ is called an *inner product space*. It can be verified that

$$\|v\| = \sqrt{\langle v, v \rangle}$$

satisfies the properties of a norm. The first two properties are straightforward; the triangle inequality can be established by first proving the Cauchy–Schwarz inequality

$$|\langle u, v \rangle| \le \|u\| \cdot \|v\|;$$

both are proved in the exercises. Thus the properties of an inner product are more restrictive than those of a norm.

We say that two elements u, v in $\mathcal{V}$ are *orthogonal* if

$$\langle u, v \rangle = 0.$$

The notation $u \perp v$ is sometimes used to indicate this relationship. It is easy to show that if $u \perp v$, then

$$\|u + v\|^2 = \|u\|^2 + \|v\|^2.$$

This is a generalization of Pythagoras' theorem.

Examples:

The standard example of an inner product space, from which much intuition about inner product spaces is gained, is Euclidean space. That is, $\mathbb{R}^n$ or $\mathbb{C}^n$ with the inner product

$$\langle x, y \rangle := x^*y := x_1^*y_1 + \cdots + x_n^*y_n,$$

where again $*$ denotes complex conjugate transpose.

The matrix generalization of this inner product is defined on $\mathbb{C}^{m \times n}$ and is given by

$$\langle A, B \rangle := \mathrm{Tr}A^*B,$$

for two elements A and B of $\mathbb{C}^{m \times n}$. Notice that this inner product induces the Frobenius norm $|A|_F$. Also if $n = 1$ then this is exactly the inner product space $\mathbb{C}^m$ from above.

We now introduce what for us will be the most important example of an inner product space. Starting with the space of functions $L_2(-\infty, \infty)$ defined before, we introduce

$$\langle x, y \rangle := \int_{-\infty}^{\infty} x^*(t)y(t)\,dt.$$

It can be shown that this integral is well defined, and is an inner product; also the corresponding norm coincides with the norm $\| \cdot \|_2$ defined above. Thus $L_2(-\infty, \infty)$ is an inner product space. Also the subspaces $L_2(-\infty, 0]$ and $L_2[0, \infty)$ of functions with support in the "past" and "future" are themselves inner product spaces. □

We are now ready to consider normed and inner product spaces which have an additional convenient property.

3.1.1 Complete spaces

We have already talked about convergence of sequences in normed and inner product spaces. We said that $\lim_{k \to \infty} v_k$ exists, if there exists an element v in the inner product space, such that the limit of real numbers $\lim_{k \to \infty} \|v - v_k\| = 0$. This is a precise definition, but means that if we are to determine whether a sequence v_k converges it is in general necessary to find such an element v. Finding v may be cumbersome or impossible, which motivates the topic of this short section. We start with the following concept.

Definition 3.3. *Suppose V is a normed space. A sequence v_k in V is a Cauchy sequence if, for each $\epsilon > 0$, there exists $M \geq 0$ such that*

$$\|v_k - v_l\| < \epsilon, \qquad \text{for all} \quad k, l \geq M.$$

This definition simply says that a sequence is Cauchy if it satisfies

$$\|v_k - v_l\| \xrightarrow{k,l \to \infty} 0.$$

A Cauchy sequence is a sequence that "appears" to be converging. It is not hard to show that every convergent sequence is necessarily a Cauchy sequence, but the converse may not be true.

A normed space is *complete* if every Cauchy sequence in it converges, and such a space is referred to as a *Banach* space. A complete inner product space is called a *Hilbert* space. Thus if we know that a space is complete, we can definitively check the convergence of any sequence v_k by simply determining whether or not it is a Cauchy sequence. We note that any closed subspace of a complete normed space, is itself a complete normed space.

Examples:

There is a plethora of Banach and Hilbert space examples. Our standard spaces $\mathbb{C}^n$ and $\mathbb{R}^n$ are complete, with the norms and inner products we have defined thus far. All the L_p spaces we have defined are complete; in particular L_2 is an example of a Hilbert space.

An example of an inner product space that is not complete is

$$\mathcal{W} = \{w \in L_2[0, \infty) : \text{there exists } T > 0 \text{ so that } w(t) = 0, \text{ for all } t \geq T\},$$

equipped with the L_2 norm. To see this consider the sequence of functions given by

$$w_k(t) := \begin{cases} e^{-t}, & \text{for } 0 \leq t \leq k, \\ 0, & \text{otherwise.} \end{cases}$$

This sequence is clearly in $\mathcal{W}$, and is a Cauchy sequence since $\|w_k - w_l\| \leq e^{-\min(k,l)}$. The sequence w_k is in $L_2[0, \infty)$ and it is easy to verify that it converges in that space to the function

$$w(t) := \begin{cases} e^{-t}, & \text{for } 0 \leq t, \\ 0, & \text{otherwise.} \end{cases}$$

But w is not in $\mathcal{W}$, and thus $\mathcal{W}$ is not complete. To summarize, there is no element $\bar{w} \in \mathcal{W}$ to which w_k converges.

$\square$

We have introduced the concept of completeness of a normed or inner product space, and throughout the course we will work with spaces of this type.

3.2 Operators

As we have mentioned in the previous examples, normed spaces can be used to characterize time domain functions, which we informally call *signals*. We now examine mappings from one normed space $\mathcal{V}$ to another normed space $\mathcal{Z}$. These are central to the course as they will eventually be used to represent *systems*. The focus of interest will be linear, bounded mappings.

Definition 3.4. *Suppose $\mathcal{V}$ and $\mathcal{Z}$ are Banach spaces. A mapping from $\mathcal{V}$ to $\mathcal{Z}$ is called a linear, bounded operator if —*

(a) *(Linearity)* $F(\alpha_1 v_1 + \alpha_2 v_2) = \alpha_1 F(v_1) + \alpha_2 F(v_2)$ *for all $v_1, v_2 \in \mathcal{V}$ and scalars α_1, α_2.*

(b) *(Boundedness) There exists a scalar $\kappa \geq 0$, such that*

$$\|Fv\|_{\mathcal{Z}} \leq \kappa \cdot \|v\|_{\mathcal{V}} \quad \text{for all} \quad v \in \mathcal{V}. \tag{3.1}$$

The space of all linear, bounded operators mapping $\mathcal{V}$ to $\mathcal{Z}$ is denoted by

$$\mathcal{L}(\mathcal{V}, \mathcal{Z}),$$

and we usually refer to *linear, bounded* operators as simply operators. We define the *induced norm*, on this space by

$$\|F\|_{\mathcal{V} \to \mathcal{Z}} = \sup_{v \in \mathcal{V}, v \neq 0} \frac{\|Fv\|_{\mathcal{Z}}}{\|v\|_{\mathcal{V}}},$$

and it can be verified that it satisfies the properties of a norm. Notice that $\|F\|_{\mathcal{V} \to \mathcal{Z}}$ is the smallest number κ that satisfies (3.1). When the spaces involved are obvious we write simply $\|F\|$. It is possible to show that $\mathcal{L}(\mathcal{V}, \mathcal{Z})$ is a complete space whenever $\mathcal{Z}$ is complete, which is always the case in this course. If $\mathcal{V} = \mathcal{Z}$ we use the abbreviation $\mathcal{L}(\mathcal{V})$ for $\mathcal{L}(\mathcal{V}, \mathcal{V})$.

We also have a natural notion of composition of operators. If $F \in \mathcal{L}(\mathcal{V}, \mathcal{Z})$ and $G \in \mathcal{L}(\mathcal{Z}, \mathcal{Y})$ and the composition GF is defined by

$$(GF)v := G(Fv), \quad \text{for each } v \text{ in } \mathcal{V}.$$

Clearly, GF is a linear mapping, and it is not difficult to show that

$$\|GF\|_{\mathcal{V} \to \mathcal{Y}} \leq \|G\|_{\mathcal{Z} \to \mathcal{Y}} \|F\|_{\mathcal{V} \to \mathcal{Z}}, \tag{3.2}$$

which implies $GF \in \mathcal{L}(\mathcal{V}, \mathcal{Y})$. The above *submultiplicative* property of induced norms has great significance in robust control theory; we shall have more to say about it below.

Examples:

As always our simplest example comes from vectors and matrices. Given a matrix $M \in \mathbb{C}^{m \times n}$, it defines a linear operator $\mathbb{C}^n \to \mathbb{C}^m$ by matrix multiplication in the familiar way:

$$z = Mv \quad \text{where } v \in \mathbb{C}^n.$$

As in Chapter 1, we will not distinguish between the matrix M and the linear operator it defines. From the singular value decomposition in Chapter 1, if we put the 2-norm on $\mathbb{C}^m$ and $\mathbb{C}^n$, then

$$\|M\|_{\mathbb{C}^m \to \mathbb{C}^n} = \bar{\sigma}(M).$$

If we put norms different from the 2-norm on $\mathbb{C}^m$ and $\mathbb{C}^n$, then it should be clear that the induced norm of M may be different from $\bar{\sigma}(M)$.

Another example that is crucial in studying systems is the case of convolution operators. Suppose f is in the space $L_1^1[0, \infty)$ of scalar functions, then it defines an operator $F : L_\infty[0, \infty) \to L_\infty[0, \infty)$ by

$$(Fu)(t) := \int_0^t f(t - \tau)u(\tau)d\tau, \quad t \geq 0,$$

for u in $L_\infty[0, \infty)$. Clearly F is linear; to see that it is a bounded mapping on $L_\infty[0, \infty)$, notice that if $u \in L_\infty[0, \infty)$ and we set $y = Fu$ then the following inequalities are satisfied for any $t \geq 0$.

$$|y(t)| = \left| \int_0^t f(t - \tau)u(\tau)d\tau \right|$$

$$\leq \int_0^t |f(t - \tau)|\,|u(\tau)|d\tau$$

$$\leq \int_0^\infty |f(\tau)|d\tau \,\|u\|_\infty = \|f\|_1 \,\|u\|_\infty.$$

Therefore we see that $\|y\|_\infty \leq \|f\|_1 \|u\|_\infty$, since t was arbitrary, and more generally that

$$\|F\|_{L_\infty \to L_\infty} \leq \|f\|_1.$$

It follows that F is a bounded operator. The next proposition says that the induced norm is exactly given by the 1-norm of f, as one might guess from our argument above.

Proposition 3.5. *With F defined on $L_\infty[0, \infty)$ as above we have*

$$\|F\|_{L_\infty \to L_\infty} = \|f\|_1.$$

Proof. We have already shown above that $\|F\|_{L_\infty \to L_\infty} \leq \|f\|_1$ and so it remains to demonstrate $\|F\|_{L_\infty \to L_\infty} \geq \|f\|_1$. We accomplish this by showing that, for every $\epsilon > 0$, there exists u, with $\|u\|_\infty = 1$ such that

$$\|Fu\|_\infty + \epsilon \geq \|f\|_1.$$

Choose $\epsilon > 0$ and t such that

$$\int_0^t |f(t - \tau)|d\tau + \epsilon > \|f\|_1.$$

Now set $u(\tau) = e^{-j\theta(t-\tau)}$ where the function $\theta(t)$ is defined to be the complex argument of $f(t)$, namely, it satisfies $f(t) = |f(t)|e^{j\theta(t)}$ for each t.

Notice that $\|u\|_\infty = 1$, and

$$(Fu)(t) = \int_0^t f(t - \tau)u(\tau)d\tau$$

$$= \int_0^t |f(t - \tau)|\, d\tau.$$

The function $(Fu)(t)$ is continuous and therefore we have

$$\|Fu\|_\infty + \epsilon \geq \|f\|_1.$$

∎

Thus this example shows us how to calculate the L_∞-induced norm of a convolution operator exactly in terms of the 1-norm of its convolution kernel.

□

3.2.1   Banach algebras

We have already seen that the set of operators $\mathcal{L}(\mathcal{V})$ on a Banach space $\mathcal{V}$ is itself a Banach space. However, the composition of operators endows it with an additional algebraic structure, which is of crucial importance. In this section we will isolate this structure and prove an important associated result.

Definition 3.6. *A Banach algebra* $\mathcal{B}$ *is a Banach space with a multiplication operation defined for elements of* $\mathcal{B}$, *mapping* $\mathcal{B} \times \mathcal{B} \to \mathcal{B}$, *satisfying the following properties:*

(a) *Algebraic properties:*

   (i) *There exists an element* $I \in \mathcal{B}$, *such that* $F \cdot I = I \cdot F = F$, *for all* $F \in \mathcal{B}$.

   (ii) $F(GH) = (FG)H$, *for all* F, G, H *in* $\mathcal{B}$.

   (iii) $F(G + H) = FG + FH$, *for all* F, G, H *in* $\mathcal{B}$.

   (iv) *For all* F, G *in* $\mathcal{B}$, *and each scalar* α, *we have* $F(\alpha G) = (\alpha F)G = \alpha FG$.

(b) *Property involving the norm: for all elements* F, G *in* $\mathcal{B}$, *we have* $\|FG\| \leq \|F\| \cdot \|G\|$.

This definition says that a Banach algebra has a multiplication operation defined between its elements, one element is the identity element, and satisfies the standard properties of being associative, distributive, and commutative with scalar multiplication. The key property of a Banach algebra is that its norm satisfies the *submultiplicative* inequality, listed in (b).

While the above definition is abstract, we should keep in mind that the main motivating example for the study of Banach algebras is the space

$\mathcal{B} = \mathcal{L}(\mathcal{V})$ of operators on a Banach space $\mathcal{V}$, equipped with the induced norm. For this example the algebraic properties are immediate, and we have already seen in (3.2) that the submultiplicative property holds.

Examples:

Of course $\mathbb{C}$ itself is a Banach algebra, but perhaps the simplest nontrivial example is the space $\mathbb{C}^{n \times n}$ of $n \times n$ complex matrices equipped with the maximum singular value norm $\bar{\sigma}(\cdot)$. Clearly this is a special case of $\mathcal{L}(\mathcal{V})$ with $\mathcal{V} = \mathbb{C}^n$.

An example of Banach algebra which does *not* correspond to a space of operators is $\mathbb{C}^{n \times n}$ endowed with the Frobenius norm $|\cdot|_F$. It is shown in the exercises that the submultiplicative inequality holds, but this is not an induced norm.   □

We now turn to the concept of an *inverse* of an element J in a Banach algebra. We use J^{-1} to denote an element of $\mathcal{B}$ which satisfies

$$J J^{-1} = J^{-1} J = I.$$

If such an element exists we say that J is *invertible* or *nonsingular*; it is not hard to see that in this case the inverse is necessarily unique.

The following result is very important in robust control theory, where it is commonly referred to as the *Small gain theorem*.

Theorem 3.7. *Suppose Q is a member of the Banach algebra $\mathcal{B}$. If $\|Q\| < 1$, then $(I - Q)^{-1}$ exists. Furthermore*

$$(I - Q)^{-1} = \sum_{k=0}^{\infty} Q^k.$$

This theorem says that if the operator Q has "gain" $\|Q\| < 1$, then $(I - Q)^{-1}$ is well defined, and a power series expression for this inverse is given. Although this norm condition is sufficient for the existence of the inverse of $I - Q$, it is not necessary as we see by example.

Examples:

If we take the Banach algebra $(\mathbb{R}^{n \times n}, \bar{\sigma}(\cdot))$ from the last example we see that

with $Q = \begin{bmatrix} 0 & \frac{1}{2} \\ \frac{1}{2} & 0 \end{bmatrix}$ the maximum singular value $\bar{\sigma}(Q) = \frac{1}{2}$.

Therefore by the theorem we know that the matrix $I - Q$ has an inverse; obviously it is easy for us to simply compute this inverse and show directly that $(I - Q)^{-1} = \frac{2}{3} \begin{bmatrix} 2 & 1 \\ 1 & 2 \end{bmatrix}$, which also agrees with the formula in the theorem.

This theorem gives sufficient conditions for the existence of $(I - Q)^{-1}$ but the following simple example shows it is not necessary: let

$$\tilde{Q} = \begin{bmatrix} 0 & 10 \\ 0 & 0 \end{bmatrix} \text{ and then } (I - \tilde{Q})^{-1} = \begin{bmatrix} 1 & -10 \\ 0 & 1 \end{bmatrix},$$

but $\bar{\sigma}(\tilde{Q}) = 10$ which does not satisfy the hypothesis of the theorem. □

Having considered some simple examples — soon we will consider infinite dimensional ones — let us turn to the proof of the theorem. We start by explaining the meaning of $\sum_{k=0}^{\infty} Q^k$. If this sum is a member of $\mathcal{B}$, this sum represents the unique element L that satisfies

$$\lim_{n \to \infty} \left\| L - \sum_{k=0}^{n} Q^k \right\| = 0;$$

that is, $L = \lim_{n \to \infty} \sum_{k=0}^{n} Q^k$. We also have the following technical lemma; we leave the proof to the reader as it is a good exercise in using the basic properties discussed so far.

Lemma 3.8. *Suppose Q and $\sum_{k=0}^{\infty} Q^k$ are both in the Banach algebra $\mathcal{B}$. Then*

$$Q \sum_{k=0}^{\infty} Q^k = \sum_{k=1}^{\infty} Q^k.$$

We are now ready to prove the theorem. The proof of the theorem relies on the completeness property of the space $\mathcal{B}$, and the submultiplicative property.

Proof of Theorem 3.7. Our first task is to demonstrate that $\sum_{k=0}^{\infty} Q^k$ is an element of $\mathcal{B}$. Since by assumption $\mathcal{B}$ is complete it is sufficient to show that the sequence $T_n := \sum_{k=0}^{n} Q^k$ is a Cauchy sequence. By the submultiplicative inequality we have that

$$\|Q^k\| \le \|Q\|^k.$$

For $m > n$ we see that

$$\|T_m - T_n\| = \| \sum_{k=n+1}^{m} Q^k \| \le \sum_{k=n+1}^{m} \|Q^k\| \le \sum_{k=n+1}^{m} \|Q\|^k,$$

where the left-hand inequality follows by the triangle inequality. It is straightforward to show that $\sum_{k=n+1}^{m} \|Q\|^k = \|Q\|^{n+1} \frac{1-\|Q\|^{m-n}}{1-\|Q\|}$, since the former is a geometric series. We conclude that

$$\|T_m - T_n\| \le \frac{\|Q\|^{n+1}}{1 - \|Q\|},$$

so T_n is a Cauchy sequence as we required.

Having established that $\sum_{k=0}^{\infty} Q^k$ is an element of $\mathcal{B}$ we now show that it is the inverse of $I - Q$. We look at the product

$$(I - Q) \sum_{k=0}^{\infty} Q^k = \sum_{k=0}^{\infty} Q^k - Q \sum_{k=0}^{\infty} Q^k = I + \sum_{k=1}^{\infty} Q^k - Q \sum_{k=0}^{\infty} Q^k = I,$$

where the right-hand side follows by invoking Lemma 3.8. Similarly we can show that $\sum_{k=0}^{\infty} Q^k \cdot (I - Q) = I$, and therefore $\sum_{k=0}^{\infty} Q^k$ is the inverse of $I - Q$.   ∎

Thus we conclude our brief incursion into the theory of Banach algebras. We now return to the special case of operators, to discuss the spectrum; here we will find a first application for the above small gain result.

3.2.2   Some elements of spectral theory

In this section we introduce the spectrum of an operator and discuss a few elementary properties. The objective is to cover only some basic facts which will be used later in the course. For an extensive treatment of this important topic in analysis, see the references at the end of the chapter.

Definition 3.9. *Let $\mathcal{V}$ be a Banach space, and $M \in \mathcal{L}(\mathcal{V})$. The spectrum of M is defined by*

$$spec(M) := \{\lambda \in \mathbb{C} : (\lambda I - M) \text{ is not invertible in } \mathcal{L}(\mathcal{V})\},$$

and the spectral radius of M by

$$\rho(M) := \max\{|\lambda| : \lambda \in spec(M)\}.$$

A non-obvious fact which is implicit in the definition of $\rho(M)$ is that the spectrum is nonempty. It is also a closed set. We will accept these facts without proof.

Examples:

As usual we begin by considering the finite dimensional case, with $M \in \mathbb{C}^{n \times n}$. It is clear that in this case the spectrum consists of the set of eigenvalues of M, and the spectral radius coincides with the largest absolute value of any eigenvalue of M.

The next example shows that there may be more to the spectrum than eigenvalues. Let $\mathcal{V} = L_2[0, \infty)$ and define the operator M as

$$M : u(t) \mapsto e^{-t} u(t).$$

It is easy to show that the operator is bounded since e^{-t} is a bounded function on $[0, \infty)$. Now we claim that $spec(M)$ is the real interval $[0, 1]$. To see this notice that

$$(\lambda I - M) : u(t) \mapsto (\lambda - e^{-t}) u(t).$$

If $\lambda \notin [0,1]$, then we can define an inverse function

$$\varphi(t) = \frac{1}{\lambda - e^{-t}} \quad \text{for } t \geq 0,$$

that is bounded on $[0, \infty)$, defining a bounded inverse operator on $L_2[0, \infty)$,

$$(\lambda I - M)^{-1} v(t) \mapsto \frac{1}{\lambda - e^{-t}} v(t).$$

So such λ is not in the spectrum, and $\text{spec}(M) \subset [0,1]$. If $\lambda \in [0,1]$, the inverse function $\varphi(t)$ will go to infinity at a $t = -\log(\lambda)$ (or $t = \infty$ when $\lambda = 0$), and it can be deduced that there is no bounded inverse to the operator $(\lambda I - M)$. Therefore $\text{spec}(M) = [0,1]$.

However, notice that

$$(\lambda - e^{-t})u(t) = 0 \quad \text{implies } u(t) = 0, \text{ for almost every } t,$$

so the operator M has no eigenvalues. $\qquad \square$

The next proposition relates the spectral radius to the norm, and in particular implies the spectral radius is always finite, or equivalently that the spectrum is a bounded set. The proof is based on the small gain result of Theorem 3.7.

Proposition 3.10. *The inequality $\rho(M) \leq \|M\|$ holds for every operator $M \in \mathcal{L}(\mathcal{V})$.*

Proof. Suppose $|\lambda| > \|M\|$. Then setting $Q = M/\lambda$ we have $\|Q\| < 1$ and we invoke Theorem 3.7 to see that $I - Q$ is invertible, with inverse

$$\left(I - \frac{M}{\lambda}\right)^{-1} = \sum_{k=0}^{\infty} \frac{M^k}{\lambda^k}.$$

Now it follows immediately that

$$\sum_{k=0}^{\infty} \frac{M^k}{\lambda^{k+1}} = \lambda^{-1} \left(I - \frac{M}{\lambda}\right)^{-1} = (\lambda I - M)^{-1}, \qquad (3.3)$$

so $\lambda \notin \text{spec}(M)$. This means that $\text{spec}(M)$ is included in the closed ball of radius $\|M\|$, which completes the proof. $\qquad \blacksquare$

The previous result generalizes the well-known relationship between spectral radius and maximum singular value of a matrix, and in general there is a gap between these two quantities. The spectral radius can, however, be characterized exactly in terms of the norms of the *powers* of M.

Proposition 3.11. *The spectral radius of an operator $M \in \mathcal{L}(\mathcal{V})$ satisfies $\rho(M) = \lim_{k \to \infty} \|M^k\|^{\frac{1}{k}}$.*

This statement implies, in particular, that the limit on the right-hand side exists for every M. We will not prove this result here, but remark that this can be done by studying the radius of convergence of the power series in (3.3), that has similar properties as scalar power series. For the case of matrices, the reader is invited to give a more direct proof based on the Jordan form.

To conclude the section we will show a result regarding the invariance of the spectral radius to changes in the order of multiplication of operators. The key observation is the following.

Lemma 3.12. *Consider the operators $M \in \mathcal{L}(\mathcal{U}, \mathcal{V})$ and $Q \in \mathcal{L}(\mathcal{V}, \mathcal{U})$, and let $I_\mathcal{U}$ and $I_\mathcal{V}$ be the identity operators. Then*

$$I_\mathcal{V} - MQ \text{ is invertible if and only if } I_\mathcal{U} - QM \text{ is invertible.}$$

Proof. Assuming $I_\mathcal{V} - MQ$ is invertible, we can explicitly construct the inverse

$$(I_\mathcal{U} - QM)^{-1} = I_\mathcal{U} + Q(I_\mathcal{V} - MQ)^{-1}M.$$

In other words, the operator on the right-hand side is well defined, and routine operations show it is the inverse of $I_\mathcal{U} - QM$. The converse implication is analogous. ∎

Having shown this, we can state the following result regarding the spectrum.

Proposition 3.13. *Consider the operators $M \in \mathcal{L}(\mathcal{U}, \mathcal{V})$ and $Q \in \mathcal{L}(\mathcal{V}, \mathcal{U})$. Then*

$$spec(MQ) \cup \{0\} = spec(QM) \cup \{0\}.$$

That is, the sets coincide except possibly for the value $0 \in \mathbb{C}$. In particular,

$$\rho(MQ) = \rho(QM).$$

Proof. Consider $0 \neq \lambda \in \mathbb{C}$. Then

$$\lambda \notin spec(MQ) \text{ if and only if } \lambda I - MQ \text{ invertible,}$$
$$\text{if and only if } I - \frac{M}{\lambda}Q \text{ invertible,}$$
$$\text{if and only if } I - Q\frac{M}{\lambda} \text{ invertible,}$$
$$\text{if and only if } \lambda I - QM \text{ invertible,}$$
$$\text{if and only if } \lambda \notin spec(QM),$$

where the only non-trivial step follows from Lemma 3.12. Therefore the spectra can differ at most in the element $\lambda = 0$; this potential difference cannot affect the spectral radius. ∎

Example:

Consider two matrices $M \in \mathbb{C}^{m \times n}$ and $Q \in \mathbb{C}^{n \times m}$, where, e.g., $m < n$. The previous result implies that the nonzero eigenvalues of $MQ \in \mathbb{C}^{m \times m}$

and $QM \in \mathbb{C}^{n\times n}$ coincide. Clearly QM will have additional eigenvalues at 0, since its rank is smaller than its dimension n.

□

So far we have developed our results for operators on Banach spaces. For the topics to follow we focus our attention on operators on Hilbert space.

3.2.3 Adjoint operators in Hilbert space

We now introduce the *adjoint* of an operator in Hilbert space.

Definition 3.14. *Suppose V and Z are Hilbert spaces, and $F \in \mathcal{L}(V, Z)$. The operator $F^* \in \mathcal{L}(Z, V)$ is the adjoint of F if*

$$\langle z, Fv\rangle_Z = \langle F^*z, v\rangle_V$$

for all $v \in V$ and $z \in Z$.

Let us look at some examples.

Examples:

The simplest example of this is where $V = \mathbb{C}^n$ and $Z = \mathbb{C}^m$, which are equipped with the usual inner product, and $F \in \mathbb{C}^{m\times n}$. Then the adjoint of the operator F is exactly equal to its conjugate transpose of the matrix F, which motivates the notation F^*.

Another example is given by convolution operators: suppose that f is a scalar function in $L_1[0, \infty)$ and the operator Q on $L_2[0, \infty)$ is again defined by

$$(Qu)(t) := \int_0^t f(t - \tau)u(\tau)d\tau, \tag{3.4}$$

for u in $L_2[0, \infty)$. Then we leave it to the reader to show that the adjoint operator Q^* is given by

$$(Q^*z)(t) = \int_t^\infty f^*(\tau - t)z(\tau)\,d\tau, \quad \text{for } z \text{ in } L_2[0, \infty),$$

where f^* denotes the complex conjugate of the function f.   □

Some basic properties of the adjoint are given in the following statement; while the existence proof is beyond the scope of this course, the remaining facts are covered in the exercises at the end of the chapter.

Proposition 3.15. *For any $F \in \mathcal{L}(V, Z)$, the adjoint exists, is unique and satisfies*

$$\|F\| = \|F^*\| = \|F^*F\|^{\frac{1}{2}}. \tag{3.5}$$

An operator F is called *self-adjoint* if $F^* = F$, which generalizes the notion of a Hermitian matrix. It is easy to show that when F is self-adjoint, the *quadratic form*

$$\phi(v) := \langle Fv, v \rangle$$

takes only real values. Therefore we have $\phi : (\mathcal{V}, \mathcal{V}) \longrightarrow \mathbb{R}$. Such quadratic forms play an important role later in the course.

As in the matrix case, it is natural to inquire about the sign of such forms. We will say that a self-adjoint operator is

- *positive semidefinite* (denoted $F \geq 0$) if $\langle Fv, v \rangle \geq 0$ for all $v \in \mathcal{V}$;

- *positive definite* (denoted $F > 0$) if there exists $\epsilon > 0$ such that $\langle Fv, v \rangle \geq \epsilon \|v\|^2$, for all $v \in \mathcal{V}$.

We remark that an operator F satisfying $\langle Fv, v \rangle > 0$ for all nonzero $v \in \mathcal{V}$, is not guaranteed to be positive as it is in the matrix case; the exercises provide examples of this. An important property is that given an operator in one of these classes, there always exists a *square root operator* $F^{\frac{1}{2}}$ of the same class, such that $(F^{\frac{1}{2}})^2 = F$.

We consider some spectral properties of self-adjoint operators.

Proposition 3.16. *Let $M \in \mathcal{L}(\mathcal{V})$, and suppose that $M = M^*$. Then $\rho(M) = \|M\|$.*

The proof is a simple exercise based on Propositions 3.15 and 3.11, and is left for the reader. The following is an immediate corollary, that generalizes the familiar matrix property that the square of $\bar{\sigma}(A)$ is equal to the maximum eigenvalue of A^*A.

Corollary 3.17. *Let $M \in \mathcal{L}(\mathcal{U}, \mathcal{V})$, where $\mathcal{U}, \mathcal{V}$ are Hilbert spaces. Then $\|M\|^2 = \rho(M^*M)$.*

We remark, without proof, that the spectrum of a self-adjoint operator is contained in the real line, and a positive semidefinite operator has spectrum in the non-negative half-line.

We now introduce another important definition involving the adjoint. An operator $U \in \mathcal{L}(\mathcal{V}, \mathcal{Z})$ is called *isometric* if it satisfies

$$U^*U = I.$$

The reason for this terminology is that these operators satisfy

$$\langle Uv_1, Uv_2 \rangle = \langle U^*Uv_1, v_2 \rangle = \langle v_1, v_2 \rangle,$$

for any $v_1, v_2 \in \mathcal{V}$, i.e., the operator preserves inner products. In particular, isometries preserve norms and therefore distances: they are "rigid" transformations. A consequence of this is that they satisfy $\|U\| = 1$, but the isometric property is clearly more restrictive.

An isometric operator is called *unitary* if $U^* = U^{-1}$; in other words,

$$U^*U = I \text{ and } UU^* = I.$$

Unitary operators are bijective mappings that preserve the all the structure of a Hilbert space; if a unitary $U \in \mathcal{L}(\mathcal{V}, \mathcal{Z})$ exists, the spaces $\mathcal{V}, \mathcal{Z}$ are *isomorphic*, they can be identified from an abstract point of view.

Example:

A matrix $U \in \mathbb{C}^{m \times n}$ whose columns $u_1, \ldots, u_n$ are orthonormal vectors in $\mathbb{C}^m$ is an isometry; if in addition $m = n$, U is unitary.

$\square$

Before leaving this section we again emphasize that we use the same notation " $*$ " to denote complex conjugate of a scalar, complex conjugate transpose of a matrix, and adjoint of an operator; a moment's thought will convince you that the former two are just special cases of the latter and thus this helps keep our notation economical.

So far in this chapter we have introduced a number of general mathematical concepts and tools. We will now turn to some more customized material.

3.3 Frequency domain spaces: Signals

In the previous sections we have introduced, by way of examples, time domain spaces of functions which will play the role of signals in our control problems. Frequently in the sequel it will be advantageous to reformulate problems in the frequency domain where they can be simpler to solve or conceptualize. In particular we will see that L_2 spaces are particularly suited for a frequency domain treatment. In this section we will introduce frequency domain spaces, and describe the various relationships they have with the time domain spaces.

3.3.1 The space $\hat{L}_2$ and the Fourier transform

We now define the complex inner product space $\hat{L}_2(j\mathbb{R})$, which consists of functions mapping the imaginary axis $j\mathbb{R}$ to $\mathbb{C}^n$ with the inner product

$$\langle \hat{u}, \hat{v} \rangle_2 := \frac{1}{2\pi} \int_{-\infty}^{\infty} \hat{u}^*(j\omega)\hat{v}(j\omega)d\omega\,.$$

Thus a function $\hat{u} : j\mathbb{R} \to \mathbb{C}^n$ is in $\hat{L}_2(j\mathbb{R})$ if $\langle \hat{u}, \hat{u} \rangle_2 = \|\hat{u}\|_2^2 < \infty$. Here we use the same notation for the norm and inner product of $\hat{L}_2$ as we did for L_2, which is intended to always be clear from the context as frequency domain objects will be denoted by " $\hat{}$ ".

The Fourier transform of a function $u : \mathbb{R} \to \mathbb{C}^n$ is defined to be

$$\hat{u}(j\omega) = \int_{-\infty}^{\infty} u(t)e^{-j\omega t}dt.$$

As usual, writing such an integral raises the issue of its convergence. It is easy to see that the integral converges absolutely for $u(t) \in L_1(-\infty, \infty)$, but we are interested in defining Fourier transforms on $L_2(-\infty, \infty)$, where this strong convergence may not hold. One approach to avoid this difficulty is to consider the "principal value" limit

$$\hat{u}(j\omega) = \lim_{T \to \infty} \int_{-T}^{T} u(t)e^{-j\omega t}dt.$$

We will adopt the following convention: $\hat{u}(j\omega)$ is defined as the above limit at values of ω where it exists, and $\hat{u}(j\omega) = 0$ otherwise. In this way the transform is defined for any function which is integrable on finite intervals. For convenience we denote this operation by the map Φ, and write simply

$$\hat{u} = \Phi u$$

to indicate the above relationship. It can be shown (see the references) that when $u(t) \in L_2(-\infty, \infty)$ this limit exists for almost all ω.

Given a function $\hat{u} : j\mathbb{R} \to \mathbb{C}^n$ we define the inverse Fourier transform of $\hat{u}$ by

$$u(t) = \frac{1}{2\pi} \int_{-\infty}^{\infty} \hat{u}(j\omega)e^{j\omega t}d\omega,$$

with an analogous convention regarding the convergence of the integral, and use $u = \Phi^{-1}\hat{u}$ to indicate this transformation.

As the notation would indicate, for certain classes of functions these maps are inverses of each other. For our purposes, it suffices to note that for $u \in L_2(-\infty, \infty)$ and $\hat{v} \in \hat{L}_2(j\mathbb{R})$, we have

$$u(t) = \Phi^{-1}(\Phi u)(t), \text{ for almost every } t,$$

and

$$\hat{v}(j\omega) = \Phi(\Phi^{-1}\hat{v})(j\omega), \text{ for almost every } \omega.$$

Furthermore, we have the following key result known as the Plancherel theorem.

Theorem 3.18. *With the Fourier transform and its inverse defined as above:*

(i) *The map* $\Phi : L_2(-\infty, \infty) \to \hat{L}_2(j\mathbb{R})$, *and given any* $u, v \in L_2(-\infty, \infty)$ *the equality*

$$\langle u, v \rangle_2 = \langle \Phi u, \Phi v \rangle_2 \quad holds.$$

(ii) The map $\Phi^{-1} : \hat{L}_2(j\mathbb{R}) \to L_2(-\infty, \infty)$, and if $\hat{u}, \hat{v} \in \hat{L}_2(j\mathbb{R})$, then

$$\langle \hat{u}, \hat{v} \rangle_2 = \langle \Phi^{-1}\hat{u}, \Phi^{-1}\hat{v} \rangle_2 \text{ is satisfied.}$$

This theorem says that the Fourier transform is an invertible map between the spaces $L_2(-\infty, \infty)$ and $\hat{L}_2(j\mathbb{R})$, and more importantly, this map preserves the inner product; thus the Fourier transform is a unitary operator which makes the two spaces isomorphic. In particular the theorem says for any $\hat{v} \in \hat{L}_2(j\mathbb{R})$ and $u \in L_2(-\infty, \infty)$ we have

$$\|\Phi u\|_2 = \|u\|_2 \quad \text{and} \quad \|\Phi^{-1}\hat{v}\|_2 = \|\hat{v}\|_2;$$

namely, norms are preserved under this mapping.

3.3.2 The spaces H_2 and $H_2^\perp$ and the Laplace transform

In this section we introduce two subspaces of $\hat{L}_2(j\mathbb{R})$ that will be important in our studies of causal systems.

Recall the definition of $L_2[0, \infty)$, the Hilbert subspace of functions in $L_2(-\infty, \infty)$ which are zero in $(-\infty, 0)$. Given $u \in L_2[0, \infty)$, we define its Laplace transform by the integral

$$\hat{u}(s) := \lim_{T \to \infty} \int_0^T e^{-st}u(t)dt = \int_0^\infty e^{-st}u(t)dt, \qquad (3.6)$$

when this limit exists, and set $\hat{u}(s) = 0$ at the divergent values of s. We will soon see, however, that the integral converges absolutely when $\text{Re}(s) > 0$. We use the notation

$$\hat{u} = \Lambda u$$

to indicate this transformation. If we evaluate the transform at $s = j\omega$, for some real number ω, we see that since $u(t) \in L_2[0, \infty)$,

$$\hat{u}(j\omega) = \int_0^\infty e^{-j\omega t}u(t)dt = \int_{-\infty}^\infty e^{-j\omega t}u(t)dt$$

is the Fourier transform of $u(t)$; therefore there is no source of ambiguity in the notation "$\hat{u}$"; depending on its argument it will denote the Laplace transform or the Fourier restriction. Notice that by the Plancherel theorem, $\hat{u}(j\omega) \in \hat{L}_2(j\mathbb{R})$, and $\|\hat{u}\|_2 = \|u\|_2$.

Next notice that for $s \in \mathbb{C}$, with $\text{Re}(s) > 0$, we have that the scalar function defined by

$$g_s(t) = \begin{cases} e^{-s^*t} & \text{for } t \geq 0, \\ 0 & \text{otherwise,} \end{cases}$$

is in $L_2[0, \infty)$. Thus we see that if $u(t)$ is scalar, the Laplace transform $\hat{u}(s)$ is given by the following inner product

$$\hat{u}(s) = \langle g_s, u \rangle_2,$$

and therefore the integral (3.6) converges for all s satisfying $\mathrm{Re}(s) > 0$. This is also true if $u(t)$ is vector valued, applying the previous argument for each component.

To state some additional properties of $\hat{u}(s)$ we introduce a new function space H_2 in the following definition.

Definition 3.19. *A function* $\hat{u} : \bar{\mathbb{C}}^+ \to \mathbb{C}^n$ *is in* H_2 *if*

(a) $\hat{u}(s)$ *is analytic in the open right half-plane* $\mathbb{C}^+$;

(b) for almost every real number ω,

$$\lim_{\sigma \to 0^+} \hat{u}(\sigma + j\omega) = \hat{u}(j\omega);$$

(c) $\sup_{\sigma \geq 0} \int_{-\infty}^{\infty} |\hat{u}(\sigma + j\omega)|_2^2 \, d\omega < \infty.$

We remark that the supremum in (c) is always achieved at $\sigma = 0$ when $\hat{u}$ is in H_2. Also that any function $\hat{u}$ which is only defined on the open half-plane $\mathbb{C}^+$ can be extended to $\bar{\mathbb{C}}^+$ so that it is in H_2 if (a) holds, and (c) is satisfied when the supremum taken over $\mathbb{C}^+$.

The key result we now state is that the Laplace transform of a function $L_2[0, \infty)$ satisfies these properties, and furthermore every function in H_2 can be obtained in this form.

Theorem 3.20.

(a) If $u \in L_2[0, \infty)$ *then* $\Lambda u \in H_2$.

(b) If $\hat{u} \in H_2$, *then there exists* $u \in L_2[0, \infty)$ *satisfying* $\Lambda u = \hat{u}$.

We will not prove this theorem, since it would take us too far afield, but only make a few remarks. Both parts are non-trivial. Part (a) says that all the requirements in Definition 3.19 are met by a Laplace transform $\hat{u}(s)$ of a function in $L_2[0, \infty)$.

Part (b) is called the Paley–Wiener theorem, and in essence says that the mapping Λ has an inverse defined on H_2. This mapping is the inverse Laplace transform and can be shown to be given explicitly by

$$u(t) = (\Lambda^{-1} \hat{u})(t) = \frac{1}{2\pi} \int_{-\infty}^{\infty} e^{\sigma t} \cdot e^{j\omega t} \hat{u}(\sigma + j\omega) d\omega,$$

where σ is any positive real number. In particular it can be shown that for $\hat{u} \in H_2$, the above integral is independent of σ and is equal to zero for $t < 0$.

A useful consequence of Theorem 3.20 is the following:

Corollary 3.21. *Let* $\hat{u}$ *and* $\hat{v}$ *be functions in* H_2, *and* $\hat{u}(j\omega) = \hat{v}(j\omega)$ *for every* $\omega \in \mathbb{R}$. *Then* $\hat{u}(s) = \hat{v}(s)$ *for every* $s \in \bar{\mathbb{C}}^+$.

Proof. From Theorem 3.20 we find $u(t)$ and $v(t)$ in $L_2[0, \infty)$ such that $\hat{u} = \Lambda u$, $\hat{v} = \Lambda v$. Then $\hat{u} - \hat{v} = \Lambda(u - v)$.

Since $\hat{u}(j\omega) - \hat{v}(j\omega) = 0$, we conclude that the Fourier transform $\Phi(u - v) = 0$. Since Φ is an isomorphism we must have $u(t) - v(t) = 0$ for almost all t. Now the Laplace transform gives $\hat{u}(s) = \hat{v}(s)$, for every $s \in \bar{\mathbb{C}}^+$.   ∎

From the previous corollary we find that there is a one-to-one correspondence between functions in H_2 and their restrictions to the imaginary axis. Thus we are justified in identifying these two functions and writing $H_2 \subset \hat{L}_2(j\mathbb{R})$, which involves some abuse of notation but no ambiguity. We say a function $\hat{u}(j\omega) \in \hat{L}_2(j\mathbb{R})$ is in H_2 if it has an *analytic continuation* $\hat{u}(s)$ to the right half-plane which satisfies the conditions of Definition 3.19.

Given this identification, we can endow H_2 with the inner product inherited from $\hat{L}_2(j\mathbb{R})$, defining for $\hat{u}$ and $\hat{v}$ in H_2

$$\langle \hat{u}, \hat{v} \rangle_2 = \frac{1}{2\pi} \int_{-\infty}^{\infty} \hat{u}^*(j\omega)\hat{v}(j\omega)\, d\omega.$$

In this way H_2 is a Hilbert subspace of $\hat{L}_2(j\mathbb{R})$.

Example:

To gain insight into Definition 3.19, consider the function

$$\hat{u}(s) = \frac{e^s}{s+1}.$$

It is easily shown that $\hat{u}(j\omega) \in \hat{L}_2(j\mathbb{R})$, and also $\hat{u}(s)$ satisfies (a) and (b) in the definition. However (c) is not satisfied, so $\hat{u}(s) \notin H_2$.

In the time domain, it is not difficult to show that

$$\Phi^{-1}\hat{u} = u(t) = \begin{cases} e^{-(t+1)}, & \text{for } t \geq -1, \\ 0, & \text{otherwise,} \end{cases}$$

that is not a function in $L_2[0, \infty)$.   □

Having identified H_2 as a subspace of $\hat{L}_2(j\mathbb{R})$, we immediately wonder about the difference between these two spaces; how do we need to complement H_2 to obtain all of $\hat{L}_2(j\mathbb{R})$?

To work towards an answer it is useful to return to the time domain; here it is very easy to characterize the difference between $L_2(-\infty, \infty)$ and $L_2[0, \infty)$. In particular, the space $L_2(-\infty, 0]$ is the *orthogonal complement* of $L_2[0, \infty)$: functions in both spaces are orthogonal to each other, and any function $u(t) \in L_2(-\infty, \infty)$ can be trivially written as a sum

$$u(t) = u_+(t) + u_-(t), \quad u_+(t) \in L_2[0, \infty), \quad u_-(t) \in L_2(-\infty, 0].$$

We denote this relationship by

$$L_2(-\infty, 0] = L_2^\perp[0, \infty).$$

Some properties of orthogonal complements in general Hilbert spaces are included in the exercises at the end of the chapter.

Returning to the frequency domain, we are led to ask the question, what is the Fourier image of $L_2(-\infty, 0]$? Not surprisingly, the answer involves the use of Laplace transforms. For $u(t) \in L_2(-\infty, 0]$, define the *left-sided* Laplace transform of u by

$$(\Lambda_- u)(s) := \int_{-\infty}^{0} e^{-st} u(t) dt,$$

with the usual convention regarding convergence. Now it is easy to see that this transform converges on the *left* hand plane $s \in \mathbb{C}^-$. Furthermore, for $s = j\omega$ it coincides with the Fourier transform.

This symmetric situation leads to introduce the space $H_2^\perp$ of functions mapping $\bar{\mathbb{C}}$ to $\mathbb{C}^n$ which satisfy: $\hat{y}$ is in $H_2^\perp$ if

$$\hat{y}(-s) \text{ is a member of } H_2.$$

In particular, functions in $H_2^\perp$ are analytic on $\mathbb{C}_-$ and have limit almost everywhere when approaching the imaginary axis horizontally from the left.

As expected we have the following result.

Proposition 3.22.

  (a) *If $u \in L_2(-\infty, 0]$ then $\Lambda_- u \in H_2^\perp$.*

  (b) *If $\hat{u} \in H_2^\perp$, then there exists $u \in L_2(-\infty, 0]$ satisfying $\Lambda_- u = \hat{u}$.*

Proof. We reduce the statement to Theorem 3.20 by a symmetry in the time axis. We have

$$u(t) \in L_2(-\infty, 0] \text{ if and only if } u(-t) \in L_2[0, \infty),$$
$$\text{if and only if } \Lambda[u(-t)] \in H_2,$$

and it is easy to show that

$$\Lambda[u(-t)](s) = \int_0^\infty u(-t) e^{-st} dt = \int_{-\infty}^0 u(t) e^{st} dt = \Lambda_-[u(t)](-s).$$

So $\Lambda[u(-t)] \in H_2$ if and only if $\Lambda_- u \in H_2^\perp$.   ∎

Analogously to Corollary 3.21 it follows that functions in $H_2^\perp$ are characterized by their values on the imaginary axis, so once again we can write $H_2^\perp \subset \hat{L}_2(j\mathbb{R})$, identifying $H_2^\perp$ with a Hilbert subspace of $\hat{L}_2(j\mathbb{R})$, consisting of functions $\hat{u}(j\omega)$ which have analytic continuation to the *left* half-plane with the corresponding boundedness property.

To complete our picture we inquire as to how do the subspaces H_2 and $H_2^\perp$ relate to each other? As the notation suggests, we have the following answer.

Proposition 3.23. *The space $H_2^\perp$ is the orthogonal complement of H_2 in $\hat{L}_2(j\mathbb{R})$. Namely, if $\hat{u} \in H_2$ and $\hat{v} \in H_2^\perp$ then $\langle \hat{u}, \hat{v} \rangle_2 = 0$, and also every* *function $\hat{u}(j\omega) \in \hat{L}_2(j\mathbb{R})$ can be written uniquely as a sum*

$$\hat{u}(j\omega) = \hat{u}_+(j\omega) + \hat{u}_-(j\omega), \quad \hat{u}_+ \in H_2, \ \hat{u}_- \in H_2^\perp.$$

The proof is a direct application of the Plancherel theorem, since we have established that as subspaces of $\hat{L}_2(j\mathbb{R})$, H_2 and $H_2^\perp$ are the Fourier images of $L_2[0,\infty)$ and $L_2(-\infty,0]$. We leave details to the reader.

3.3.3 Summarizing the big picture

It is useful now to summarize what we have learned about frequency domain signal spaces. The main conclusions are represented in the following diagram, which includes all the spaces and relevant transforms.

$$
\begin{array}{ccccc}
L_2(-\infty,0] & \subset & L_2(-\infty,\infty) & \supset & L_2[0,\infty) \\
\Lambda_- \downarrow \uparrow \Lambda_-^{-1} & & \Phi \downarrow \uparrow \Phi^{-1} & & \Lambda \downarrow \uparrow \Lambda^{-1} \\
H_2^\perp & \subset & \hat{L}_2(j\mathbb{R}) & \supset & H_2
\end{array}
$$

The time domain space $L_2(-\infty,\infty)$ is isomorphic to the frequency domain space $\hat{L}_2(j\mathbb{R})$ by means of the Fourier transform, which is invertible and preserves inner products. In the time domain we can identify the natural subspaces $L_2(-\infty,0]$ and $L_2[0,\infty)$, which are mutually orthogonal. In the frequency domain, these are mapped by the Fourier transform to the subspaces $H_2^\perp$ and H_2 of $\hat{L}_2(j\mathbb{R})$; functions in these spaces are defined over the imaginary axis, but also admit analytic continuations respectively to each half-plane, with bounded norms over vertical lines. The Laplace transforms Λ_- and Λ extend the Fourier transform to each analytic domain.

Examples:

To gain more insight into the frequency domain spaces that have been introduced, it is useful to discuss which *rational functions* belong to each of them. Recall from §2.6 that a scalar rational function is a ratio of polynomials

$$\hat{u}(s) = \frac{p(s)}{q(s)},$$

and the concept extends to vectors or matrices with components of this form. We will denote by $R\hat{L}_2$, RH_2, and $RH_2^\perp$ the sets of rational functions which belong, respectively, to $\hat{L}_2(j\mathbb{R})$, H_2, and $H_2^\perp$. For example, $\hat{u}(s)$ is in $R\hat{L}_2$ if $\hat{u}(j\omega) \in \hat{L}_2(j\mathbb{R})$.

It is not difficult to show that a vector rational function is

(i) in $R\hat{L}_2$ if it is strictly proper and has no poles on the imaginary axis;

(ii) in RH_2 if it is strictly proper and has no poles on the closed right half-plane $\bar{\mathbb{C}}^+$;

(iii) in $RH_2^\perp$ if it is strictly proper and has no poles on the closed left half-plane $\bar{\mathbb{C}}^-$.

The subspaces $R\hat{L}_2$, RH_2, and $RH_2^\perp$ are themselves inner product spaces, but are not complete. In particular it can be shown that they are dense in their respective Hilbert spaces $\hat{L}_2(j\mathbb{R})$, H_2 and $H_2^\perp$; this means that for every element in H_2 is the limit of a sequence of functions in RH_2, and analogously for the other spaces.

It follows directly from the partial fraction expansion that every function in $R\hat{L}_2$ can be written as a sum of functions RH_2, and $RH_2^\perp$; what is not so immediate is that these subspaces are perpendicular; i.e., given a rational function $\hat{u}(s)$ with no poles in $\bar{\mathbb{C}}^+$, and another one $\hat{v}(s)$ with no poles in $\bar{\mathbb{C}}^-$,

$$\frac{1}{2\pi} \int_{-\infty}^{\infty} \hat{u}^*(j\omega)\hat{v}(j\omega)\,d\omega = 0.$$

This latter fact follows from the preceding theory.   □

Our next goal is to move from alternate representations for signals to equivalent system representations.

3.4   Frequency domain spaces: Operators

Having studied the structure of the L_2 and $\hat{L}_2$ Hilbert spaces, we now turn to the study of *operators*. Since the above spaces are isomorphic it is clear that so must $\mathcal{L}(L_2)$ and $\mathcal{L}(\hat{L}_2)$. Namely, with each $M \in \mathcal{L}(L_2)$ we can associate $\tilde{M} = \Phi M \Phi^{-1} \in \mathcal{L}(\hat{L}_2)$, where Φ is the Fourier transform. Thus we can choose to study operators in whichever domain is most convenient. In particular, we will see in this section that the frequency domain provides a very elegant, exact characterization of an important class of operators on L_2: those which are linear and time invariant.

Our study begins by introducing the class of *multiplication* operators on $\hat{L}_2(j\mathbb{R})$. First define the space $\hat{L}_\infty(j\mathbb{R})$ of matrix valued functions $j\mathbb{R} \to \mathbb{C}^{m \times n}$ such that

$$\|\hat{G}\|_\infty = \operatorname*{ess\,sup}_{\omega \in \mathbb{R}} \bar{\sigma}(\hat{G}(j\omega)) < \infty.$$

The subscript ∞ in this norm corresponds to the essential supremum over frequency, analogously to the time domain L_∞ space introduced before.

Notice, however, that we are using matrix functions and taking a maximum singular value norm at each frequency [1].

The next result shows that $\hat{L}_\infty(j\mathbb{R})$ is a representation for a set of linear bounded operators on $\hat{L}_2(j\mathbb{R})$.

Proposition 3.24. *Every function $\hat{G} \in \hat{L}_\infty(j\mathbb{R})$ defines a bounded linear operator*

$$M_{\hat{G}} : \hat{L}_2(j\mathbb{R}) \to \hat{L}_2(j\mathbb{R}),$$

via the relationship $(M_{\hat{G}}\hat{u})(j\omega) = \hat{G}(j\omega)\hat{u}(j\omega)$. Also

$$\|M_{\hat{G}}\|_{\hat{L}_2 \to \hat{L}_2} = \|\hat{G}\|_\infty.$$

The operator $M_{\hat{G}}$ is called, for obvious reasons, a multiplication operator; we see that the induced norm of this operator is exactly $\|\hat{G}\|_\infty$, which is left as an exercise. We also have that $\hat{G}$ defines an operator on $L_2(-\infty, \infty)$, denoted by G, and defined by

$$G = \Phi^{-1} M_{\hat{G}} \Phi.$$

This operator has some very special properties. We call $\hat{G}(j\omega)$ the *frequency response* of G, when G satisfies the last equation. The diagram below summarizes the relationships between operators and spaces defined so far.

$$
\begin{array}{ccc}
L_2(-\infty, \infty) & \xrightarrow{\ \ G\ \ } & L_2(-\infty, \infty) \\
\Phi \downarrow & & \downarrow \Phi \\
\hat{L}_2(j\mathbb{R}) & \xrightarrow{\ \ M_{\hat{G}}\ \ } & \hat{L}_2(j\mathbb{R})
\end{array}
$$

Having defined the concept of a multiplication operator, we proceed to study some of its features.

3.4.1 Time invariance and multiplication operators

In this section we investigate some of the temporal properties of multiplication operators. Our discussion is brought into focus by defining the shift operator S_τ, for $\tau \geq 0$, on time domain vector valued functions. If u is a function $\mathbb{R} \to \mathbb{C}^n$, the action of the shift operator is given by

$$(S_\tau u)(t) = u(t - \tau).$$

---

[1] Bear in mind that, even in the scalar case, $\hat{L}_\infty(j\mathbb{R})$ is *not* related to $L_\infty(-\infty, \infty)$ via the Fourier transform.

The inverse S_τ^{-1} is also defined on these functions, and satisfies

$$(S_\tau^{-1}u)(t) = u(t+\tau).$$

Suppose $u \in L_2(-\infty, \infty)$, and let $y = S_\tau u$. Then taking Fourier transforms we have

$$\hat{y}(j\omega) = e^{-j\tau\omega} \cdot \hat{u}(j\omega),$$

so S_τ is represented in the frequency domain by the multiplication by $e^{-j\tau\omega}$ in $\hat{L}_\infty(j\mathbb{R})$.

This brings us to the definition of *time invariance*: an operator Q mapping $L_2(-\infty, \infty) \to L_2(-\infty, \infty)$ is *time invariant* if

$$S_\tau Q = QS_\tau, \quad \text{for all} \quad \tau \geq 0.$$

From this definition it is immediate to see that every multiplication operator $M_{\hat{G}}$ defines a time invariant $G = \Phi^{-1}M_{\hat{G}}\Phi$. In fact, $S_\tau G$ is represented by

$$e^{-j\tau w}\hat{G}(j\omega) = \hat{G}(j\omega)e^{-j\tau w},$$

where the right-hand expression is a representation for GS_τ. Thus G is necessarily time invariant.

A more surprising fact is the converse: *all* time invariant operators on $L_2(-\infty, \infty)$ can be represented in this way.

Theorem 3.25. *An operator $G : L_2(-\infty, \infty) \to L_2(-\infty, \infty)$ is time invariant if and only if it can be represented in $\hat{L}_2(j\mathbb{R})$ by a multiplication by a function*

$$\hat{G} \in \hat{L}_\infty(j\mathbb{R}).$$

Thus this theorem says that the set of time invariant operators is in exact correspondence with the set of functions $\hat{L}_\infty(j\mathbb{R})$. In mathematical terminology these spaces are isomorphic.

3.4.2 Causality with time invariance

We now formalize the notion of causality, which informally means that the present is only determined by the past and present. It turns out that the space of functions $\hat{L}_\infty(j\mathbb{R})$ gives us representations for multiplication operators that are not always causal. One of our goals is to classify those operators that are both time invariant and causal.

To begin, define the truncation operator P_τ, for $\tau \in \mathbb{R}$, on vector valued functions by

$$(P_\tau u)(t) = \begin{cases} u(t), & t \leq \tau, \\ 0, & t > \tau, \end{cases}$$

where $u : \mathbb{R} \to \mathbb{C}^n$. The function $(P_\tau u)(t)$ has support on the interval $(-\infty, \tau]$, which means it is zero outside this interval.

We define an operator $G : L_2(-\infty, \infty) \to L_2(-\infty, \infty)$ to be *causal* if

$$P_\tau G P_\tau = P_\tau G, \tag{3.7}$$

for all $\tau \in \mathbb{R}$. If we apply this identity to some element $u \in L_2(-\infty, \infty)$, we get

$$P_\tau G P_\tau u = P_\tau G u. \tag{3.8}$$

Notice that $(P_\tau u)(t)$ is zero for $t > \tau$. Thus the above equation says that $(Gu)(t)$, for $t \le \tau$, only depends on $u(t)$ for $t \le \tau$. That is, the past only depends on the past.

The above definition of causality depends on the real parameter τ, and we now attempt to simplify this condition in the special case of G being a time invariant operator. Suppose that G is time invariant and satisfies the above causality condition for the special case where $\tau = 0$; precisely,

$$P_0 G (I - P_0) = 0 \quad \text{holds.} \tag{3.9}$$

We now show that this in fact guarantees that the causality condition holds for all τ, and therefore that G is causal. To see this notice that the following relationship holds, for any $\tau \ge 0$, immediately from the definitions of S_τ and P_τ:

$$P_\tau S_\tau = S_\tau P_0. \tag{3.10}$$

That is, if we truncate a function over positive time and shift it by τ, this is the same as shifting it by τ and then truncating it after time τ.

Turning back to (3.9) we fix a positive τ and see it holds if and only if

$$S_\tau P_0 G (I - P_0) = 0$$

is satisfied, since S_τ is nonsingular. Now applying (3.10) we obtain the equivalent conditions

$$0 = S_\tau P_0 G (I - P_0) = P_\tau S_\tau G (I - P_0)$$
$$= P_\tau G S_\tau (I - P_0) = P_\tau G (I - P_\tau) S_\tau,$$

where the latter equalities follow from (3.10) and the fact that G is time invariant. Finally, since the shift operator is invertible we see this latter equality holds exactly when $P_\tau G (I - P_\tau) = 0$. Therefore condition (3.8) holds for every positive τ whenever Equation (3.9) does. For $\tau < 0$ we can use the same argument starting with $P_\tau S_{-\tau}^{-1} = S_{-\tau}^{-1} P_0$ instead of (3.10), to show that (3.9) implies (3.8). Clearly (3.9) is a special case of (3.8), and so we have the following result.

Proposition 3.26. *A time invariant operator G on $L_2(-\infty, \infty)$ is causal if and only if*

$$P_0 G (I - P_0) = 0 \quad \text{is satisfied.} \tag{3.11}$$

We can also write the following equivalent characterization, which follows by noticing that P_0 and $I - P_0$ are the orthogonal projection operators taking $L_2(-\infty, \infty)$ to $L_2(-\infty, 0]$ and $L_2[0, \infty)$, respectively. We leave details to the reader.

Corollary 3.27. *A time invariant operator G mapping $L_2(-\infty, \infty)$ to $L_2(-\infty, \infty)$ is causal if and only if it maps $L_2[0, \infty)$ to $L_2[0, \infty)$.*

In plain language this corollary says that a time invariant operator on $L_2(-\infty, \infty)$ is causal exactly when it maps every function that is zero for negative time to a function which also is zero on the negative time axis. More compactly, the space $L_2[0, \infty)$ must be G-invariant.

3.4.3   Causality and H_∞

Having characterized causality in the time domain, we now return to frequency domain spaces. The key question is, which multiplication operators $M_{\hat{G}}$ on $\hat{L}_2(j\mathbb{R})$ correspond to causal time invariant operators in the time domain?

A partial answer is given by translating Corollary 3.27 to the frequency domain. From our discussion of signal spaces, it follows that

$$G = \Phi^{-1} M_{\hat{G}} \Phi \quad \text{is causal if and only if } M_{\hat{G}} \text{ maps } H_2 \text{ to } H_2.$$

Recall that H_2 is made of functions $\hat{u}(j\omega)$ which have analytic continuation to the right half-plane, with a certain boundedness condition. Which property of $\hat{G}(j\omega)$ is required for the product $\hat{G}(j\omega)\hat{u}(j\omega)$ to have a similar continuation? Clearly if $\hat{G}$ itself can be analytically extended to $\hat{G}(s)$, analytic and bounded on $\mathbb{C}^+$, then $\hat{G}(s)\hat{u}(s)$ would be in H_2 whenever $\hat{u}(s)$ is. Now, is this a necessary requirement for $\hat{G}$? The following example suggests that this is indeed the case.

Example:

Suppose G is a causal, time invariant operator between scalar L_2 spaces. Choose

$$u(t) = \begin{cases} e^{-t} & \text{for } t \geq 0, \\ 0 & \text{otherwise.} \end{cases}$$

Then $\hat{u}(s) = \frac{1}{s+1}$ and we have

$$\hat{z}(j\omega) = \hat{G}(j\omega) \frac{1}{j\omega + 1},$$

for a scalar function $\hat{G}$. Thus we see that $\hat{G}(j\omega) = (j\omega + 1)\hat{z}(j\omega)$. Since $\hat{z} \in H_2$, we can extend the domain of $\hat{G}$, setting

$$\hat{G}(s) := (s + 1)\hat{z}(s)$$

so that $\hat{G}(s)$ is defined, and analytic, for all $s \in \bar{\mathbb{C}}^+$. It is also possible to show that

$$\|\hat{G}\|_\infty = \sup_{s \in \bar{\mathbb{C}}^+} |\hat{G}(s)| \,.$$

So we see from this generic scalar example that causality seems to imply that $\hat{G}(s)$ is naturally defined on $\bar{\mathbb{C}}^+$ rather than just $j\mathbb{R}$.   □

This discussion leads us to define the following important class of functions

Definition 3.28. *A function $\hat{G} : \bar{\mathbb{C}}^+ \to \mathbb{C}^{n \times m}$ is in H_∞ if*

(a) $\hat{G}(s)$ is analytic in $\mathbb{C}^+$;

(b) for almost every real number ω,

$$\lim_{\sigma \to 0^+} \hat{G}(\sigma + j\omega) = \hat{G}(j\omega);$$

(c) $\displaystyle\sup_{s \in \bar{\mathbb{C}}^+} \bar{\sigma}(\hat{G}(s)) < \infty$.

As with H_2, it can be shown that H_∞ functions are determined by their values on the imaginary axis. Thus we can regard H_∞ as a subspace of $\hat{L}_\infty(j\mathbb{R})$, with the same norm

$$\|\hat{G}\|_\infty = \operatorname*{ess\,sup}_{\omega \in \mathbb{R}} \bar{\sigma}(\hat{G}(j\omega)) \,.$$

We remark that the above quantity also coincides with the supremum in Definition 3.28(c). This maximum principle is discussed in the exercises. We also point out that if a function $\hat{G}$ is only defined on $\mathbb{C}^+$, but satisfies (a) and the supremum condition in (c) restricted to $\mathbb{C}^+$, then one can show that its domain of definition can be extended to include the imaginary axis so that it is in H_∞.

We are now ready to characterize causal, time invariant operators in the frequency domain.

Theorem 3.29.

(a) Every $\hat{G} \in H_\infty$ defines a causal, time invariant operator G on $L_2(-\infty, \infty)$, where $z = Gu$ is defined by $\hat{z}(j\omega) = \hat{G}(j\omega)\hat{u}(j\omega)$.

(b) For each causal, time invariant operator G on $L_2(-\infty, \infty)$ there exists a function $\hat{G} \in H_\infty$ such that $z = Gu$ satisfies $\hat{z}(j\omega) = \hat{G}(j\omega)\hat{u}(j\omega)$ for all u in $L_2(-\infty, \infty)$.

The following diagram complements the one given before, focusing exclusively on *causal*, time invariant operators, or equivalently for multiplication functions $\hat{G} \in H_\infty$.

$$
\begin{array}{ccc}
L_2[0,\infty) & \xrightarrow{\;\;G\;\;} & L_2[0,\infty) \\
\Big\downarrow{\scriptstyle\Lambda} & & \Big\downarrow{\scriptstyle\Lambda} \\
H_2 & \xrightarrow{\;\;M_{\hat{G}}\;\;} & H_2
\end{array}
$$

Examples:

As a final illustrative example, we turn once more to the case of rational functions. We define the sets $R\hat{L}_\infty$ and RH_∞ to consist of rational, matrix functions that belong, respectively, to $\hat{L}_\infty$ and H_∞. It is not hard to see that a matrix rational function is:

(i) in $R\hat{L}_\infty$ if it is proper and has no poles on the imaginary axis;

(ii) in RH_∞ if it is proper and has no poles on the closed right half-plane $\bar{\mathbb{C}}^+$.

It follows from §2.6 that every function in $R\hat{L}_\infty$ can be expressed in the form

$$
\hat{G}(j\omega) = C(j\omega I - A)^{-1}B + D,
$$

where A has no eigenvalues on the imaginary axis, and every function in RH_∞ can be expressed in the form

$$
\hat{G}(s) = C(sI - A)^{-1}B + D,
$$

where A is Hurwitz. In the latter case, the time domain operator $G = \Phi^{-1}M_{\hat{G}}\Phi$, restricted to $L_2[0,\infty)$, has the convolution form

$$
u(t) \mapsto \int_0^t Ce^{A(t-\tau)}Bu(\tau)d\tau + Du(t),
$$

that was studied in Chapter 2 when we discussed minimal realizations.

A remark is in order regarding the $R\hat{L}_\infty$ case. When A has a right half-plane eigenvalue, rather than the previous convolution, which would give a causal, but unbounded map, the multiplication $M_{\hat{G}}$ corresponds to a non-causal, yet bounded operator. More insight into this distinction is provided in the exercises.   □

To conclude the chapter we focus our attention on operators on the subspace $L_2[0,\infty)$; these will play a significant role in the sequel. Clearly an operator on $L_2(-\infty,\infty)$ can be restricted to $L_2[0,\infty)$ provided that it leaves this subspace *invariant*. For instance, the shift operator S_τ for $\tau \geq 0$ has this property. Notice, however, that its inverse does not, so in fact the restricted operator $S_\tau : L_2[0,\infty) \to L_2[0,\infty)$ is *not* invertible.

Causality and time invariance are defined in the same way as before for operators on $L_2[0, \infty)$. The following theorem provides a complete characterization of such operators. Part (a) below follows directly from Theorem 3.29, since we have already seen that a causal, time invariant operator on $L_2(-\infty, \infty)$ leaves the subspace $L_2[0, \infty)$ invariant. For part (b) see the exercises.

Theorem 3.30.

(a) Every $\hat{G} \in H_\infty$ defines a causal, time invariant operator G on $L_2[0, \infty)$, where $z = Gu$ is defined by $\hat{z}(j\omega) = \hat{G}(j\omega)\hat{u}(j\omega)$.

(b) If the operator $G \in \mathcal{L}(L_2[0, \infty))$ is time invariant, then there exists a function $\hat{G} \in H_\infty$ such that $z = Gu$ satisfies $\hat{z}(j\omega) = \hat{G}(j\omega)\hat{u}(j\omega)$, for all u in $L_2[0, \infty)$.

This theorem states that all LTI operators on $L_2[0, \infty)$ are represented by functions in H_∞. Notice this means that an LTI operator on $L_2[0, \infty)$ is *necessarily* causal. This is in contrast to the LTI operators on $L_2(-\infty, \infty)$ which we saw need not be causal.

We have thus concluded our survey of function spaces and basic operator theory, and are now ready to put these tools to use for the study of controlled systems. In particular we will measure signals in terms of L_2 norms; since this is the only norm employed we will drop the norm subscript and use $\| \cdot \|$ to denote the L_2 norm. Another important comment as we transition these tools, is that from this point onwards in the course our signals will be *complex* valued. This convention helps simplify some of the proofs to follow, particularly when dealing with transforms. We note, however, that unless otherwise indicated, all the results extend to systems that operate with real valued signals.

Exercises

3.1. Let $\langle\ ,\ \rangle$ be the inner product on a Hilbert space. Use the fact that $\langle u + \lambda v, u + \lambda v \rangle \geq 0$ for every scalar λ to derive the Cauchy–Schwarz inequality. Deduce from it the triangle inequality.

3.2. Prove, for $v \in \mathbb{C}^n$, that

$$\lim_{p \to \infty} |v|_p = |v|_\infty.$$

3.3. Prove the submultiplicative inequality (3.2) for operators.

3.4. Assuming existence of the adjoint has been established, prove the remainder of Proposition 3.15.

3.5. The inner product space ℓ_2 is defined to be the set of elements $x = (x_0, x_1, x_2, \ldots)$ formed from the complex sequences, which satisfy

$$\sum_{k=0}^{\infty} |x_k|^2 < \infty.$$

The inner product on this space is $\langle x, y \rangle = \sum_{k=0}^{\infty} x_k^* y_k$. Verify that ℓ_2 is indeed an inner product space, and from the completeness of $\mathbb{C}$, prove that it is complete.

3.6. Let $\mathcal{H}$ be a Hilbert space, and $\mathcal{S}$ a linear subspace. We define the *orthogonal complement* of $\mathcal{S}$ by

$$\mathcal{S}^{\perp} = \{v \in \mathcal{H} : \langle v, s \rangle = 0 \ \text{ for all } s \in \mathcal{S}\}.$$

Recall, a subspace $\mathcal{S}$ is *closed* if every convergent sequence with elements in $\mathcal{S}$ has limit in $\mathcal{S}$. The Projection theorem states that if $\mathcal{S}$ is closed, every $v \in \mathcal{H}$ can be written in a unique way as

$$v = v_1 + v_2, \quad v_1 \in \mathcal{S}, \quad v_2 \in \mathcal{S}^{\perp}.$$

The element v_1 is called the *projection* of v in $\mathcal{S}$, and $P_{\mathcal{S}} \in \mathcal{L}(\mathcal{H})$ defined by $P_{\mathcal{S}} : v \mapsto v_1$ is the orthogonal projection operator.

   (a) If $\mathcal{S}$ is closed, show that $(\mathcal{S}^{\perp})^{\perp} = \mathcal{S}$. What happens if $\mathcal{S}$ is not closed?
   (b) Show that $P_{\mathcal{S}} = P_{\mathcal{S}}^2 = P_{\mathcal{S}}^*$.
   (c) For any operator $M \in \mathcal{L}(\mathcal{H})$, show that $\text{Im}(M)^{\perp} = \text{Ker}(M^*)$.
   (d) Consider the multiplication operator $M_{\hat{G}}$ on H_2, defined by $\hat{G}(s) = \frac{s}{s+1}$. Describe the subspaces in part (c) for this case. Is $\text{Im}(M)$ closed?

3.7. Consider the space $\mathbb{C}^{n \times n}$ of square matrices. Show that the *Frobenius* norm $|M|_F$ makes this space a Banach algebra, but that this norm cannot be induced by any vector norm in $\mathbb{C}^n$.

3.8. Consider the scalar spaces $L_p(-\infty, \infty)$; for which value(s) of p does the pointwise multiplication $(fg)(t) = f(t)g(t)$ give the structure of a Banach algebra?

3.9. Use the Jordan form to show that for any $M \in \mathbb{C}^{n \times n}$, satisfying $\rho(M) < 1$, the limit $M^k \to 0$, as $k \to \infty$, holds. Use this fact to show that $\rho(M) = \lim_{k \to \infty} \bar{\sigma}(M^k)^{\frac{1}{k}}$.

3.10. Prove Proposition 3.16.

3.11. Given an operator M on a Hilbert space $\mathcal{V}$, show that

$$\rho(M) = \inf_{D \in \mathcal{L}(\mathcal{V}), \text{ invertible}} \|DMD^{-1}\|,$$

by proving the equivalence between

(i) $\rho(M) < 1$;

(ii) there exists $D \in \mathcal{L}(\mathcal{V})$, invertible such that $\|DMD^{-1}\| < 1$;

(iii) there exists $P \in \mathcal{L}(\mathcal{V})$, $P > 0$ such that

$$M^*PM - P < 0.$$

Hint: consider the series $\sum_{k=0}^{\infty}(M^*)^k M^k$.

3.12. Prove Proposition 3.24; in particular show that the induced norm is exactly $\|\hat{G}\|_{\infty}$. You can assume that $\hat{G}(j\omega)$ is a continuous function. *Hint:* choose an input that concentrates its energy around the peak frequency response.

3.13. Consider the system $\dot{x} = Ax + Bu$, where A is Hurwitz. Show that if $u(t) \in L_2[0, \infty)$, then

$$\lim_{t \to \infty} x(t) = 0.$$

Hint: Express the entries $x_i(t)$ as inner products of the form $\langle S_t f, g \rangle$ for appropriately chosen L_2 functions f, g.

3.14. Consider the multiplication operator $M_{\hat{G}}$, where

$$\hat{G}(j\omega) = C(j\omega I - A)^{-1}B + D \in \hat{L}_{\infty}(j\mathbb{R}).$$

(a) Characterize the adjoint $M_{\hat{G}}^*$ as a multiplication operator, and give a state space realization for the corresponding function.

(b) Give conditions in terms of $\hat{G}(j\omega)$ that make
(i) $M_{\hat{G}}$ self-adjoint; (ii) $M_{\hat{G}} \geq 0$; (iii) $M_{\hat{G}} > 0$.

(c) Assuming $M_{\hat{G}}$ is self-adjoint, determine when $\langle \hat{u}, M_{\hat{G}}\hat{u} \rangle > 0$ holds, for all nonzero $\hat{u} \in \hat{L}_2(j\mathbb{R})$. Are these conditions equivalent to $M_{\hat{G}} > 0$?

(d) What is the spectrum of $M_{\hat{G}}$?

3.15. *Maximum modulus principle.* It follows from analytic function theory that given a scalar function $f(s)$ in H_{∞} (analytic and bounded in $Re(s) > 0$), its maximum modulus is achieved at the boundary; i.e.,

$$\sup_{Re(s)>0} |f(s)| = \text{ess} \sup_{\omega} |f(j\omega)| = \|f\|_{\infty}.$$

Using this fact, extend it to matrix valued functions in H_{∞} norm.

3.16. Decompose the function $\hat{u}(s) = \frac{e^s}{s+1}$ as a sum of functions in H_2 and $H_2^{\perp}$.

3.17. (a) Given two functions u, v in $L_2(-\infty, \infty)$ with Fourier transforms satisfying

$$|\hat{u}(j\omega)| \geq |\hat{v}(j\omega)| \quad \text{for almost all } \omega \in \mathbb{R}. \qquad (3.12)$$

Show that there exists a linear time invariant operator G, with $\|G\| \leq 1$, satisfying $Gu = v$. Can G always be chosen to be causal?

(b) Let $Gu = v$, $\|G\| \leq 1$, but G not necessarily time invariant. Is (3.12) satisfied?

3.18. Let $\hat{G}(j\omega) = \frac{1}{j\omega-1}$. Express $G = \Phi^{-1}M_{\hat{G}}\Phi$, as a convolution in $L_2(-\infty, \infty)$, giving the corresponding convolution function $g(t)$. Is G causal?

3.19. Prove part (b) of Theorem 3.30. To do this show that the domain of G can be extended to $L_2(-\infty, \infty)$ by setting

$$Gu := \lim_{\tau \to \infty} S_\tau^{-1} G S_\tau (I - P_{-\tau})u,$$

and that G on this extended domain is LTI. Now use Proposition 3.26 and Theorem 3.29.

3.20. Consider the operator Q on $L_2[0, \infty)$ defined in (3.4). Is it both time invariant and causal? Derive the explicit formula for its adjoint Q^*. Is Q^* time invariant?

Notes and references

For additional background material on linear functional analysis consult [18] for an excellent introduction; also see [181] for more basic information on Hilbert space. More advanced work on Hilbert space operator theory, and a main source of references on this topic, is the book [71].

Proofs of all the results in this chapter (excepting Theorems 3.25 and 3.29) on the function spaces L_p, H_2, and H_∞ can be found in [136]. These proofs are for the scalar valued case, but easily generalize to the vector or matrix valued case. In particular proofs of the Plancherel and Paley–Wiener theorems are given. We point out that some aspects of our presentation are nonstandard, and different from [136]. In particular the definition of the Fourier transform, pointwise as a principal value limit, is based on [87]; see also [105]. Also our definitions for H_2 and H_∞ are made on the closed, rather than open, half-plane. These modifications to the usual definitions have been done for clarity of exposition, and do not affect any results.

Theorem 3.25 is proved in [17, Thms71-72] for the scalar case; the extension to the vector case given here is routine. Theorem 3.29 follows from Theorem 3.25 and Proposition 3.26, and standard analytic continuation theory; see [136]. See also [169] for more information on time invariant operators and functions. Reference [53] has many results on the discrete time operator valued case. This book also contains much additional information on the connections between function theory and operator theory, for both the matrix and operator valued cases, and is a particularly good reference for such results as related to control theory. For a complete mathematical treatment of causal operators, and more generally triangular operators, see reference [32].

Classic references for work on H_p and related spaces are [47, 74]; another important reference is [59]. Additional advanced information on these spaces in the operator valued case can be located in [135, 157].

4

Model Realizations and Reduction

In this chapter we start our investigation of quantitative input and output properties of systems. To do this we will require the state space systems theory of Chapter 2 combined with the new viewpoint and framework gained in the preceding chapter. We first consider issues related to the relative controllability and observability of system states, and their relationships with the overall input–output characteristics of a system. We then turn to the important question of systematically finding reduced order approximations to systems. We will develop a powerful technique to accomplish this, and the operator perspective of the previous chapter will play a central role.

4.1 Lyapunov equations and inequalities

In this section we describe a basic tool for the study of stability, controllability, and observability in linear systems. This topic is closely related to the material in Chapter 2, but we present it in this chapter, where it will be extensively applied.

Lyapunov equations come in the two dual forms

$$A^*X + XA + Q = 0$$

and

$$AX + XA^* + Q = 0.$$

In both cases A, Q are given square matrices, and X is the unknown. For concreteness we will focus on the first case, but by the trivial substitution of A by A^*, all the following properties have their dual counterpart. Notice in particular that the Hurwitz property is preserved by the "$*$" operation.

We now see that the Lyapunov equation can always be solved when A is Hurwitz.

Theorem 4.1. *Suppose A and Q are square matrices and that A is Hurwitz. Then*

$$X = \int_0^\infty e^{A^*\tau} Q e^{A\tau} d\tau \qquad (4.1)$$

is the unique solution to the Lyapunov equation $A^*X + XA + Q = 0$.

Proof. We first remark that the above integral converges because A is Hurwitz. Next, we verify that the given X is a solution, observing that

$$A^*X + XA = \int_0^\infty \frac{d}{d\tau}\{e^{A^*\tau} Q e^{A\tau}\} d\tau = e^{A^*\tau} Q e^{A\tau}\big|_0^\infty = -Q.$$

It remains to show that it is the unique solution. For this purpose, define the linear mapping $\Pi : \mathbb{R}^{n \times n} \to \mathbb{R}^{n \times n}$ by

$$\Pi(X) := A^*X + XA\,,$$

where $X \in \mathbb{R}^{n \times n}$. Now for any matrix $Q \in \mathbb{R}^{n \times n}$ we have seen that the equation

$$\Pi(X) = -Q$$

has a solution. This means that the dimension of the image of Π is n^2; but the dimension of the domain of Π is also n^2 and therefore

$$\mathrm{Ker}\,\Pi = 0\,.$$

This means $\Pi(X) = -Q$ has a unique solution for each Q. ∎

In terms of computation, we emphasize that the Lyapunov equation is linear in the matrix variable X, and therefore it is in essence no more than a system of n^2 linear equations in the n^2 entries of X, which can be solved by standard techniques. In particular when A is Hurwitz we are guaranteed to have a unique solution; direct computation of the integral in (4.1) is never required.

Another useful property of the Lyapunov solution is that

$$\begin{bmatrix} I & 0 \\ X & I \end{bmatrix} \begin{bmatrix} A & 0 \\ Q & -A^* \end{bmatrix} \begin{bmatrix} I & 0 \\ -X & I \end{bmatrix} = \begin{bmatrix} A & 0 \\ A^*X + XA + Q & -A^* \end{bmatrix} = \begin{bmatrix} A & 0 \\ 0 & -A^* \end{bmatrix}\,.$$

Thus X can also be found by performing an eigenvalue decomposition.

From now on we focus on the case where $Q = Q^*$, for which one always seeks solutions $X = X^*$ to the Lyapunov equations. The following result shows that the Hurwitz nature of A is directly related to the sign definiteness of the solutions.

Proposition 4.2. *Suppose $Q > 0$. Then A is Hurwitz if and only if there exists a solution $X > 0$ to the Lyapunov equation $A^*X + XA + Q = 0$.*

Proof. We first establish the "only if" direction. Since A is Hurwitz we already know that the unique solution is

$$X = \int_0^\infty e^{A^*\tau} Q e^{A\tau} d\tau.$$

Now since $Q > 0$ it is straightforward to verify that $X > 0$. We leave details as an exercise.

For the converse, suppose $X > 0$ solves the equation. If λ is an eigenvalue of A, with eigenvector $v \neq 0$, we have

$$0 = v^*(A^*X + XA + Q)v = \lambda^* v^* X v + \lambda v^* X v + v^* Q v.$$

Since $v^* X v > 0$ we have

$$2\text{Re}(\lambda) = -\frac{v^* Q v}{v^* X v} < 0,$$

so A is Hurwitz.   ■

We make the following remarks:

- The preceding proof is purely algebraic; an alternative argument which helps explain our terminology is to consider the *Lyapunov function* $V(x) = x^* X x$ as a tool to show that $\dot{x} = Ax$ is asymptotically stable. This approach can be found in standard references on differential equations.

- The preceding result remains true if the hypothesis $Q > 0$ is weakened to $Q = C^*C > 0$, with (C, A) observable. The verification of this stronger theorem is covered in the exercises at the end of the chapter.

In that case, the Lyapunov solution is denoted by Y_o, satisfying

$$A^*Y_o + Y_o A + C^*C = 0, \tag{4.2}$$

and is called the *observability gramian* of (C, A); this is the topic of the next section.

- It is sometimes convenient to rewrite Proposition 4.2 as an LMI:

Corollary 4.3. *The matrix A is Hurwitz if and only if there exists $X > 0$ satisfying*

$$A^*X + XA < 0.$$

We finish the section elaborating some more on such *Lyapunov inequalities*. In general, these have the form

$$A^*X + XA + Q \leq 0$$

or the strict version

$$A^*X + XA + Q < 0.$$

Focusing on the case where A is Hurwitz, it is clear that these two LMIs always have solutions. To see this notice that the left-hand side of the inequality in the above corollary can be made less than any matrix simply by scaling any solution X. Now we ask, how do the solutions to these inequalities relate to the corresponding Lyapunov equation solution? We have the following result, which is left as an exercise.

Proposition 4.4. *Suppose that A is Hurwitz, and X_0 satisfies $A^*X_0 + X_0 A + Q = 0$, where Q is a symmetric matrix. If X satisfies $A^*X + XA + Q \leq 0$, then*

$$X \geq X_0.$$

This concludes our general study of Lyapunov equations. We now focus on the problem of quantifying the degree of controllability and observability of state space realizations.

4.2   Observability operator and gramian

We begin our study by focusing on the autonomous system given by

$$\dot{x}(t) = Ax(t), \quad x(0) = x_0 \in \mathbb{C}^n$$
$$y(t) = Cx(t),$$

where A is a Hurwitz matrix. This is our usual state space system with the input set to zero, and a nonzero initial condition on the state. The solution to this system is $Ce^{At}x_0$, for $t \geq 0$. Define the observability operator $\Psi_o :$ $\mathbb{C}^n \to L_2[0, \infty)$ by

$$x_0 \mapsto \begin{cases} Ce^{At}x_0, & \text{for } t \geq 0, \\ 0, & \text{otherwise.} \end{cases}$$

This operator is analogous to the map Ψ encountered in Chapter 2; here, however, we exploit the fact that A is Hurwitz. In particular, we can easily show there exist positive constants α and β satisfying

$$|Ce^{At}x_0| \leq \beta \cdot e^{-\alpha t}|x_0| \, ,$$

for all initial conditions x_0 and times t. If we set $y = \Psi_o x_0$ this gives

$$\|y\| \leq \frac{\beta}{\sqrt{2\alpha}} \, |x_0| \, ,$$

so we see that $y \in L_2[0, \infty)$ and Ψ_o is a bounded operator. Let us now take a more careful look at the energy of $y = \Psi_o x_0$, for $x_0 \in \mathbb{C}^n$. It is given by

$$\|y\|^2 = \langle \Psi_o x_0, \Psi_o x_0 \rangle = \langle x_0, \Psi_o^* \Psi_o x_0 \rangle \, ,$$

where $\Psi_o^* : L_2[0, \infty) \to \mathbb{C}^n$ and is the adjoint of Ψ_o. An exact expression for the adjoint operator can be found from its definition and the following equations. For each $z \in L_2[0, \infty)$ and $x_0 \in \mathbb{C}^n$

$$\langle \Psi_o^* z, x_0 \rangle_{\mathbb{C}^n} = \langle z, \Psi_o x_0 \rangle_2 = \int_0^\infty z^*(\tau) C e^{A\tau} x_0 d\tau$$

$$= \left(\int_0^\infty e^{A^*\tau} C^* z(\tau) d\tau \right)^* x_0$$

$$= \left\langle \int_0^\infty e^{A^*\tau} C^* z(\tau) d\tau, x_0 \right\rangle_{\mathbb{C}^n} .$$

Thus we see that Ψ_o^* is given by

$$\Psi_o^* z = \int_0^\infty e^{A^*\tau} C^* z(\tau) d\tau ,$$

for $z \in L_2[0, \infty)$. Now the 2-norm of $y = \Psi_o x_0$ is given by $\langle x_0, \Psi_o^* \Psi_o x_0 \rangle$ and we have

$$\Psi_o^*(\Psi_o x_0) = \int_0^\infty e^{A^*\tau} C^* C e^{A\tau} x_0 d\tau$$

$$= \left(\int_0^\infty e^{A^*\tau} C^* C e^{A\tau} d\tau \right) x_0 = (\Psi_o^* \Psi_o) x_0 .$$

Therefore the operator $\Psi_o^* \Psi_o$ is given by the matrix

$$Y_o := \Psi_o^* \Psi_o = \int_0^\infty e^{A^*\tau} C^* C e^{A\tau} d\tau ,$$

which is the *observability gramian* of (C, A), already encountered in the previous section. In fact, using Theorem 4.1 it is clear that Y_o satisfies the Lyapunov equation (4.2).

This gramian is positive semidefinite

$$Y_o \geq 0,$$

and as mentioned before is positive definite when (C, A) is observable. However, the previous calculation shows that the gramian carries quantitative information about observability; this is in addition to simply indicating whether the matrix pair is observable. We have shown that the energy of the output $y = \Psi_o x_0$, starting from an initial condition $x_0 \in \mathbb{C}^n$, is given by

$$\|y\|^2 = x_0^* Y_o x_0 . \tag{4.3}$$

In particular, if we only consider states that satisfy $|x_0| = 1$, then clearly some of these states will yield higher output norms $\|y\|$ than others. Therefore the gramian measures "how observable" a given initial condition is.

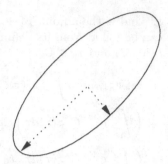

Figure 4.1. Observability ellipsoid

This idea can be described geometrically in a very intuitive way. Consider the vector

$$Y_o^{\frac{1}{2}} x_0,$$

where x_0 has unit length. This vector has length equal to the 2-norm of y, from (4.3), and as we will see contains directional information as well. Define the set

$$\mathcal{E}_o := \{ Y_o^{\frac{1}{2}} x_0 : \ x_0 \in \mathbb{C}^n \text{ and } |x_0| = 1 \} \, ,$$

which is the collection of all such vectors. Since Y_o is positive semidefinite this set is an ellipsoid, and is depicted two dimensionally in Figure 4.1. This shape is called the *observability ellipsoid*, and tells us the output norm associated with a particular direction in state space. Let

$$\eta_1 \geq \eta_2 \geq \cdots \geq \eta_n \geq 0 \, ,$$

be the eigenvalues of $Y_o^{\frac{1}{2}}$ and

$$v_1, \ldots, v_n \, ,$$

their respective unit-norm eigenvectors. Then the v_k give the directions of the principal axes of the ellipsoid, and the η_k the length of each axis.

If $\eta_k = 0$ for some k, the corresponding v_k produces no output energy and is therefore unobservable. In fact the span of all the vectors v_k with $\eta_k = 0$ is precisely the unobservable subspace $\mathcal{N}_{CA}$ from Chapter 2. Geometrically, the ellipsoid degenerates into a subspace of lower dimension, orthogonal to this space.

If η_k, η_l are both nonzero but, for example, $\eta_k \gg \eta_l$, then the output energy resulting from initial state v_l is much smaller than that observed when the initial condition is v_k. Thus both states are observable, but intuitively state v_k is "more observable" than state v_l. We conclude that the observability gramian provides us with a way to assess the relative observability of various directions in the state space. When a state corresponds to a large eigenvalue η_k we say it is *strongly* observable; if instead $\eta_k > 0$

is small compared with the other eigenvalues we say the state is *weakly observable*.

As an aside let us see the significance of this gramian for determining the observability of a pair (C, A) where A is Hurwitz. In Chapter 2 we saw how this could be done by checking whether the matrix

$$Q^* = \begin{bmatrix} C^* & A^*C^* & \cdots & (A^*)^{n-1}C^* \end{bmatrix}$$

has rank n. Another alternative, also based on rank, is the PBH test. This raises the question of numerically determining the rank of a matrix, in particular the question of numerical tolerance for this determination; perhaps one would check the eigenvalues of Q^*Q, but we have no interpretation assigned to the size of these eigenvalues. In particular, you will show in the exercises that these are unrelated to the eigenvalues of Y_o. In short, the above rank test can only provide a yes or no answer. In contrast, the eigenvalues of Y_o have a clear association with the degree of observability and thus provide a much more sound numerical method for observability assessment.

In line with this reasoning, it seems natural that these eigenvalues could be used as a basis of a *model reduction* scheme, based on the elimination of weakly observable states. We will return to this idea later, but first we study the dual situation of controllability.

4.3   Controllability operator and gramian

Here our focus is the dual idea of that pursued in the preceding section. Given a matrix pair (A, B) define the *controllability operator* Ψ_c : $L_2(-\infty, 0] \to \mathbb{C}^n$ by

$$u \mapsto \int_{-\infty}^{0} e^{-A\tau} B u(\tau) d\tau .$$

Notice this can be thought of as the response of the system described by

$$\dot{x}(t) = Ax(t) + Bu(t), \quad x(-\infty) = 0 \tag{4.4}$$

to an input function $u \in L_2(-\infty, 0]$, where the output is the vector $x(0)$.

To see the significance of this operator, we ask the following question:

Given the final condition $x(0) \in \mathbb{C}^n$ with unit norm, $|x_0| = 1$, what $u \in L_2(-\infty, 0]$ solves

$$\Psi_c u = x_0$$

with the smallest norm $\|u\|$?

In system terms this question is, what is the input with smallest energy $\|u\|$ which drives the state in (4.4) to satisfy $x(0) = x_0$ at time zero? Clearly if the matrix pair (A, B) is not controllable it may not be possible even to satisfy $\Psi_c u = x_0$, so for the next result we focus on the case where the pair is controllable.

Proposition 4.5. *Suppose that (A, B) is controllable. Then*

(i) *the matrix $\Psi_c \Psi_c^* =: X_c$ is nonsingular;*

(ii) *for any $x_0 \in \mathbb{C}^n$, the input $u_{opt} := \Psi_c^* X_c^{-1} x_0$ is the element of minimum norm in the set*

$$\{u \in L_2(-\infty, 0], \ \Psi_c u = x_0\}. \tag{4.5}$$

Proof. First, we find an expression for X_c. Proceeding analogously to the observability case, the definition of the adjoint implies that $\Psi_c^* : \mathbb{C}^n \to L_2(-\infty, 0]$ is given by

$$\Psi_c^* : \xi \mapsto \begin{cases} B^* e^{-A^* t} \xi & \text{for } t \leq 0, \\ 0 & \text{otherwise}; \end{cases}$$

and therefore

$$X_c = \Psi_c \Psi_c^* = \int_{-\infty}^{0} e^{-At} BB^* e^{-A^* t} dt = \int_{0}^{\infty} e^{At} BB^* e^{A^* t} dt,$$

called the *controllability gramian* of the matrix pair (A, B). Notice that this is an infinite horizon version of the definition given in Chapter 2, and also that X_c satisfies the Lyapunov equation

$$AX_c + X_c A^* + BB^* = 0,$$

dual to the observability case. It follows from here that (A, B) is controllable if and only if $X_c > 0$, which establishes (i).

Next, we verify that given $x_0 \in \mathbb{C}^n$,

$$\Psi_c u_{opt} = \Psi_c \Psi_c^* X_c^{-1} x_0 = x_0 ,$$

and thus $u = u_{opt}$ belongs to the set of allowable inputs described in (4.5). So it remains to show, for a general u in this set, that $\|u\| \geq \|u_{opt}\|$. Define the operator

$$P = \Psi_c^* X_c^{-1} \Psi_c$$

on $L_2(-\infty, 0]$, and observe that $P^2 = P = P^*$. An operator satisfying these identities is called an orthogonal projection. In particular, it satisfies

$$\langle Pu, (I - P)u \rangle = \langle u, P(I - P)u \rangle = 0,$$

and consequently

$$\|u\|^2 = \|Pu\|^2 + \|(I - P)u\|^2 \geq \|Pu\|^2,$$

for any $u \in L_2(-\infty, 0]$. This is called the Bessel inequality.

Now let $u \in L_2(-\infty, 0]$ be any function satisfying $\Psi_c u = x_0$. Applying $\Psi_c^* X_c^{-1}$ on both sides we get

$$Pu = \Psi_c^* X_c^{-1} \Psi_c u = \Psi_c^* X_c^{-1} x_0 = u_{opt},$$

so the Bessel inequality gives $\|u\|^2 \geq \|u_{opt}\|^2$. ∎

The proposition says that if we want to reach a state x_0, then

$$u_{opt} = \Psi_c^* X_c^{-1} x_0$$

is the most economical input in terms of energy. This input energy is given by

$$\begin{aligned}
\|u_{opt}\|^2 &= \langle \Psi_c^* X_c^{-1} x_0, \Psi_c^* X_c^{-1} x_0 \rangle \\
&= \langle X_c^{-1} x_0, \Psi_c \Psi_c^* X_c^{-1} x_0 \rangle \\
&= x_0^* X_c^{-1} x_0.
\end{aligned}$$

We now provide a geometric interpretation of the controllability gramian, similar to that for Y_o. The key question is, what are all the final states

$$x_0 = \Psi_c u$$

that can result from an input $u \in L_2(-\infty, 0]$ of unit norm? The answer is given in the following proposition, which holds even if the pair (A, B) is not controllable. We will, however, concentrate here on the controllable case; the general case is covered in the exercises at the end of the chapter.

Proposition 4.6. *The following sets are equal:*

(a) $\{\Psi_c u : u \in L_2(-\infty, 0] \text{ and } \|u\| \leq 1\}$;

(b) $\{X_c^{\frac{1}{2}} x_c : x_c \in \mathbb{C}^n \text{ and } |x_c| \leq 1\}$.

Proof. We first show that set (a) is contained in set (b): choose any element $u \in L_2(-\infty, 0]$ with $\|u\| \leq 1$. Set

$$x_c = X_c^{-\frac{1}{2}} \Psi_c u,$$

which has norm

$$\begin{aligned}
|x_c|^2 &= \langle X_c^{-\frac{1}{2}} \Psi_c u, X_c^{-\frac{1}{2}} \Psi_c u \rangle = \langle u, \Psi_c^* X_c^{-1} \Psi_c u \rangle \\
&= \langle u, Pu \rangle = \|Pu\|^2 \leq \|u\|^2 \leq 1.
\end{aligned}$$

This means that $\Psi_c u = X_c^{\frac{1}{2}} x_c$ is in the set (b).

We now demonstrate that set (b) is contained in set (a). Let x_c be any vector that has length less than or equal to one. We choose the input of minimum norm which gives $\Psi_c u = X_c^{\frac{1}{2}} x_c$. From the previous results this input exists and its norm is

$$\|u_{opt}\|^2 = (X_c^{\frac{1}{2}} x_c)^* X_c^{-1} (X_c^{\frac{1}{2}} x_c) = |x_c|^2 \leq 1.$$

Therefore $X_c^{\frac{1}{2}} x_c$ is in the set (a).

■

This result says that all the states reachable with u satisfying $\|u\| \leq 1$ are given by

$$X_c^{\frac{1}{2}} x_c \ ,$$

where $|x_c| \leq 1$. Notice the norm squared of any such state is given by $x_c^* X_c x_c$. We define the controllability ellipsoid by

$$\mathcal{E}_c = \{X_c^{\frac{1}{2}} x_c : \ x_c \in \mathbb{C}_n \text{ and } |x_c| = 1\} \ .$$

This gives us the boundary of the set given in the proposition. Let

$$\mu_1 \geq \mu_2 \geq \cdots \geq \mu_n \geq 0 \ ,$$

be the eigenvalues of $X_c^{\frac{1}{2}}$ and

$$v_1, \ldots, v_n$$

their corresponding orthonormal eigenvectors. Then clearly the principal axes of the ellipsoid $\mathcal{E}_c$ are given by

$$\mu_k v_k \ .$$

The unit vectors $v_1, \ldots, v_n$ give a basis for the state space and the values $\mu_1, \ldots, \mu_n$ tell us that given an input $\|u\| = 1$, the largest we can make a state in the direction v_k is μ_k. Thus we conclude that if $\mu_k = 0$, then v_k is an unreachable state. In the same vein, if $\mu_k \gg \mu_l$ then state direction v_k is "more" controllable than direction v_l. As with observability this gives rise to the terms *strong* and *weak* controllability.

4.4   Balanced realizations

The preceding two sections have given us geometric ways, in terms of the gramians introduced, to assess which directions in the state space are strongly or weakly controllable and observable. We now look at our usual system given by

$$\dot{x}(t) = Ax(t) + Bu(t)$$
$$y(t) = Cx(t),$$

for $t \geq 0$. Here the direct feed-through term given by the D-matrix is not included, since it makes no difference to our investigation. We are interested in finding a natural basis for this state space that gives some idea as to which states dominate the system behavior. In our study so far we concluded that states corresponding to small eigenvalues of the observability gramian are not very observable. Does this mean such states do not contribute much to the input–output behavior of the above system? The answer is, not necessarily, since such states may be very controllable. This phenomenon is easily appreciated by looking at Figure 4.2, which shows both the observability and controllability ellipsoids on the same plot.

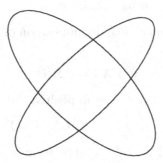

Figure 4.2. Unbalanced system ellipsoids

As this picture shows, these two ellipsoids need not be aligned, and therefore we can have a situation as drawn here, where the major and minor axes of the ellipsoids are nearly opposite. Therefore we reason that the most intuition would be gained about the system if the controllability and observability ellipsoids were exactly aligned, and is certainly the natural setting for this discussion. This raises the question of whether or not it is possible to arrange such an alignment by means of a state transformation; answering this is our next goal.

A change of basis to the state space of the above system yields the familiar transformed realization

$$\tilde{A} = TAT^{-1}, \quad \tilde{B} = TB, \quad \text{and} \quad \tilde{C} = CT^{-1},$$

where T is a state transformation. The controllability gramian associated with this new realization is

$$\tilde{X}_c = \int_0^\infty e^{\tilde{A}\tau} \tilde{B}\tilde{B}^* e^{\tilde{A}^*\tau} d\tau$$

$$= \int_0^\infty Te^{A\tau}T^{-1}TBB^*T^*(T^*)^{-1}e^{A^*\tau}T^* d\tau$$

$$= TX_cT^*$$

Similarly the observability gramian in the transformed basis is

$$\tilde{Y}_o = (T^*)^{-1}Y_oT^{-1}.$$

The above are called *congruence* transformations, which typically arise from quadratic forms under a change of basis. The following result concerns *simultaneous diagonalization* of positive definite matrices by this kind of transformation.

Proposition 4.7. *Given positive definite matrices X and Y, there exists a nonsingular matrix T such that*

$$TXT^* = (T^*)^{-1}YT^{-1} = \Sigma$$

where Σ is a diagonal, positive definite matrix.

Proof. Perform a singular value decomposition of the matrix $X^{\frac{1}{2}}YX^{\frac{1}{2}}$ to get

$$X^{\frac{1}{2}}YX^{\frac{1}{2}} = U\Sigma^2 U^* \,,$$

where U is unitary and Σ is diagonal, positive definite. Therefore we get

$$\Sigma^{-\frac{1}{2}}U^*X^{\frac{1}{2}}YX^{\frac{1}{2}}U\Sigma^{-\frac{1}{2}} = \Sigma.$$

Now set $T^{-1} = X^{\frac{1}{2}}U\Sigma^{-\frac{1}{2}}$ and the above states that $(T^{-1})^*YT^{-1} = \Sigma$. Also

$$TXT^* = (\Sigma^{\frac{1}{2}}U^*X^{-\frac{1}{2}})\,X\,(X^{-\frac{1}{2}}U\Sigma^{\frac{1}{2}}) = \Sigma.$$

∎

Applying this proposition to the gramians, the following conclusion can be drawn:

Corollary 4.8. *Suppose (A, B, C) is a controllable and observable realization. Then there exists a state transformation T such that the equivalent realization $(\tilde{A}, \tilde{B}, \tilde{C}) = (TAT^{-1}, TB, CT^{-1})$ satisfies*

$$\tilde{X}_c = \tilde{Y}_o = \Sigma$$

with $\Sigma > 0$ diagonal.

A state space realization such that the controllability and observability gramians are equal and diagonal, is called a *balanced realization*. The previous corollary implies that there always exists a balanced realization for a transfer function in RH_∞; in fact, starting with any minimal realization (which will necessarily have A Hurwitz), a balanced one can be obtained from the above choice of state transformation.

Clearly the controllability and observability ellipsoids for the system are exactly aligned when a system is balanced. Thus the states which are the least controllable are also the least observable. Balanced realizations play a key role in the model reduction studies of the rest of this chapter.

Before leaving this section, we state an extension of Proposition 4.7 which covers the general case, where the original realization is not necessarily controllable or observable. For a proof, see the chapter references.

Proposition 4.9. *Given positive semidefinite matrices X and Y there exists a nonsingular matrix T such that*

(a) $TXT^* = \begin{bmatrix} \Sigma_1 & & & \\ & \Sigma_2 & & \\ & & 0 & \\ & & & 0 \end{bmatrix}$,

(b) $(T^*)^{-1}YT^{-1} = \begin{bmatrix} \Sigma_1 & & & \\ & 0 & & \\ & & \Sigma_3 & \\ & & & 0 \end{bmatrix}$,

where the matrices Σ_k are diagonal and positive definite.

When applied to the gramians X_c and Y_o, we find that if the system is either uncontrollable or unobservable, then each of the Σ_k blocks of $\tilde{Y}_o$ and $\tilde{X}_c$ have the following interpretation:

- Σ_1 captures controllable and observable states;
- Σ_2 captures controllable and unobservable states;
- Σ_3 captures observable and uncontrollable states;
- $\Sigma_4 = 0$ fp captures unobservable and uncontrollable states.

Under such a transformation the state matrix $\tilde{A}$ has the form

$$\tilde{A} = \begin{bmatrix} \tilde{A}_1 & 0 & \tilde{A}_6 & 0 \\ \tilde{A}_2 & \tilde{A}_3 & \tilde{A}_4 & \tilde{A}_5 \\ 0 & 0 & \tilde{A}_7 & 0 \\ 0 & 0 & \tilde{A}_8 & \tilde{A}_9 \end{bmatrix},$$

which is the so-called Kalman decomposition; this can be verified from the various invariance properties of the controllability and observability subspaces. See the Chapter 2 exercises. fp

4.5 Hankel operators

In the previous sections we have introduced an input-to-state operator Ψ_c and a state-to-output operator Ψ_o that where naturally related to the degree of controllability and observability of a state space system. It seems natural, then, that these pieces should come together to study input–output properties of a system. These ideas are now explored.

We begin with a causal, bounded input–output operator G mapping $L_2(-\infty, \infty)$ to $L_2(-\infty, \infty)$ described by the state space convolution

$$(Gu)(t) := \int_{-\infty}^{t} Ce^{A(t-\tau)}Bu(\tau)d\tau,$$

where A is a Hurwitz matrix, and (A, B, C) is a *minimal* realization. In the language of Chapter 3, this operator is $G = \Phi^{-1}M_{\hat{G}}\Phi$, where $\hat{G}(s) =$

$C(sI - A)^{-1}B$ is a transfer function in RH_∞. We focus on the strictly proper case; the presence of a D term would make no difference to the discussion in this section.

Now we define the Hankel operator $\Gamma_G : L_2(-\infty, 0] \to L_2[0, \infty)$ of G by

$$\Gamma_G := P_+ G\big|_{L_2(-\infty, 0]},$$

where $G\big|_{L_2(-\infty, 0]}$ denotes the restriction of G to the subspace $L_2(-\infty, 0]$, and P_+ is the operator that projects a signal in $L_2(-\infty, \infty)$ into $L_2[0, \infty)$ by truncation. In other words, Γ_G takes an input supported in the "past," and maps it to the "future" output, as illustrated in Figure 4.3 for scalar inputs and outputs.

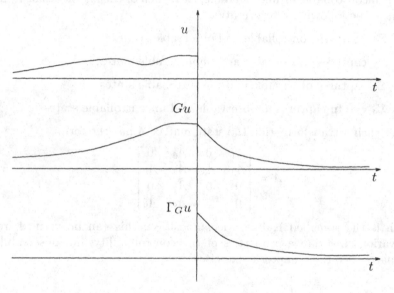

Figure 4.3. Results of G and Γ_G operating on a given $u \in L_2(-\infty, 0]$

The Hankel operator is closely related to the controllability and observability operators, owing to the causality of G; in fact, if $u(t)$ is supported in the past, the future output $P_+ y$ depends on u only through the initial state $x(0)$; thus we can think of composing two stages:

- Ψ_c which maps $u \in L_2(-\infty, 0]$ to $x(0)$.

- Ψ_o which maps $x(0)$ to $y(t), t \geq 0$, with no input applied for $t \geq 0$.

It is not difficult to formalize this idea and obtain

$$\Gamma_G = \Psi_o \Psi_c,$$

using the definitions of these operators; we leave details to the reader. Thus the commutative diagram in Figure 4.4 is satisfied.

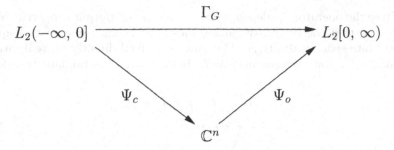

Figure 4.4. Hankel, controllability, and observability operators

Notice that Γ_G has at most rank n for a state space system of order n; namely, the dimension of $\mathrm{Im}(\Gamma_G)$ is at most n.

Our first result on the Hankel operator tells us how to compute its $L_2(-\infty, 0] \rightarrow L_2[0, \infty)$ induced norm. Below $\lambda_{\max}$ denotes the largest eigenvalue of the argument matrix.

Proposition 4.10. *The norm of Γ_G is given by*

$$\|\Gamma_G\|_{L_2 \rightarrow L_2} = (\lambda_{\max}(Y_o X_c))^{\frac{1}{2}},$$

where Y_o and X_c are the gramians of the minimal realization (A, B, C).

Proof. We begin with the following identity established in Corollary 3.17:

$$\|\Gamma_G\|^2 = \rho(\Gamma_G^* \Gamma_G) = \rho(\Psi_c^* \Psi_o^* \Psi_o \Psi_c),$$

where $\rho(\cdot)$ denotes spectral radius. We recall as well that the spectral radius is invariant under commutation, and obtain

$$\|\Gamma_G\|^2 = \rho(\Psi_o^* \Psi_o \Psi_c \Psi_c^*) = \rho(Y_o X_c).$$

Now, $Y_o X_c$ is a matrix, and has only positive eigenvalues; for this latter point, observe that the eigenvalues of $Y_o X_c$ coincide with those of $X_c^{\frac{1}{2}} Y_o X_c^{\frac{1}{2}} > 0$.

Therefore $\rho(Y_o X_c) = \lambda_{\max}(Y_o X_c)$, which completes the proof. ∎

We remark that it can also be shown that $\Gamma_G^* \Gamma_G$, a finite rank operator, has only eigenvalues in its spectrum, namely, the eigenvalues of $Y_o X_c$ in addition to the eigenvalue zero. For this reason, in analogy to the matrix case, the square roots of the eigenvalues of $Y_o X_c$ are called the *singular values* of Γ_G or the *Hankel singular values* of the system G. We order these and denote them by

$$\sigma_1 \geq \sigma_2 \geq \cdots \geq \sigma_n > 0.$$

Clearly $\sigma_1 = \|\Gamma_G\|$. Note that we do not include zero as an allowable Hankel singular value in our definition.

Since the operator Γ_G depends only on the input–output properties of G, it is clear that the Hankel singular values must be the same for two equivalent state space realizations. This can be verified directly for realizations related by a state transformation T. Recall that the gramians transform via

$$\tilde{Y}_o = (T^{-1})^* Y_o T^{-1},$$
$$\tilde{X}_c = T X_c T^*,$$

and therefore

$$\tilde{Y}_o \tilde{X}_c = (T^{-1})^* Y_o X_c T^*.$$

Thus we have $\tilde{Y}_o \tilde{X}_c$ and $Y_o X_c$ are related by a similarity transformation, and therefore their eigenvalues coincide. Notice that if instead we had a non-minimal realization, the non-zero eigenvalues of $Y_o X_c$ would be exactly the Hankel singular values of the input–output operator.

In the special case of a balanced realization $X_c = Y_0 = \Sigma$, we see that the diagonal of Σ is formed precisely by the Hankel singular values.

We now turn to the question of relating the Hankel singular values to the norm of the original operator G. While these are different objects we will see that important bounds can be established. The first one is the following.

Proposition 4.11. *With the above definitions,*

$$\sigma_1 = \|\Gamma_G\| \le \|G\| = \|\hat{G}\|_\infty.$$

Proof. Since the projection P_+ has induced norm 1, then

$$\|\Gamma_G\| = \|P_+ G\big|_{L_2(-\infty,0]}\| \le \|G\big|_{L_2(-\infty,0]}\| \le \|G\|.$$

For the last step, notice that the norm of an operator cannot increase when restricting it to a subspace. In fact, it is not difficult to show that this step is an equality. ∎

Thus the largest Hankel singular value provides us with a lower bound for the H_∞ norm of the transfer function $\hat{G}$. The next result provides an upper bound in terms of the Hankel singular values.

Proposition 4.12. *Suppose $\hat{G}(s) = C(Is - A)^{-1} B$ and that A is a Hurwitz matrix. Then*

$$\|\hat{G}\|_\infty \le 2(\sigma_1 + \cdots + \sigma_n),$$

where the σ_k are the Hankel singular values of G.

A proof with elementary tools is given in the exercises at the end of the chapter. In the next section we will see a slight refinement of this bound using more advanced methods.

The Hankel operator plays an important part in systems theory, as well as pure operator theory, and is intimately related to the question of approximation in the $\| \cdot \|_\infty$ norm. The latter has many connections to robust

control theory; in this chapter we will use it in connection to model reduction. For further discussion consult the references at the end of the
chapter.

4.6 Model reduction

We are now ready to consider approximation of systems. Frequently in
modeling or control synthesis it is of value to use a simple model that
approximates a more complex one. In our context, complexity is measured
by the dimensionality of state space realizations. Given a system, we would
like to reduce the order of a state space realization, while keeping the system
input–output properties approximately the same.

An extreme case of model reduction we have already studied is the elimination of uncontrollable or unobservable states. As shown in Chapter 2,
this can be done without affecting the input–output transfer function, i.e.,
with zero error. We are now ready to take the next step, that is, to allow
some error, and study the following problem: given a transfer function $\hat{G}(s)$
of order n, find a transfer function $\hat{G}_r(s)$ of order $r < n$, such that $\hat{G}$ and
$\hat{G}_r$ are "close."

Before this problem has a precise meaning, a notion of error between
transfer functions must be chosen; indeed there are many metrics that can
be used, all based on the input–output behavior of the systems. We will
focus on the H_∞ model reduction problem where

$$\text{error}(\hat{G}, \hat{G}_r) = \|\hat{G} - \hat{G}_r\|_\infty = \|G - G_r\|_{L_2 \to L_2}.$$

This choice deserves some comment.

In the first place, with this choice we are restricting our model reduction
to stable systems G, that is, systems which correspond to bounded operators on L_2. Later on in the course we will remark on how these techniques
can be extended to study errors between unstable systems; see Exercise 5.6.

Secondly, even under the stability restriction, there are other system
norms which could be candidates for model approximation. The reason
for our preference towards an *induced* norm can be illustrated as follows.
Suppose our system G is to be connected in cascade with another system
H; if we are going to use the approximation G_r instead of G, we would
like to be able to quantify the error easily between the composed systems
HG and HG_r. For any norm satisfying the submultiplicative inequality we
have

$$\|HG - HG_r\| \leq \|H\| \cdot \|G - G_r\|,$$

which tells us how to propagate our error to the cascaded system. Thus we
see the convenience of operator norms, in particular the H_∞ norm.

We recapitulate our discussion by stating more explicitly the H_∞ model
reduction problem:

Given: transfer function $\hat{G}(s) = \left[\begin{array}{c|c} A & B \\ \hline C & D \end{array}\right]$, where $A \in \mathbb{R}^{n \times n}$
is Hurwitz.

Find: lower-order function $\hat{G}_r(s) = \left[\begin{array}{c|c} A_r & B_r \\ \hline C_r & D_r \end{array}\right]$, where $A_r \in \mathbb{R}^{r \times r}$
and is Hurwitz, such that

$$\|\hat{G} - \hat{G}_r\|_\infty \text{ is minimized.}$$

Here we use the transfer function notation from §2.6, defined in (2.16). Again, throughout this section we have the assumption that (A, B, C, D) is a minimal realization; clearly, this is made without loss of generality, since we can otherwise reduce to a minimal realization with no error.

An immediate observation is that we can assume $D = 0$, since

$$\text{error}(\hat{G}, \hat{G}_r) = \left[\begin{array}{c|c} A & B \\ \hline C & D \end{array}\right] - \hat{G}_r = \left[\begin{array}{c|c} A & B \\ \hline C & 0 \end{array}\right] - (\hat{G}_r - D)$$

and feed-through terms do not affect the order of the realization.

What is not so clear is whether D_r can be set to zero; in other words, is the best approximation to a strictly proper system, itself strictly proper? Indeed there may be advantages in using $D_r \neq D$, as is shown in the exercises at the end of the chapter; in this section, however, we concentrate on the case where both $\hat{G}$ and $\hat{G}_r$ are strictly proper.

As stated, the H_∞ model reduction problem is open, in the sense that there is no known computationally tractable method to obtain the optimal approximation of a given order. In the absence of this optimal solution, we will study model reduction methods in terms of *bounds* on the approximation error. This includes both lower bounds on the error achievable by any model reduction method, and upper bounds on the error achieved by a specific scheme, in our case the *balanced truncation* method.

4.6.1 Limitations

Faced with the approximation problem above it is natural for us to ask whether there are some fundamental limits to how well we can approximate a given system with a lower-order one. To work towards the answer, as well as to build intuition, we begin by studying a *matrix* approximation problem that is strongly related to the system approximation problem: given a square matrix $N \in \mathbb{C}^{n \times n}$, how well can it be approximated, in the maximum singular value norm, by a matrix of rank r? We have the following result:

Lemma 4.13. *Suppose that $N \in \mathbb{C}^{n \times n}$, with singular values $\sigma_1 \geq \sigma_2 \geq \cdots \geq \sigma_n$. Then for any $R \in \mathbb{C}^{n \times n}$, rank$(R) \leq r < n$,*

$$\bar{\sigma}(N - R) \geq \sigma_{r+1} .$$

Proof. We start by taking the singular value decomposition of N:

$$N = U\Sigma V^*, \quad \text{where} \quad \Sigma = \begin{bmatrix} \sigma_1 & & \\ & \ddots & \\ & & \sigma_n \end{bmatrix},$$

and matrices U, V are both unitary.

Now define the column vectors v_k by

$$[v_1 \cdots v_n] = V ,$$

and consider the $(r+1)$-dimensional subspace $\text{span}\{v_1, \ldots, v_{r+1}\}$. Since the subspace $\text{Ker}(R)$ has dimension at least $n - r$, these two subspaces must intersect non-trivially. Therefore $\text{span}\{v_1, \ldots, v_{r+1}\}$ contains an element x, with $|x| = 1$, such that

$$Rx = 0 .$$

The vectors v_k are orthonormal since V is unitary and so x can be expressed by

$$x = \sum_{k=1}^{r+1} \alpha_k v_k \quad \text{with} \quad \sum_{k=1}^{r+1} |\alpha_k|^2 = 1,$$

for appropriately chosen scalars α_k.

We now let the matrix $N - R$ act on this vector x to get

$$(N - R)x = Nx = \sum_{k=1}^{r+1} \alpha_k N v_k = \sum_{k=1}^{r+1} \sigma_k \alpha_k u_k,$$

where we define u_k by $[u_1 \cdots u_n] = U$. The u_k are orthonormal and so we have

$$|(N - R)x|^2 = \sum_{k=1}^{r+1} \sigma_k^2 |\alpha_k|^2.$$

To complete the proof notice that since the singular values are ordered we have

$$\sum_{k=1}^{r+1} \sigma_k^2 |\alpha_k|^2 \geq \sigma_{r+1}^2 \sum_{k=1}^{r+1} |\alpha_k|^2 = \sigma_{r+1}^2,$$

and so

$$\bar{\sigma}(N - R) \geq |(N - R)x| \geq \sigma_{r+1} .$$

∎

The preceding lemma shows us that there are fundamental limits to the matrix approximation problem in the $\bar{\sigma}(\cdot)$ norm, and that singular values play a key role in characterizing these limits. What does this have to do with the system model reduction problem? The key connection is given by the Hankel operator, since as we saw the dimension of a state space realization is the same as the rank of the Hankel operator. Not surprisingly, then, we can express the fundamental limitations to model reduction in terms of the Hankel singular values:

Theorem 4.14. *Let* $\sigma_1 \geq \sigma_2 \geq \cdots \geq \sigma_r \geq \sigma_{r+1} \cdots \geq \sigma_n > 0$ *be the Hankel singular values associated with the transfer function* $\hat{G} \in RH_\infty$. *Then for any* $\hat{G}_r$ *of order* $r < n$, *the inequality*

$$\|\hat{G} - \hat{G}_r\|_\infty \geq \sigma_{r+1} \text{ holds.}$$

This result says that if we are seeking a reduced order model of state dimension r, then we cannot make the approximation error smaller than the $(r+1)$th Hankel singular value of the original system.

Proof. To begin we have that

$$\|\hat{G} - \hat{G}_r\|_\infty \geq \|\Gamma_{G-G_r}\| = \|\Gamma_G - \Gamma_{G_r}\| .$$

It therefore suffices to demonstrate that

$$\|\Gamma_G - \Gamma_{G_r}\| \geq \sigma_{r+1} .$$

As we have noted before these Hankel operators satisfy

$$\text{rank}\,(\Gamma_G) \leq n \quad \text{and} \quad \text{rank}\,(\Gamma_{G_r}) \leq r ,$$

which is a fact we will now exploit.

We recall the definitions of the observability and controllability operators Ψ_o and Ψ_c associated with a minimal realization (A, B, C), and the identity

$$\Gamma_G = \Psi_o \Psi_c .$$

Now we define the maps $P_o : L_2[0, \infty) \to \mathbb{C}^n$ and $P_c : \mathbb{C}^n \to L_2(-\infty, 0]$ by

$$P_o = Y_o^{-\frac{1}{2}} \Psi_o^* \quad \text{and} \quad P_c = \Psi_c^* X_c^{-\frac{1}{2}} ,$$

and verify that $\|P_o\| = \|P_c\| = 1$. We therefore have

$$\|\Gamma_G - \Gamma_{G_r}\| = \|P_o\| \cdot \|\Gamma_G - \Gamma_{G_r}\| \cdot \|P_c\| \geq \bar{\sigma}(P_o \Gamma_G P_c - P_o \Gamma_{G_r} P_c) , \tag{4.6}$$

where the submultiplicative inequality is used. We further have that

$$P_o \Gamma_G P_c = Y_o^{-\frac{1}{2}} \Psi_o^* \Psi_o \Psi_c \Psi_c^* X_c^{-\frac{1}{2}} = Y_o^{\frac{1}{2}} X_c^{\frac{1}{2}} ,$$

which has rank equal to n since its singular values are $\sigma_1, \ldots, \sigma_n$ and are all positive. But matrix $P_o \Gamma_{G_r} P_c$ has rank at most equal to r since rank $(\Gamma_{G_r}) \leq r$. Invoke Lemma 4.13 to see that

$$\bar{\sigma}(P_o \Gamma_G P_c - P_o \Gamma_{G_r} P_c) \geq \sigma_{r+1} .$$

Now apply (4.6) to arrive at the conclusion we seek.

■

4.6.2 Balanced truncation

We now move on to one of the major methods for finding a reduced order approximation. Recall we showed that a balanced realization for a transfer function can always be found.

Formally this means a realization such that the gramians satisfy

$$X_c = Y_o = \begin{bmatrix} \sigma_1 & & & \\ & \sigma_2 & & \\ & & \ddots & \\ & & & \sigma_n \end{bmatrix}.$$

Also we saw that if a given σ_k was small compared with the other eigenvalues, the corresponding state had the interpretation of being weakly controllable and observable. The technique for model reduction we now consider, essentially amounts to discarding or *truncating* such states.

Here we necessarily assume that the realization (A, B, C) is a balanced realization and order the Hankel singular values

$$\sigma_1 \geq \sigma_2 \geq \cdots \geq \sigma_n.$$

Let us suppose we want to find a realization of order $r < n$. We assume that the *strict* inequality

$$\sigma_{r+1} < \sigma_r \quad \text{is satisfied;}$$

that is, we only attempt to truncate the system at an order where the Hankel singular values have a clear separation. Then we compatibly partition our realization as

$$A = \begin{bmatrix} A_{11} & A_{12} \\ A_{21} & A_{22} \end{bmatrix}, \quad B = \begin{bmatrix} B_1 \\ B_2 \end{bmatrix}, \quad \text{and} \quad C = [C_1 \; C_2],$$

where $A_{11} \in \mathbb{R}^{r \times r}$. The reduced order model is chosen to be

$$\hat{G}_r(s) = \left[\begin{array}{c|c} A_{11} & B_1 \\ \hline C_1 & 0 \end{array} \right](s).$$

The minimum requirement on $\hat{G}_r$ is that A_{11} be Hurwitz, and this is indeed the case.

Proposition 4.15.

(a) The matrix A_{11}, as defined above, is Hurwitz.

(b) The realization (A_{11}, B_1, C_1) is balanced with Hankel singular values $\sigma_1, \ldots, \sigma_r$.

Proof. We begin by writing down the Lyapunov equations of the balanced realization (A, B, C), highlighting the relevant partition:

$$\begin{bmatrix} A_{11}^* & A_{21}^* \\ A_{12}^* & A_{22}^* \end{bmatrix} \begin{bmatrix} \Sigma_1 & 0 \\ 0 & \Sigma_2 \end{bmatrix} + \begin{bmatrix} \Sigma_1 & 0 \\ 0 & \Sigma_2 \end{bmatrix} \begin{bmatrix} A_{11} & A_{12} \\ A_{21} & A_{22} \end{bmatrix} + \begin{bmatrix} C_1^* \\ C_2^* \end{bmatrix} [C_1 \; C_2] = 0, \quad (4.7)$$

$$\begin{bmatrix} A_{11} & A_{12} \\ A_{21} & A_{22} \end{bmatrix} \begin{bmatrix} \Sigma_1 & 0 \\ 0 & \Sigma_2 \end{bmatrix} + \begin{bmatrix} \Sigma_1 & 0 \\ 0 & \Sigma_2 \end{bmatrix} \begin{bmatrix} A_{11}^* & A_{21}^* \\ A_{12}^* & A_{22}^* \end{bmatrix} + \begin{bmatrix} B_1 \\ B_2 \end{bmatrix} [B_1^* \; B_2^*] = 0. \quad (4.8)$$

Here Σ_1 and Σ_2 are diagonal matrices with, respectively, $\sigma_1, \ldots, \sigma_r$, and $\sigma_{r+1}, \ldots, \sigma_n$ in the diagonal. Notice that, by assumption, they have no eigenvalues in common.

From here we extract the top left blocks and obtain

$$A_{11}^* \Sigma_1 + \Sigma_1 A_{11} + C_1^* C_1 = 0, \quad (4.9)$$
$$A_{11} \Sigma_1 + \Sigma_1 A_{11}^* + B_1 B_1^* = 0. \quad (4.10)$$

We see then that if A_{11} is Hurwitz, we immediately have that Σ_1 is a balanced gramian for the truncated system, proving (b). Thus we concentrate on part (a).

Fix λ to be an eigenvalue of A_{11}, and V be a full column rank matrix that generates the eigenvector space; therefore,

$$\mathrm{Ker}\,(A_{11} - \lambda I) = \mathrm{Im}(V).$$

Multiplying (4.9) on the left by V^* and on the right by V leads to

$$(\lambda^* + \lambda)V^* \Sigma_1 V + V^* C_1^* C_1 V = 0.$$

Since $V^* \Sigma_1 V > 0$ we see that $\mathrm{Re}(\lambda) \leq 0$. Thus it only remains to rule out the possibility of a purely imaginary eigenvalue; we make this assumption from now on.

Setting $\lambda = j\omega$, the preceding equation implies that the corresponding V must satisfy

$$C_1 V = 0.$$

Now we multiply (4.9) only on the right by V; we conclude that

$$A_{11}^*(\Sigma_1 V) = -j\omega \Sigma_1 V.$$

We have now an invariant subspace of A_{11}^*, so we turn to an analogous study of the dual equation (4.10). First we multiply it on the left by $V^* \Sigma_1$ and on the right by $\Sigma_1 V$ to show

$$B_1^* \Sigma_1 V = 0;$$

then we return to (4.10) and perform only the right multiplication by $\Sigma_1 V$ to yield

$$A_{11}(\Sigma_1^2 V) = j\omega \Sigma_1^2 V.$$

So we are back with an eigenspace of A_{11}; in particular, we have

$$\mathrm{Im}(\Sigma_1^2 V) \subset \mathrm{Ker}\,(A_{11} - j\omega I) = \mathrm{Im}(V);$$

namely, $\mathrm{Im}(V)$ is *invariant* under Σ_1^2. We can now finish this exercise in invariant subspaces by concluding that Σ_1^2 must have an eigenvector v in $\mathrm{Im}(V)$, satisfying

$$\Sigma_1^2 v = \sigma^2 v.$$

We have thus found $v \neq 0$ that is simultaneously an eigenvector of A_{11} with eigenvalue $j\omega$, and of Σ_1^2, with eigenvalue σ^2. In addition, from the previous development we have

$$C_1 v = 0 \quad \text{and} \quad B_1^* \Sigma_1 v = 0.$$

We claim that

$$\begin{bmatrix} A_{11} & A_{12} \\ A_{21} & A_{22} \end{bmatrix} \begin{bmatrix} v \\ 0 \end{bmatrix} = j\omega \begin{bmatrix} v \\ 0 \end{bmatrix}. \tag{4.11}$$

This contradicts the fact that A is Hurwitz, and thus the assumption that $\lambda = j\omega$ is an eigenvalue of A_{11} would be ruled out.

Thus, to complete the proof it is sufficient for us to show that $A_{21}v = 0$. To do this we return to the original Lyapunov equations (4.7) and (4.8) and multiply them, respectively, by

$$\begin{bmatrix} v \\ 0 \end{bmatrix} \quad \text{and} \quad \begin{bmatrix} \Sigma_1 v \\ 0 \end{bmatrix}.$$

Looking only at the *second* row of equations and using the above properties of v, we get

$$A_{12}^* \Sigma_1 v + \Sigma_2 A_{21} v = 0,$$
$$A_{21} \sigma^2 v + \Sigma_2 A_{12}^* \Sigma_1 v = 0,$$

which lead to the key relation

$$\Sigma_2^2 A_{21} v = \sigma^2 A_{21} v.$$

At this point (only!) we bring in the assumption that Σ_1 and Σ_2 have no common eigenvalues; thus σ^2 cannot be an eigenvalue of Σ_2^2, so $A_{21}v$ must be zero, which proves (4.11). ∎

In the exercises you will show, by counterexample, that the requirement of distinct eigenvalues for Σ_1 and Σ_2 ($\sigma_r > \sigma_{r+1}$) is indeed essential to ensure the stability of the truncation.

Having discussed stability, we turn to the key model reduction question, what is the H_∞ norm error achieved by this method? We will see that the model reduction error can be quantified by a special purpose argument based on so-called *allpass dilations*; to develop it requires a short diversion into the topic of inner transfer functions, after which we can return to the balanced truncation error.

4.6.3 Inner transfer functions

We begin by recalling the notion of an isometric operator between Hilbert spaces, characterized by the identity

$$U^* U = I,$$

where U^* is the usual adjoint. Such operators preserve norms and hence have induced norm $\|U\| = 1$.

Focusing now on linear-time invariant operators characterized by a transfer function $\hat{U}(s) \in RH_\infty$, we inquire which of these are isometric. Converting to frequency domain we see that U is an isometry if and only if

$$\langle M_{\hat{U}}\hat{x}, \, M_{\hat{U}}\hat{x} \rangle = \langle \hat{x}, \, \hat{x} \rangle, \text{ holds for all } x \in L_2.$$

Therefore we conclude that a system defines an isometric operator if and only if its transfer function satisfies

$$\hat{U}(j\omega)^* \hat{U}(j\omega) = I, \text{ for all } \omega.$$

Rational transfer functions satisfying the above identity are called *inner* functions. The term *allpass* function is also used sometimes, since in the scalar case these transfer functions have unit gain at every frequency.

The next step is to obtain a state space test for inner functions. For this purpose, notice that if

$$\hat{U}(s) = \left[\begin{array}{c|c} A & B \\ \hline C & D \end{array} \right].$$

then the realization

$$\hat{U}^\sim(s) = \left[\begin{array}{c|c} -A^* & -C^* \\ \hline B^* & D^* \end{array} \right]$$

satisfies $\hat{U}^\sim(j\omega) = \hat{U}(j\omega)^*$. The rational function $\hat{U}^\sim(s)$ is called the *para-Hermitian conjugate* of $\hat{U}(s)$. If $\hat{U}(s) \in RH_\infty$, $\hat{U}^\sim(s)$ will be analytic on the closed *left* half-plane.

Composing $\hat{U}^\sim$ and $\hat{U}$, we obtain the joint realization

$$\hat{U}^\sim(s)\hat{U}(s) = \left[\begin{array}{cc|c} -A^* & -C^*C & -C^*D \\ 0 & A & B \\ \hline B^* & D^*C & D^*D \end{array} \right]. \tag{4.12}$$

Clearly, the only way the above rational function can be constant across frequency, as required for $\hat{U}$ to be inner, is that all states must be uncontrollable or unobservable. An important case where this can be guaranteed is given in the next lemma.

Lemma 4.16. *Consider the transfer function* $\hat{U}(s) = \left[\begin{array}{c|c} A & B \\ \hline C & D \end{array} \right]$ *where* A *is Hurwitz. If the observability gramian* Y_o *of* (C, A) *satisfies the additional property* $C^*D + Y_o B = 0$, *then*

$$\hat{U}^\sim(s)\hat{U}(s) = D^*D.$$

Proof. Starting with the realization (4.12), we introduce the state transformation

$$T = \begin{bmatrix} I & -Y_o \\ 0 & I \end{bmatrix},$$

which leads to the equivalent realization

$$\hat{U}^{\sim}(s)\hat{U}(s) = \left[\begin{array}{cc|c} -A^* & -(A^*Y_o + Y_oA + C^*C) & -(C^*D + Y_oB) \\ 0 & A & B \\ \hline B^* & D^*C + B^*Y_o & D^*D \end{array}\right].$$

Now the Lyapunov equation for Y_o, and the additional property in the hypothesis give

$$\hat{U}^{\sim}(s)\hat{U}(s) = \left[\begin{array}{cc|c} -A^* & 0 & 0 \\ 0 & A & B \\ \hline B^* & 0 & D^*D \end{array}\right] = D^*D,$$

since all states in this latter realization are uncontrollable or unobservable.
∎

We are now ready to return to the model reduction problem.

4.6.4 Bound for the balanced truncation error

In this section we develop a bound for the H_∞ norm error associated with balanced truncation. This bound will show that if the sum of $\sigma_{r+1}, \ldots, \sigma_n$ is small compared with the remaining Hankel singular values, then G_r is a near approximation to G.

We first consider the case where the $n-r$ smallest Hankel singular values are equal.

Lemma 4.17. *Suppose the Hankel singular values of G satisfy*

$$\sigma_1 \geq \sigma_2 \geq \cdots \geq \sigma_r > \sigma_{r+1} = \sigma_{r+2} \cdots = \sigma_n .$$

If $\hat{G}_r$ is obtained by r-th order balanced truncation, then $\|\hat{G} - \hat{G}_r\|_\infty \leq 2\sigma_{r+1}$.

In the proof we use standard manipulations of transfer functions; see Exercise 2.18 for relevant formulas.

Proof. Start by defining the error system

$$\hat{E}_{11}(s) = \hat{G}(s) - \hat{G}_r(s) ,$$

which has state space realization

$$\hat{E}_{11}(s) = \left[\begin{array}{ccc|c} A_{11} & 0 & 0 & B_1 \\ 0 & A_{11} & A_{12} & B_1 \\ 0 & A_{21} & A_{22} & B_2 \\ \hline -C_1 & C_1 & C_2 & 0 \end{array}\right].$$

Notice that by Proposition 4.15, this realization has a Hurwitz A-matrix. Our technique of proof will be to construct a so-called *allpass dilation* of $\hat{E}_{11}(s)$, by finding

$$\hat{E}(s) = \begin{bmatrix} \hat{E}_{11}(s) & \hat{E}_{12}(s) \\ \hat{E}_{21}(s) & \hat{E}_{22}(s) \end{bmatrix},$$

which contains $\hat{E}_{11}$ as a sub-block and is, up to a constant, inner. If this can be done, the error norm can be bounded by the norm of $\hat{E}(s)$, which is easy to compute.

We first use the state transformation

$$T = \begin{bmatrix} I & I & 0 \\ I & -I & 0 \\ 0 & 0 & I \end{bmatrix},$$

to get

$$\hat{E}_{11}(s) = \left[\begin{array}{ccc|c} A_{11} & 0 & A_{12}/2 & B_1 \\ 0 & A_{11} & -A_{12}/2 & 0 \\ A_{21} & -A_{21} & A_{22} & B_2 \\ \hline 0 & -2C_1 & C_2 & 0 \end{array}\right].$$

Next, we define the augmentation as

$$\hat{E}(s) = \left[\begin{array}{ccc|cc} A_{11} & 0 & A_{12}/2 & B_1 & 0 \\ 0 & A_{11} & -A_{12}/2 & 0 & \sigma_{r+1}\Sigma_1^{-1}C_1^* \\ A_{21} & -A_{21} & A_{22} & B_2 & -C_2^* \\ 0 & -2C_1 & C_2 & 0 & 2\sigma_{r+1}I \\ \hline -2\sigma_{r+1}B_1^*\Sigma_1^{-1} & 0 & -B_2^* & 2\sigma_{r+1}I & 0 \end{array}\right]$$

$$=: \left[\begin{array}{c|c} \bar{A} & \bar{B} \\ \hline \bar{C} & \bar{D} \end{array}\right].$$

Here we are using the notation

$$\Sigma = \begin{bmatrix} \Sigma_1 & 0 \\ 0 & \sigma_{r+1}I \end{bmatrix}$$

for the balanced gramian of the original system G.

While this construction is as yet unmotivated, the underlying aim is to be able to apply Lemma 4.16. Indeed, it can be verified by direct substitution that the observability gramian of the above realization is

$$\bar{Y}_o = \begin{bmatrix} 4\sigma_{r+1}^2\Sigma_1^{-1} & 0 & 0 \\ 0 & 4\Sigma_1 & 0 \\ 0 & 0 & 2\sigma_{r+1}I \end{bmatrix},$$

and that it satisfies the additional restriction

$$\bar{C}^*\bar{D} + \bar{Y}_o\bar{B} = 0.$$

While somewhat tedious, this verification is based only the Lyapunov equations satisfied by the gramian Σ.

Therefore we can apply Lemma 4.16 to conclude that

$$\hat{E}^*(j\omega)\hat{E}(j\omega) = \bar{D}^*\bar{D} = 4\sigma_{r+1}^2 I \quad \text{for all } \omega,$$

and therefore we have

$$\|\hat{E}_{11}\|_\infty \leq \|\hat{E}\|_\infty = 2\sigma_{r+1}.$$

■

We now turn to the general case, where the truncated eigenvalues $\sigma_{r+1},\ldots,\sigma_n$ are not necessarily all equal. Still, there may be repetitions among them, so we introduce the notation $\sigma_1^t,\ldots\sigma_k^t$ to denote the distinct values of this tail. More precisely, we assume that

$$\sigma_1^t > \sigma_2^t > \cdots > \sigma_k^t$$

and $\{\sigma_{r+1},\ldots,\sigma_n\} = \{\sigma_1^t,\ldots\sigma_k^t\}$. Equivalently, the balanced gramian of the full order system is given by

$$\Sigma = \left[\begin{array}{ccc|ccc} \sigma_1 & & & & & \\ & \ddots & & & & \\ & & \sigma_r & & & \\ \hline & & & \sigma_{r+1} & & \\ & & & & \ddots & \\ & & & & & \sigma_n \end{array}\right] = \left[\begin{array}{c|ccc} \Sigma_1 & & & \\ \hline & \sigma_1^t I & & \\ & & \ddots & \\ & & & \sigma_k^t I \end{array}\right]$$

where the block dimensions of the last expression correspond to the number of repetitions of each σ_i^t

We are now ready to state the main result:

Theorem 4.18. *With the above notation for the Hankel singular values of $\hat{G}$, let $\hat{G}_r$ be obtained by r-th order balanced truncation. Then the following inequality is satisfied:*

$$\|\hat{G} - \hat{G}_r\|_\infty \leq 2(\sigma_1^t + \cdots \sigma_k^t).$$

Proof. The idea is to apply successively the previous lemma, truncating one block of repeated Hankel singular values at every step. By virtue of Proposition 4.15, in this process the realizations we obtain remain balanced and stable, and the Hankel singular values are successively removed. Therefore at each step we can apply the previous lemma and obtain an error bound of twice the corresponding σ_i^t. Finally, the triangle inequality implies that the overall error bound is the sum of these terms.

More explicitly, consider the transfer functions $\hat{G}^{(0)},\ldots,\hat{G}^{(k)}$, where $\hat{G}^{(k)} = \hat{G}$, and for each $i \in \{1,\ldots,k\}$, the transfer function $\hat{G}^{(i-1)}$ is obtained by balanced truncation of the repeated Hankel singular value σ_i^t

of $\hat{G}^{(i)}$. By induction, each $\hat{G}^{(i)}$ has a stable, balanced realization with gramian

$$\begin{bmatrix} \Sigma_1 & & & \\ & \sigma_1^t I & & \\ & & \ddots & \\ & & & \sigma_i^t I \end{bmatrix}.$$

Therefore Lemma 4.17 applies at each step and gives

$$\|\hat{G}^{(i)} - \hat{G}^{(i-1)}\|_\infty \le 2\sigma_i^t.$$

Also $\hat{G}^{(0)} = \hat{G}_r$, so we have

$$\|\hat{G} - \hat{G}_r\| = \|\sum_{i=1}^k \left(\hat{G}^{(i)} - \hat{G}^{(i-1)} \right) \|_\infty$$

$$\le \sum_{i=1}^k \|\hat{G}^{(i)} - \hat{G}^{(i-1)}\|_\infty \le 2(\sigma_1^t + \cdots \sigma_k^t).$$

∎

As a comment, notice that by specializing the previous result to the case $r = 0$ we can bound $\|\hat{G}\|_\infty$ by twice the sum of its Hankel singular values; since repeated singular values are only counted once, this is actually a tighter version of Proposition 4.12.

We have therefore developed a method by which to reduce the dynamic size of a state space model predictably, with the H_∞ norm as the quality measure on the error.

Example:

We now give a numerical example of the uses of this method. Consider the 7th order transfer function in RH_∞,

$$\hat{G}(s) = \frac{(s+10)(s-5)(s^2+2s+5)(s^2-0.5s+5)}{(s+4)(s^2+4s+8)(s^2+0.2s+100)(s^2+5s+2000)}.$$

The magnitude Bode plot, i.e. the plot of $|\hat{G}(j\omega)|$ in a log-log scale, is depicted with solid lines in each graph of Figure 4.5.

The Hankel singular values of $\hat{G}$ can be computed to be

$$\begin{bmatrix} \sigma_1 \\ \sigma_2 \\ \sigma_3 \\ \sigma_4 \\ \sigma_5 \\ \sigma_6 \\ \sigma_7 \end{bmatrix} = \begin{bmatrix} 0.17926 \\ 0.17878 \\ 0.10768 \\ 0.10756 \\ 0.00076 \\ 0.00008 \\ 0.00003 \end{bmatrix}.$$

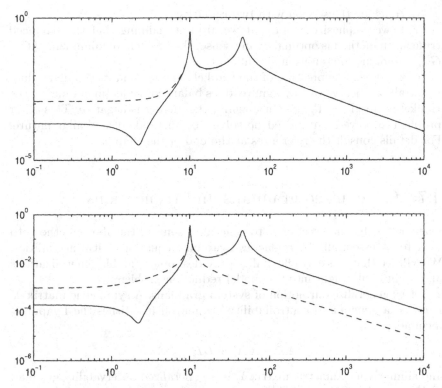

Figure 4.5. Log-log plots for the example. Top: $|\hat{G}(j\omega)|$ (solid) and $|\hat{G}_4(j\omega)|$ (dashed). Bottom: $|\hat{G}(j\omega)|$ (solid) and $|\hat{G}_2(j\omega)|$ (dashed).

Given these values, it is apparent that a natural model reduction would be to $r = 4$ states, since the remaining states are clearly very weakly observable and controllable. We may also wish to study a more drastic reduction to $r = 2$ states. The magnitude Bode plots of $\hat{G}_4$ and $\hat{G}_2$, obtained by balanced truncation, are shown respectively in the top and bottom plots of Figure 4.5, with dashed lines. We see that $\hat{G}_4$ "picks-up" with great accuracy the high-frequency behavior of the system (where it has its greatest gain), but commits errors at low frequency. $\hat{G}_2$ focuses its approximation on the largest resonant peak, around $\omega = 10$.

We can compute the model reduction bounds from Theorem 4.18, and also (e.g., by frequency gridding, or with methods to be covered later in the course) the actual H_∞ errors. We find

$$0.0014 = \|\hat{G} - \hat{G}_4\|_\infty \le 2(\sigma_5 + \sigma_6 + \sigma_7) = 0.0017,$$
$$0.2153 = \|\hat{G} - \hat{G}_2\|_\infty \le 2(\sigma_3 + \sigma_4 + \sigma_5 + \sigma_6 + \sigma_7) = 0.4322.$$

$\square$

We conclude the section with two remarks:

First, we emphasize again that we are not claiming that the balanced truncation method is optimal in any sense; the problem of minimizing $\|\hat{G} - \hat{G}_r\|_\infty$ remains computationally difficult.

Second, given the presence of the Hankel operator and its singular values in the above theory, we may inquire, does balanced truncation minimize the Hankel norm $\|\Gamma_{\hat{G}} - \Gamma_{\hat{G}_r}\|$? Once again, the answer is negative. This other problem, however, can indeed be solved by methods of a similar nature. For details consult the references at the end of the chapter.

4.7 Generalized gramians and truncations

The goal of this final section is to generalize some of the above methods to LMI tools, essentially by replacing Lyapunov equations with inequalities. We will see that this extra flexibility provides some valuable generalizations and additional insight into the model reduction problem.

Let us generalize our notion of system gramians. A symmetric matrix X is called a *generalized* controllability gramian if it satisfies the Lyapunov inequality

$$AX + XA^* + BB^* \leq 0,$$

and similarly a symmetric matrix Y is a *generalized* observability gramian if $A^*Y + YA + C^*C \leq 0$ holds.

The system gramians X_c and Y_o are both generalized gramians, but are not unique in this regard; in general there is a set of generalized gramians, even when (A, B, C) is a minimal realization. What special properties do the actual gramians enjoy? From Proposition 4.4 we know, in fact, they are the minimal elements among the generalized gramians, so

$$X \geq X_c, \quad Y \geq Y_o, \tag{4.13}$$

for any generalized gramians X, Y.

As with X_c and Y_o we say that two generalized gramians X and Y are *balanced* when they are diagonal and equal:

$$X = Y = \begin{bmatrix} \gamma_1 & & \\ & \ddots & \\ & & \gamma_n \end{bmatrix}.$$

Generalized gramians can be balanced in the exact same way as before; when (A, B, C) is minimal, X and Y are positive definite and we can simply apply the simultaneous diagonalization result of Proposition 4.7.

The entries γ_k are called the *generalized* Hankel singular values. It is shown in the exercises, based on the inequalities (4.13), that with the order

$\gamma_1 \geq \gamma_2 \geq \cdots \geq \gamma_n$, we have as expected

$$\gamma_i \geq \sigma_i, \quad i = 1, \ldots, n.$$

Generalized gramians can also be used for model reduction by truncation; we have the following generalization to Proposition 4.15 and Theorem 4.18:

Proposition 4.19. If $\hat{G} = \left[\begin{array}{c|c} A & B \\ \hline C & 0 \end{array} \right]$ is minimal, with A Hurwitz, and

$\hat{G}_r = \left[\begin{array}{c|c} A_r & B_r \\ \hline C_r & 0 \end{array} \right]$ is formed by truncating according to the balanced generalized gramians, then A_r is Hurwitz and

$$\|\hat{G} - \hat{G}_r\|_\infty \leq 2(\gamma_1^t + \cdots \gamma_q^t), \qquad (4.14)$$

where γ_i^t denote the distinct generalized Hankel singular values which satisfy $\gamma_i^t < \gamma_r$.

The proof of this result follows readily from our earlier balanced truncation result, and is included in the exercises.

At first sight, the previous generalization may not appear to have a particular application, because we have seen that the generalized Hankel singular values are always larger than the σ_i. Therefore it seems that Theorem 4.18 would always give a smaller error bound.

Notice, however, that there is the issue of repetition in the singular values. Suppose, for example, that a third-order system has Hankel singular values $(\sigma_1, \sigma_2, \sigma_3) = (1, 0.4, 0.3)$. If by some means we managed to obtain generalized Hankel singular values $(\gamma_1, \gamma_2, \gamma_3) = (1, 0.4, 0.4)$, then in a truncation to $r = 1$ states the generalized method will have a smaller error bound.

Can such repetition always be arranged? The interesting fact is that the answer is affirmative, and can be used to characterize exactly the H_∞ model reduction problem. This is stated in the following result.

Theorem 4.20. Given $\hat{G} = \left[\begin{array}{c|c} A & B \\ \hline C & D \end{array} \right]$, with $A \in \mathbb{R}^{n \times n}$ Hurwitz, the following are equivalent:

(a) There exists $\hat{G}_r = \left[\begin{array}{c|c} A_r & B_r \\ \hline C_r & D_r \end{array} \right]$ of order r such that $\|\hat{G} - \hat{G}_r\|_\infty < \epsilon$.

(b) There exist $X > 0$ and $Y > 0$ satisfying

   (i) $AX + XA^* + BB^* < 0$,
   (ii) $A^*Y + YA + C^*C < 0$,
   (iii) $\lambda_{\min}(XY) = \epsilon^2$, with $\text{rank}(XY - \epsilon^2 I) \leq r$.

We will not prove this theorem here, since it will follow as a corollary of a general result on H_∞ control, to be proved in Chapter 7. We will also see there how $\hat{G}_r$ can be constructed if we have the appropriate X, Y.

This theorem says that if there exists an H_∞ model reduction with error less than ϵ, then we can always find (strict) generalized gramians such that the smallest generalized Hankel singular value is ϵ, repeated $n - r$ times. In particular, there would be only one term in the error bound from truncation (4.14), although truncation is not the method we will use to obtain the above result. Notice also that the factor of 2 does not appear in the H_∞ norm bound, which is related to the use of a term $D_r \neq D$ (see the exercises).

At this point, it might appear that we can solve the H_∞ model reduction problem, since we have an exact finite dimensional characterization of when a reduction of order r with a certain error can be achieved. The difficulty is that the above conditions, while easy to state, are not computationally tractable. In particular the condition on the minimum eigenvalue is not easy to enforce in a convex way. This means that the above theorem cannot truly be considered a solution. Nevertheless, it is a valuable addition to our insight on this problem, since it provides a very compact description of where the computational difficulty lies.

There are other important reasons to work with generalized gramians, perhaps the most compelling one is that these generalize to multidimensional and uncertain systems, as discussed in Chapter 11. In addition, we remark that for *discrete time* systems, generalized gramians appear naturally since the actual gramians are not preserved by truncation; for more details see the exercises and the references.

Exercises

4.1. Suppose A, X and C satisfy $A^*X + XA + C^*C = 0$. Show that any two of the following implies the third:

(i) A Hurwitz.
(ii) (C, A) observable.
(iii) $X > 0$.

4.2. Prove Proposition 4.4.

4.3. Use the controllability canonical form to prove Proposition 4.6 in the general case of uncontrollable (A, B).

4.4. Controllability gramian vs. controllability matrix. We have seen that the singular values of the controllability gramian X_c can be used to determine "how controllable" the states are. In this problem you will show that the controllability matrix

$$M_c = [B \ AB \ A^2B \ \cdots \ A^{n-1}B]$$

cannot be used for the same purpose, since its singular values are unrelated to those of X_c. In particular, construct examples ($A \in \mathbb{C}^{2\times 2}, B \in \mathbb{C}^{2\times 1}$ suffices) such that

(a) $X_c = I$, but $\underline{\sigma}(M_c)$ is arbitrarily small.
(b) $M_c = I$, but $\underline{\sigma}(X_c)$ is arbitrarily small.

4.5. Proof of Proposition 4.12.

(a) For an integrable matrix function $M(t)$, show that

$$\bar{\sigma}\left(\int_a^b M(t)\, dt\right) \le \int_a^b \bar{\sigma}(M(t))\, dt.$$

You can use the fact that the property holds in the scalar case.

(b) Let $\hat{G} = \left[\begin{array}{c|c} A & B \\ \hline C & 0 \end{array}\right]$. Using (a) and the fact that $\hat{G}(j\omega)$ is the Fourier transform of $Ce^{At}B$, derive the inequality

$$\|\hat{G}\|_\infty \le \int_0^\infty \bar{\sigma}(Ce^{\frac{At}{2}})\,\bar{\sigma}(e^{\frac{At}{2}}B)dt.$$

(c) If X_c, Y_o are the gramians of (A, B, C), show that

$$\int_0^\infty \bar{\sigma}(Ce^{\frac{At}{2}})^2 \, dt \le 2\mathrm{Tr}(Y_o),$$

$$\int_0^\infty \bar{\sigma}(e^{\frac{At}{2}}B)^2 \, dt \le 2\mathrm{Tr}(X_c).$$

(d) Combine (b) and (c) for a balanced realization (A, B, C), to show that

$$\|\hat{G}\|_\infty \le 2(\sigma_1 + \cdots + \sigma_n) .$$

4.6. In Proposition 4.15 the strict separation $\sigma_r > \sigma_{r+1}$ of Hankel singular values was used to ensure the stability of the truncation (i.e., that A_{11} is Hurwitz). Show that indeed this is a necessary requirement, by constructing an example (e.g., with $n = 2, r = 1$) where the truncated matrix is not Hurwitz.

4.7. Consider the transfer function $\hat{G}(s) = \frac{1}{s+1}$.

(a) Find the (only) Hankel singular value of $\hat{G}$. Compare the error bound and actual error when truncating to 0 states (clearly the truncation is $\hat{G}_0 = 0$).

(b) Show that by allowing a nonzero d_r term, the previous error can be reduced by a factor of 2. Do this by solving explicitly

$$\min_{d\in\mathbb{R}} \|\hat{G} - d\|_\infty.$$

4.8. (a) Let X and Y be two positive definite matrices in $\mathbb{C}^{n \times n}$. Show that the following are equivalent:

(i) $X \geq Y^{-1}$;   (ii) $\lambda_{\min}(XY) \geq 1$;   (iii) $\begin{bmatrix} X & I \\ I & Y \end{bmatrix} \geq 0$.

(b) Under the conditions of part (a), show that: $\lambda_{\min}(XY) = 1$, with multiplicity k, if and only if $\begin{bmatrix} X & I \\ I & Y \end{bmatrix} \geq 0$ and has rank $2n - k$.

(c) Let (A, B, C) be a state space realization of $\hat{G}(s)$ with A Hurwitz. Assume there exist X_0, Y_0 in $\mathbb{C}^{r \times r}$ such that

$$A \begin{bmatrix} X_0 & 0 \\ 0 & \epsilon I_{n-r} \end{bmatrix} + \begin{bmatrix} X_0 & 0 \\ 0 & \epsilon I_{n-r} \end{bmatrix} A^* + BB^* \;<\; 0$$

$$A^* \begin{bmatrix} Y_0 & 0 \\ 0 & \epsilon I_{n-r} \end{bmatrix} + \begin{bmatrix} Y_0 & 0 \\ 0 & \epsilon I_{n-r} \end{bmatrix} A + C^*C \;<\; 0$$

$$\begin{bmatrix} X_0 & \epsilon I_r \\ \epsilon I_r & Y_0 \end{bmatrix} \;\geq\; 0$$

Using Theorem 4.20 prove that there exists a r-th order realization $\hat{G}_r(s)$ with $\|\hat{G} - \hat{G}_r\|_\infty < \epsilon$. Is the above problem convex?

4.9. Discrete time gramians and truncation. This problem concerns the discrete time system

$$\begin{aligned} x_{k+1} &= Ax_k + Bw_k, \\ z_k &= Cx_k. \end{aligned}$$

(a) The discrete Lyapunov equations are

$$\begin{aligned} A^*L_oA - L_o + C^*C = 0, \\ AJ_cA^* - J_c + BB^* = 0. \end{aligned}$$

Assume that each eigenvalue in the set $\text{eig}(A)$ has magnitude less than one (stable in the discrete sense). Explicitly express the solutions J_c and L_o in terms of A, B, and C; prove these solutions are unique.

A realization is said to be balanced if J_c and L_o are diagonal and equal. Is it always possible to find a balanced realization?

(b) In this part we assume, in contrast to (a), that $J > 0$ and $L > 0$ are generalized gramians that are solutions to the strict Lyapunov inequalities

$$\begin{aligned} A^*LA - L + C^*C < 0; \\ AJA^* - J + BB^* < 0. \end{aligned}$$

Show that $L_o < L$ and $J_c < J$. Can generalized gramians be balanced?

(c) Suppose $J = L = \begin{bmatrix} \Sigma_1 & 0 \\ 0 & \sigma I \end{bmatrix}$ are given generalized gramians, where $\Sigma_1 > 0$ is diagonal. Partition the state space accordingly:

$$A = \begin{bmatrix} A_{11} & A_{12} \\ A_{21} & A_{22} \end{bmatrix}; \quad B = \begin{bmatrix} B_1 \\ B_2 \end{bmatrix};$$
$$C = \begin{bmatrix} C_1 & C_2 \end{bmatrix};$$

and define

$$\hat{G}_{11} = \left[\begin{array}{c|c} A_{11} & B_1 \\ \hline C_1 & 0 \end{array} \right].$$

Show that (A_{11}, B_1, C_1) is a balanced realization in the sense of generalized gramians, and that all the eigenvalues of A_{11} have absolute value less than one.

4.10. Recall, from Exercise 1.16, the Courant-Fischer formula for the ordered eigenvalues of an $n \times n$ Hermitian matrix.

(a) Use this result to deduce that if $A \geq B$, then $\lambda_k(A) \geq \lambda_k(B)$, for every $k = 1, \dots, n$.

(b) Notice that $\lambda_k(AC) = \lambda_k(CA)$, for positive semidefinite matrices A and C. Use this fact and (a) to show that the *generalized* Hankel singular values of a system are always greater than the corresponding Hankel singular values. That is,

$$\gamma_k \geq \sigma_k, \text{ for all } k = 1, \dots, n.$$

4.11. Prove Proposition 4.19. Use the following steps:

(a) Use the balanced Lyapunov inequalities to show that there exists an augmented system

$$\hat{G}_a = \left[\begin{array}{c|cc} A & B & B_a \\ \hline C & 0 & 0 \\ C_a & 0 & 0 \end{array} \right]$$

whose gramians are X and Y.

(b) Apply Theorem 4.18 directly to $\hat{G}_a$ to give a reduced system $\hat{G}_{ar}$. Show that $\hat{G}_r$ of Proposition 4.19 is embedded in $\hat{G}_{ar}$ and must therefore satisfy the proposition.

Notes and references

Balanced realizations were first introduced into the control theory literature in [106], motivated by principal component analysis. In this paper weak and strong controllability and observability were introduced, and balanced truncation model reduction was suggested. For a proof of the generalization in Proposition 4.9, see [188].

A proof that balanced truncation yields a stable system was subsequently given in [120]. The initial proof of the error bound in Theorem 4.18 appeared in [48], and an independent but subsequent proof in [63]. The concise proof given here of the key Lemma 4.17 is due to [63]. For computational methods on obtaining balanced realizations see [95] and [141].

The books [119] and [127] can be consulted for in depth treatments of Hankel operators.

Another major approach to model reduction, also involving Hankel singular values is that of so-called optimal Hankel norm approximation. The main results in this area were presented in [63], based on scalar results appearing in [1]. A more compact treatment of this approach is given in [64]. An interesting comment is that this approach, in conjunction with a nonzero D_r term, can be used to improve the error bound in Theorem 4.18 by a factor of two.

The discrete time version of balanced truncation model reduction was initiated in [2] using discrete gramians. The use of gramians in discrete time only yields an analog of Lemma 4.17, but surprisingly Theorem 4.18 does not hold in general: the induction argument breaks down since balancing is not preserved by truncation. In [73] it was shown that if *strict* generalized gramians are used in lieu of gramians, then a discrete version of Proposition 4.19 does hold.

The use of generalized gramians to characterize exactly the H_∞ model reduction problem, is essentially from [86], although our version of Theorem 4.20 emphasizes the LMI methods we will use later for H_∞ control synthesis. A discrete time version, based on LMI methods, was obtained in [12]. In this latter work, notions of balanced realizations and model reduction are extended to multi-dimensional and uncertain systems. We will give an outline of these generalizations in Chapter 11.

5
Stabilizing Controllers

We begin here our study of feedback design, which will occupy our attention in the next three chapters. In these chapters we will consider *systematic* design methods where objectives are first specified, and one can then exactly characterize when the specifications can be met, as well as find suitable controllers. In other words, design is based solely on clearly formulated specifications, rather than on a specific strategy chosen a priori.

The only a priori structure will be a very general feedback arrangement which is described below. Once introduced, we will focus in this chapter on a first necessary specification for any feedback system: that it be stable in some appropriate sense. In particular we will precisely define feedback stability and then proceed to parametrize *all* controllers that stabilize the feedback system.

The general feedback setup we are concerned with is shown in Figure 5.1, which was already introduced in a descriptive manner in Chapter 0. As depicted the so-called closed-loop system has one external input and one output, given by w and z, respectively; each of these may, however, consist of multiple channels. The signal or function w captures the effects of the environment on the feedback system; for instance, noise, disturbances, and commands. The signal z contains all characteristics of the feedback system that are to be controlled. The maps G and K represent linear subsystems where G is a given system, frequently called the *plant*, which is fixed, and K is the controller or control law whose aim is to ensure that the mapping from w to z has the desired characteristics. To accomplish this task the control law utilizes signal y, and chooses an action u which directly affects the behavior of G.

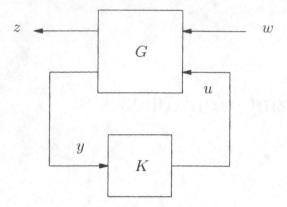

Figure 5.1. General feedback arrangement

Here G and K are state space systems, with G evolving according to

$$\dot{x}(t) = \quad A\,x(t) \;+\; \begin{bmatrix} B_1 & B_2 \end{bmatrix} \begin{bmatrix} w(t) \\ u(t) \end{bmatrix},$$

$$\begin{bmatrix} z(t) \\ y(t) \end{bmatrix} = \begin{bmatrix} C_1 \\ C_2 \end{bmatrix} x(t) + \begin{bmatrix} D_{11} & D_{12} \\ D_{21} & D_{22} \end{bmatrix} \begin{bmatrix} w(t) \\ u(t) \end{bmatrix},$$

and K being described by

$$\dot{x}_K(t) = A_K x_K(t) + B_K y(t),$$
$$u(t) = C_K x_K(t) + D_K y(t).$$

Throughout this chapter we have the standing assumption that the matrix triples (C, A, B) and (C_K, A_K, B_K) are both stabilizable and detectable.

As shown in the figure, G is naturally partitioned with respect to its two inputs and two outputs. We therefore partition the transfer function of G as

$$\hat{G}(s) = \left[\begin{array}{c|cc} A & B_1 & B_2 \\ \hline C_1 & D_{11} & D_{12} \\ C_2 & D_{21} & D_{22} \end{array} \right] = \begin{bmatrix} \hat{G}_{11}(s) & \hat{G}_{12}(s) \\ \hat{G}_{21}(s) & \hat{G}_{22}(s) \end{bmatrix},$$

so that we can later refer to these constituent transfer functions.

At first we must determine under what conditions this interconnection of components makes sense. That is, we need to know when these equations have a solution for an arbitrary input w. The system of Figure 5.1 is *well-posed* if unique solutions exist for $x(t)$, $x_K(t)$, $y(t)$ and $u(t)$, for all initial conditions $x(0)$, $x_K(0)$, and all input functions[1] $w(t)$; and further, such unique solutions must exist if the realizations are perturbed in some neighborhood of (A, B, C, D) and (A_K, B_K, C_K, D_K).

---

[1] Inputs are assumed to be sufficiently regular.

Proposition 5.1. *The connection of G and K in Figure 5.1 is well-posed if and only if $I - D_{22}D_K$ is nonsingular.*

Proof. Writing out the state equations of the overall system, we have

$$\dot{x}(t) = Ax(t) + B_1 w(t) + B_2 u(t) \qquad (5.1)$$
$$\dot{x}_K(t) = A_K x_K(t) + B_K y(t),$$

and

$$\begin{bmatrix} I & -D_K \\ -D_{22} & I \end{bmatrix} \begin{bmatrix} u(t) \\ y(t) \end{bmatrix} = \begin{bmatrix} 0 & C_K \\ C_2 & 0 \end{bmatrix} \begin{bmatrix} x(t) \\ x_K(t) \end{bmatrix} + \begin{bmatrix} 0 \\ D_{21} \end{bmatrix} w(t). \qquad (5.2)$$

Now it is easily seen that the left-hand side matrix is invertible if and only if $I - D_{22}D_K$ is nonsingular. If this holds, clearly one can substitute u, y into (5.1) and find a unique solution to the state equations, and this can always be done for a sufficiently small neighborhood of D_{22} and D_K.

Conversely if $I - D_{22}D_K$ is singular, from (5.2) we can always perturb C_2 and C_K slightly to find a linear combination of $x(t)$, $x_K(t)$, and $w(t)$ which must be zero, which means that $x(0)$, $x_K(0)$, $w(0)$ cannot be chosen arbitrarily. ∎

Notice in particular that if either $D_{22} = 0$ or $D_K = 0$ (strictly proper $\hat{G}_{22}$ or $\hat{K}$), then the interconnection in Figure 5.1 is well-posed.

We are now ready to talk about stability.

5.1   System stability

In this section we discuss the notion of internal stability, and discuss its relation to the boundedness of input–output maps. Internal stability was already introduced in Chapter 2 for an autonomous system; we now specialize the definition to our feedback arrangement.

Definition 5.2. *The system in Figure 5.1 is internally stable if it is well-posed, and for every initial condition $x(0)$ of G, and $x_K(0)$ of K, the limits*

$$x(t), x_K(t) \xrightarrow{t \to \infty} 0 \quad hold,$$

when $w = 0$.

The following is an immediate test for internal stability.

Proposition 5.3. *The system of Figure 5.1 is internally stable if and only if $I - D_{22}D_K$ is invertible and*

$$A_{cl} = \begin{bmatrix} A & 0 \\ 0 & A_K \end{bmatrix} + \begin{bmatrix} B_2 & 0 \\ 0 & B_K \end{bmatrix} \begin{bmatrix} I & -D_K \\ -D_{22} & I \end{bmatrix}^{-1} \begin{bmatrix} 0 & C_K \\ C_2 & 0 \end{bmatrix} \qquad (5.3)$$

is Hurwitz.

Proof. Given the well-posedness condition as in Proposition 5.1, it is easy to solve from (5.1) and (5.2) to show that A_{cl} is the A-matrix of the closed-loop; therefore the result follows from our work in Chapter 2.   ∎

As defined, internal stability refers to the autonomous system dynamics in the absence of an input w; in this regard it coincides with the standard notion of asymptotic stability of dynamical systems. However, it has immediate implications on the input–output properties of the system. In particular, the transfer function from w to z, denoted $\hat{T}(s)$, is proper and will have as poles a subset of the eigenvalues of A_{cl}; for example, when $D_K = 0$ we have

$$\hat{T}(s) = \begin{bmatrix} C_1 & D_{12}C_K \end{bmatrix} (sI - A_{cl})^{-1} \begin{bmatrix} B_1 \\ B_K D_{21} \end{bmatrix} + D_{11}.$$

If A_{cl} is Hurwitz, this function has all its poles in the left half plane: in the language of Chapter 3, it is an element of RH_∞. Equivalently, the closed-loop map $w \mapsto z$ is a bounded, causal operator on $L_2[0, \infty)$; this is termed *input–output* stability.

The question immediately arises as to whether the two notions are interchangeable, that is, whether the boundedness of $w \mapsto z$ implies internal stability. Clearly, the answer is negative: an extreme example would be to have C_1, D_{11}, D_{12} all be zero, which gives $\hat{T}(s) = 0$ but clearly says nothing about A_{cl}. In other words, the internal dynamics need not be reflected in the external map: they could be unobservable or uncontrollable.

We are still interested, however, in seeking an external characterization of internal stability; this will allow us, in a later section, to describe *all* possible stabilizing controllers. Such characterization should only impose non-restrictive requirements on the internal dynamics. From this point of view, the variables w and z are not relevant to the discussion, since the conditions of Proposition 5.3 depend exclusively on the components in u and y of the state space system G.

These considerations lead us to discuss a second interconnection diagram, represented in Figure 5.2. This is precisely the same as the bottom loop in Figure 5.1, except that we have injected *interconnection noise* into the feedback loop. In particular, the controller K has the same description as before, and the system G_{22} is the lower block of G, described by the state space equations

$$\dot{x}_{22}(t) = Ax_{22}(t) + B_2 v_1(t)$$
$$v_2(t) = C_2 x_{22}(t) + D_{22} v_1(t),$$

where (C_2, A, B_2, D_{22}) are the same matrices as in the state space description of G.

The only addition to the diagram are the external inputs d_1 and d_2, introduced at the interconnection between G_{22} and K.

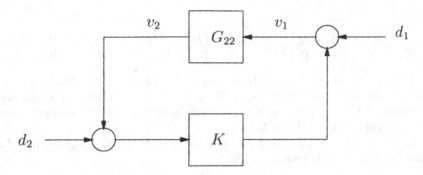

Figure 5.2. Input–output stability

As with our more general system, we say that this new system is well-posed if there exist unique solutions for $x_{22}(t)$, $x_K(t)$, $v_1(t)$ and $v_2(t)$ for all inputs $d_1(t)$ and $d_2(t)$ and initial conditions $x_{22}(0)$, $x_K(0)$. We say it is internally stable if it is well-posed and for $d_i = 0$ we have

$$x_{22}(t),\; x_K(t) \overset{t \to \infty}{\longrightarrow} 0 \quad \text{holds}$$

for every initial condition.

It is an easy exercise to see that the system is well-posed if and only if $I - D_{22}D_K$ is nonsingular; this is the same well-posedness condition we have for Figure 5.1. Also noticing that *all* the states in the description of G are included in the equations for G_{22}, it follows immediately that internal stability of one is equivalent to internal stability of the other.

Lemma 5.4. *Given a controller K, Figure 5.1 is internally stable if and only if Figure 5.2 is internally stable.*

The next result shows that with this new set of inputs, internal stability can be characterized by the boundedness of an input–output map. When this map is bounded, we say the system is *input–output stable*.

Lemma 5.5. *Suppose that (A, B_2, C_2) is stabilizable and detectable. Then Figure 5.2 is internally stable if and only if the transfer function of $\begin{bmatrix} d_1 \\ d_2 \end{bmatrix} \mapsto$*

$\begin{bmatrix} v_1 \\ v_2 \end{bmatrix}$ *is in RH_∞.*

Proof. We begin by finding an expression for the transfer function. For convenience denote

$$\bar{D} = \begin{bmatrix} I & -D_K \\ -D_{22} & I \end{bmatrix};$$

then routine calculations lead to the following relationship:

$$\begin{bmatrix} \hat{v}_1(s) \\ \hat{v}_2(s) \end{bmatrix} = \hat{W}(s) \begin{bmatrix} \hat{d}_1(s) \\ \hat{d}_2(s) \end{bmatrix},$$

where

$$\hat{W}(s) := \bar{D}^{-1} \begin{bmatrix} 0 & C_K \\ C_2 & 0 \end{bmatrix} (sI - A_{cl})^{-1} \begin{bmatrix} B_2 & 0 \\ 0 & B_K \end{bmatrix} \bar{D}^{-1} + \bar{D}^{-1} + \begin{bmatrix} 0 & 0 \\ 0 & -I \end{bmatrix},$$

and A_{cl} is the closed-loop matrix from (5.3). Therefore the "only if" direction follows immediately, since the poles of this transfer function are a subset of the eigenvalues of A_{cl}, which is by assumption Hurwitz; see Proposition 5.3 and Lemma 5.4.

To prove "if": assume that the transfer function has no poles in $\bar{\mathbb{C}}^+$; therefore the same is true of

$$\underbrace{\begin{bmatrix} 0 & C_K \\ C_2 & 0 \end{bmatrix}}_{\bar{C}} (sI - A_{cl})^{-1} \underbrace{\begin{bmatrix} B_2 & 0 \\ 0 & B_K \end{bmatrix}}_{\bar{B}}.$$

We need to show that A_{cl} is Hurwitz; it is therefore sufficient (see Exercise 2.19) to show that $(\bar{C}, A_{cl}, \bar{B})$ is a stabilizable and detectable realization. Let

$$\bar{F} = \begin{bmatrix} F & 0 \\ 0 & F_K \end{bmatrix} - \bar{D}^{-1} \begin{bmatrix} 0 & C_K \\ C_2 & 0 \end{bmatrix},$$

where F and F_K are chosen so that $A + B_2 F$ and $A_K + B_K F_K$ are both Hurwitz. It is routine to show that

$$A_{cl} + \bar{B}\bar{F} = \begin{bmatrix} A + B_2 F & 0 \\ 0 & A_K + B_K F_K \end{bmatrix},$$

and thus $(A_{cl}, \bar{B})$ is stabilizable.

A formally similar argument shows that $(\bar{C}, A_{cl})$ is detectable. ∎

The previous characterization will be used later on to develop a parametrization of all stabilizing controllers. As for the stabilizability and detectability assumption on (A, B_2, C_2), we will soon see that it is non-restrictive — it is necessary for the existence of any stabilizing controller.

5.2 Stabilization

In the previous section we have discussed the analysis of stability of a given feedback configuration; we now turn to the question of design of a stabilizing controller. The following result explains when this can be achieved.

Proposition 5.6. *A necessary and sufficient condition for the existence of an internally stabilizing K for Figure 5.1 is that (A, B_2, C_2) is stabilizable* *and detectable. In that case, one such controller is given by*

$$\hat{K}(s) = \left[\begin{array}{c|c} A + B_2F + LC_2 + LD_{22}F & -L \\ \hline F & 0 \end{array} \right],$$

where F and L are matrices such that $A + B_2F$ and $A + LC_2$ are Hurwitz.

Proof. If the stabilizability or detectability of (C_2, A, B_2) is violated, we can choose an initial condition which excites the unstable hidden mode. It is not difficult to show that the state will not converge to zero, regardless of the controller. Details are left as an exercise. Consequently no internally stabilizing K exists, which proves necessity.

For the sufficiency side, it is enough to verify that the given controller is indeed internally stabilizing. Notice that this is precisely the observer-based controller encountered in Chapter 2. In particular, well-posedness is ensured since $D_K = 0$, and it was shown in Chapter 2 that the closed loop eigenvalues are exactly the eigenvalues of $A + B_2F$ and $A + LC_2$. ∎

The previous result is already a solution to the stabilization problem, since it provides a constructive procedure for finding a stabilizer, in this case with the structure of a state feedback combined with an asymptotic observer. However, there are other aspects to the stabilization question, in particular:

- Can we find a stabilizer with lower order? The above construction provides a controller of the order of the plant.

- What are *all* the stabilizing controllers? This is an important question since one is normally interested in other performance properties beyond internal stability.

We will address the first question using purely LMI techniques, starting with the state feedback problem, and later considering the general case. The second problem will then be addressed in §5.3.

5.2.1 Static state feedback stabilization via LMIs

We begin our search for stabilizing controllers in terms of LMIs by considering the simplest type of control, *static state feedback*. A state feedback controller has direct access to the state of G, namely,

$$y = x.$$

This means that $C_2 = I$, $D_{22} = 0$. The feedback is static if it has no dynamics, and it is given simply by

$$u(t) = Fy(t),$$

where for convenience we have set $D_K = F$. As a beginning for controller parametrization we would like to find all such F that are stabilizing. That is, referring to Proposition 5.3, all matrices F such

$$A_{cl} := A + B_2 F$$

is Hurwitz. We already know that such a F exists exactly when (A, B_2) is stabilizable, and we now seek a constructive way to find such F's in terms of an LMI. The natural path to follow is to impose that A_{cl} is Hurwitz by means of a Lyapunov inequality of the type discussed in Chapter 4. Here we will impose that the inequality

$$A_{cl}X + XA_{cl}^* < 0$$

has a solution $X > 0$ (dual to the test in Corollary 4.3). Substituting for A_{cl} we get

$$(A + B_2 F)X + X(A + B_2 F)^* < 0,$$

that can be expanded out as

$$AX + B_2(FX) + XA^* + (XF^*)B_2^* < 0.$$

The above is not an LMI in F and X since their product appears. However, it is natural to introduce the new variable $Z = FX$, which makes the above an LMI in X and Z; once this LMI is solved, the desired F can be easily reconstructed.

Thus we find that all stabilizing feedback laws can be found from a convex feasibility problem. We summarize the conclusions in the following statement.

Theorem 5.7. *A static state feedback* $D_K = F$ *stabilizes Figure 5.1 if and only if there exist matrices* $X > 0$ *and* Z *such that*

$$F = ZX^{-1}$$

and

$$\begin{bmatrix} A & B_2 \end{bmatrix} \begin{bmatrix} X \\ Z \end{bmatrix} + \begin{bmatrix} X & Z^* \end{bmatrix} \begin{bmatrix} A^* \\ B_2^* \end{bmatrix} < 0 \quad \text{is satisfied.} \tag{5.4}$$

We are now ready to move on to a more general LMI characterization of fixed order controllers.

5.2.2   *An LMI characterization of the stabilization problem*

Given a fixed controller order, a direct approach to the stabilization problem is to write down a generic controller realization (A_K, B_K, C_K, D_K), and impose that the closed-loop matrix A_{cl} in (5.3) is Hurwitz by means of the Lyapunov inequality

$$X_{cl} > 0, \qquad A_{cl}^* X_{cl} + X_{cl} A_{cl} < 0.$$

The question is whether the above condition, with unknowns X_{cl}, A_K, B_K, C_K, and D_K can be computed tractably. The following is a result in this direction, stated for simplicity in the case of a strictly proper plant.

Theorem 5.8. *Consider the system of Figure 5.1, with $D_{22} = 0$. There exists a controller of order n_K, which internally stabilizes the system if and only if there exist $n \times n$ matrices $X > 0$, $Y > 0$, such that*

$$N_o^*(A^*X + XA)N_o \; < \; 0, \tag{5.5}$$

$$N_c^*(AY + YA^*)N_c \; < \; 0, \tag{5.6}$$

$$\begin{bmatrix} X & I \\ I & Y \end{bmatrix} \; \geq \; 0, \quad and \tag{5.7}$$

$$\mathrm{rank} \begin{bmatrix} X & I \\ I & Y \end{bmatrix} \; \leq \; n + n_K \, , \tag{5.8}$$

where N_o and N_c are full column rank matrices such that

$$\mathrm{Im}N_o = \mathrm{Ker}C_2,$$
$$\mathrm{Im}N_c = \mathrm{Ker}B_2^*.$$

We make the following remarks: the LMIs (5.5) and (5.6) are equivalent, respectively, to the detectability and stabilizability of the realization (A, B_2, C_2). In fact the feasibility of (5.6) is equivalent to that of (5.4).

Since (5.5) and (5.6) are homogeneous in the unknowns, the solutions can always be scaled up to satisfy (5.7). This means that if there is no constraint in the order n_k, the existence of a stabilizer is equivalent to stabilizability and detectability of (A, B_2, C_2), consistently with the previous analysis.

Furthermore, clearly the rank constraint (5.8) is not felt when $n_K \geq n$, which implies that we never need an order greater than n to stabilize, once again consistently with Proposition 5.6. In this case, the set of LMIs (5.5–5.7) is a convex problem that can be solved for X, Y. For $n_K < n$, stabilizability and detectability may not suffice to obtain a stabilizer; furthermore, the rank constraint is not convex and makes the computation of X, Y more difficult; still, the above proposition provides useful insight into the role of controller order.

Once X, Y are obtained, algebraic operations can be used to compute X_{cl}, and from it an additional LMI leads to the controller. We will not extend ourselves on this construction here, or on the proof of Theorem 5.8, since these topics will all be covered in Chapter 7, where we extend the LMI method for the solution of the H_∞ control problem. After going through Chapter 7, the reader is invited to prove Theorem 5.8 as an exercise.

5.3   Parametrization of stabilizing controllers

In this section we pursue the objective of fully characterizing the family of all state space controllers which stabilize a given plant.

This is a very important tool since in many problems one wishes to optimize a performance objective with stability as a constraint. A parametrization of the stabilizing controllers reduces these problems to unconstrained optimization over the parameter.

The parametrization will be strongly based on transfer function methods, for which we will exploit the input–output characterization of internal stability, summarized in the following corollary of Lemmas 5.4 and 5.5 and Proposition 5.6.

Corollary 5.9. *A necessary and sufficient condition for the existence of an internally stabilizing controller for Figure 5.1, is that $(A, B_2.C_2)$ is stabilizable and detectable. In that case, the internally stabilizing controllers are exactly those for which the configuration of Figure 5.2 makes the transfer function of*

$$\begin{bmatrix} d_1 \\ d_2 \end{bmatrix} \mapsto \begin{bmatrix} v_1 \\ v_2 \end{bmatrix}$$

an element of RH_∞.

It is important to note that given any two appropriately dimensioned functions $\hat{Q}_1$ and $\hat{Q}_2$ in RH_∞, their product $\hat{Q}_1\hat{Q}_2$ will be in RH_∞, as will be their sum $\hat{Q}_1 + \hat{Q}_2$; namely, RH_∞ has the structure of an algebra. Similarly, the set RP of proper rational functions, introduced in §2.6, also has this structure of being closed under addition and multiplication. The following subsection introduces tools to analyze these algebras which are our key to the controller parametrization.

5.3.1   Coprime factorization

Throughout this section we work mostly with transfer functions. So far we have used the "hat" convention to denote a transfer function, for instance, $\hat{T}(s)$. We will now also use the "wave" notation to denote functions, as in $\tilde{Q}(s)$. We will work exclusively with proper rational functions from the set RP, defined in (2.19), which is the set of all transfer functions. Also, given $\hat{Q} \in RP$, the inverse $\hat{Q}^{-1}$ will refer to the inverse as a rational function, which we note might *not* be proper.

Given two matrix valued functions $\hat{M}$ and $\hat{N}$ in RH_∞, they are *right coprime* if there exist $\tilde{X}, \tilde{Y} \in RH_\infty$ such that

$$\tilde{X}\hat{M} - \tilde{Y}\hat{N} = I \ .$$

Similarly, $\tilde{M}$ and $\tilde{N}$ are *left coprime* if there exist $\hat{X}$ and $\hat{Y}$ in RH_∞, such that

$$\tilde{M}\hat{X} - \tilde{N}\hat{Y} = I .$$

These equations are called Bezout identities.

A *right coprime factorization* of a function $\hat{P} \in RP$ is a factorization

$$\hat{P} = \hat{N}\hat{M}^{-1} ,$$

where

$$\hat{N}, \hat{M} \in RH_\infty, \quad \hat{M}^{-1} \text{ is proper,}$$

and $\hat{N}$ and $\hat{M}$ are right coprime. A *left coprime* factorization for $\hat{P}$ would be

$$\hat{P} = \tilde{M}^{-1}\tilde{N} ,$$

with similar restrictions imposed on $\tilde{N}$ and $\tilde{M}$ as in a right factorization.

Example:

To gain insight into the definitions, consider the case of a scalar $\hat{p}(s)$. In this case a (right or left) coprime factorization is

$$\hat{p}(s) = \frac{\hat{n}(s)}{\hat{m}(s)},$$

where $\hat{n}(s)$, $\hat{m}(s)$ are in RH_∞, and

$$\hat{x}(s)\hat{m}(s) - \hat{y}(s)\hat{n}(s) = 1 \quad \text{for some} \quad \hat{x}(s), \hat{y}(s) \in RH_\infty.$$

It is not difficult to show that coprimeness is equivalent to the property that $\hat{n}(s)$ and $\hat{m}(s)$ have no common zeros on the closed right half-plane $\bar{\mathbb{C}}^+$, or at $s = \infty$.

Take, for instance, $\hat{p}(s) = \frac{s-1}{s-2}$; then

$$\hat{n}(s) = \frac{s-1}{s+a}, \quad \hat{m}(s) = \frac{s-2}{s+a}$$

defines a coprime factorization for any $a > 0$, but

$$\hat{n}(s) = \frac{s-1}{(s+a)^2}, \quad \hat{m}(s) = \frac{s-2}{(s+a)^2}$$

does not. □

Do these factorizations exist in the general matrix case? The following result shows constructively that this is the case for any proper rational matrix.

Proposition 5.10. *Given a function $\hat{P}$ in the set RP, there exist both right and left coprime factorizations*

$$\hat{P} = \hat{N}\hat{M}^{-1} = \tilde{M}^{-1}\tilde{N} ,$$

satisfying

$$\begin{bmatrix} \tilde{X} & -\tilde{Y} \\ -\tilde{N} & \tilde{M} \end{bmatrix} \begin{bmatrix} \hat{M} & \hat{Y} \\ \hat{N} & \hat{X} \end{bmatrix} = I \qquad (5.9)$$

for appropriate functions $\tilde{X}$, $\tilde{Y}$, $\hat{X}$, and $\hat{Y}$ in RH_∞.

Here $\tilde{X}$, $\tilde{Y}$ $\hat{X}$, and $\hat{Y}$ satisfy an additional identity, and (5.9) is therefore termed a *doubly coprime* factorization of $\hat{P}$.

Proof. Our proof is by direct construction. Let

$$\hat{P}(s) = \left[\begin{array}{c|c} A & B \\ \hline C & D \end{array} \right] ,$$

where (C, A, B) is a stabilizable and detectable realization. Introduce the state space system

$$\dot{x}(t) = Ax(t) + Bu(t) \qquad x(0) = 0$$
$$y(t) = Cx(t) + Du(t) .$$

Now choose a matrix F such that $A^F := A + BF$ is Hurwitz. Further define

$$v(t) = u(t) - Fx(t)$$
$$C^F = C + DF .$$

Then we define the system

$$\dot{x}(t) = A^F x(t) + Bv(t) \qquad x(0) = 0$$
$$u(t) = Fx(t) + v(t)$$
$$y(t) = C^F x(t) + Dv(t) ,$$

whose solutions have a one-to-one correspondence with the solutions to the first system. The transfer function from v to u is

$$\hat{M}(s) = \left[\begin{array}{c|c} A^F & B \\ \hline F & I \end{array} \right] ,$$

and the mapping from v to y has transfer function

$$\hat{N}(s) = \left[\begin{array}{c|c} A^F & B \\ \hline C^F & D \end{array} \right] .$$

Thus we see that

$$\hat{P}(s) = \hat{N}(s)\hat{M}^{-1}(s) ,$$

a fact that can also be checked via direct state space manipulations.
   Similarly a factorization

$$\hat{P}(s) = \tilde{M}^{-1}(s)\tilde{N}(s)$$

can be found, with

$$\tilde{M}(s) = \left[\begin{array}{c|c} A^L & L \\ \hline C & I \end{array}\right], \qquad \tilde{N}(s) = \left[\begin{array}{c|c} A^L & B^L \\ \hline C & D \end{array}\right],$$

where L is chosen such that $A^L = A + LC$ is Hurwitz, and $B^L = B + LD$. A quick way to derive these is to transpose[2] the problem to

$$\hat{P}'(s) = \left[\begin{array}{c|c} A' & C' \\ \hline B' & D' \end{array}\right] = \tilde{N}'(s)(\tilde{M}'(s))^{-1},$$

and reproduce the steps of the right coprime case.

The other four transfer functions are given by

$$\hat{X}(s) = \left[\begin{array}{c|c} A^F & -L \\ \hline C^F & I \end{array}\right] \qquad \hat{Y}(s) = \left[\begin{array}{c|c} A^F & -L \\ \hline F & 0 \end{array}\right]$$

and

$$\tilde{X}(s) = \left[\begin{array}{c|c} A^L & -B^L \\ \hline F & I \end{array}\right] \qquad \tilde{Y}(s) = \left[\begin{array}{c|c} A^L & -L \\ \hline F & 0 \end{array}\right].$$

Observe now that the terms in (5.9) can be very compactly represented by

$$\begin{bmatrix} \hat{M} & \hat{Y} \\ \hat{N} & \hat{X} \end{bmatrix} = \left[\begin{array}{c|cc} A^F & B & -L \\ \hline F & I & 0 \\ C^F & D & I \end{array}\right]$$

and

$$\begin{bmatrix} \tilde{X} & -\tilde{Y} \\ -\tilde{N} & \tilde{M} \end{bmatrix} = \left[\begin{array}{c|cc} A^L & -B^L & L \\ \hline F & I & 0 \\ C & -D & I \end{array}\right].$$

A notable feature of these realizations is that, by exploiting the common matrices between the different transfer functions, we do not require a higher order than n.

That these transfer functions multiply out to the identity matrix can be verified by routine state space manipulations; this is left as an exercise. ∎

We now move to our main purpose and apply these new tools.

5.3.2 Controller parametrization

We are now ready to parametrize all stabilizing controllers for Figure 5.1. From Corollary 5.9, we may focus on the input–output properties of the arrangement of Figure 5.2. To motivate what follows, we start by looking again at the case of scalar functions.

---

[2]This is a pure matrix transpose at each s, with no conjugation.

Example:

Let $\hat{G}_{22}$ be a scalar transfer function with coprime factorization $\hat{G}_{22} = \frac{\hat{n}(s)}{\hat{m}(s)}$. Let $\hat{x}(s), \hat{y}(s) \in RH_\infty$ satisfy

$$\hat{x}(s)\hat{m}(s) - \hat{y}(s)\hat{n}(s) = 1. \tag{5.10}$$

Now we claim that $\hat{K}(s) = \frac{\hat{y}(s)}{\hat{x}(s)}$ is input–output stabilizing for Figure 5.2. To see this, it is routine to check that the transfer function of $\begin{bmatrix} d_1 \\ d_2 \end{bmatrix} \mapsto \begin{bmatrix} v_1 \\ v_2 \end{bmatrix}$ is given by

$$\frac{1}{1 - \hat{K}\hat{G}_{22}} \begin{bmatrix} 1 & \hat{K} \\ \hat{G}_{22} & \hat{K}\hat{G}_{22} \end{bmatrix} = \frac{1}{\hat{x}\hat{m} - \hat{y}\hat{n}} \begin{bmatrix} \hat{x}\hat{m} & \hat{y}\hat{m} \\ \hat{x}\hat{n} & \hat{y}\hat{n} \end{bmatrix} = \begin{bmatrix} \hat{x}\hat{m} & \hat{y}\hat{m} \\ \hat{x}\hat{n} & \hat{y}\hat{n} \end{bmatrix}.$$

Notice that the relationship (5.10) played a key role in making the common denominator disappear. The last transfer function is in RH_∞, which proves stability.

In the same vein, noticing that

$$(\hat{x} - \hat{n}\hat{q})\hat{m} - (\hat{y} - \hat{m}\hat{q})\hat{n} = 1$$

for *any* $\hat{q}$ in RH_∞, it follows analogously that

$$\hat{K}(s) = \frac{\hat{y}(s) - \hat{m}(s)\hat{q}(s)}{\hat{x}(s) - \hat{n}(s)\hat{q}(s)}$$

is a stabilizing controller for *any* $\hat{q}$ in RH_∞ that makes the above transfer function proper. Thus we see that a coprime factorization generates a family of stabilizing controllers over the parameter $\hat{q}(s)$.

The striking fact, which is proved below for the general matrix case, is that indeed *all* stabilizing controllers can be expressed in the latter form.

□

We now proceed to extend the idea in the example to the matrix setting, and also build up the theory necessary to show the ensuing parametrization is complete. Let

$$\begin{bmatrix} \tilde{X} & -\tilde{Y} \\ -\tilde{N} & \tilde{M} \end{bmatrix} \begin{bmatrix} \hat{M} & \hat{Y} \\ \hat{N} & \hat{X} \end{bmatrix} = I$$

be a doubly coprime factorization of $\hat{G}_{22}$, and

$$\hat{K} = \hat{U}\hat{V}^{-1} = \tilde{V}^{-1}\tilde{U}$$

be coprime factorizations for $\hat{K}$.

The following gives us a new condition for stability.

Lemma 5.11. *Given the above definitions the following are equivalent:*

(a) The controller K input–output stabilizes G_{22} in Figure 5.2;

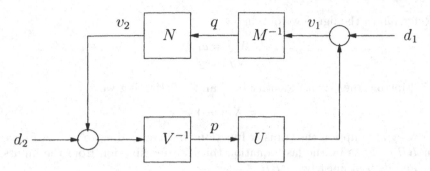

Figure 5.3. Coprime factorizations and stability

(b) $\begin{bmatrix} \hat{M} & \hat{U} \\ \hat{N} & \hat{V} \end{bmatrix}$ is invertible in RH_∞;

(c) $\begin{bmatrix} \tilde{V} & -\tilde{U} \\ -\tilde{N} & \tilde{M} \end{bmatrix}$ is invertible in RH_∞.

Proof. Note that

$$\begin{bmatrix} \hat{M} & \hat{U} \\ \hat{N} & \hat{V} \end{bmatrix} = \begin{bmatrix} -I & 0 \\ 0 & I \end{bmatrix} \begin{bmatrix} \hat{M} & -\hat{U} \\ -\hat{N} & \hat{V} \end{bmatrix} \begin{bmatrix} -I & 0 \\ 0 & I \end{bmatrix},$$

and therefore condition (b) is equivalent to the statement

$$\begin{bmatrix} \hat{M} & -\hat{U} \\ -\hat{N} & \hat{V} \end{bmatrix}^{-1} \in RH_\infty. \tag{5.11}$$

We will work with this equivalent condition.

First we demonstrate that condition (5.11) implies condition (a). This proof centers around Figure 5.3, that is exactly Figure 5.2 redrawn with the factorizations for G_{22} and K. Clearly

$$\begin{bmatrix} \hat{M} & -\hat{U} \\ -\hat{N} & \hat{V} \end{bmatrix} \begin{bmatrix} \hat{q} \\ \hat{p} \end{bmatrix} = \begin{bmatrix} \hat{d}_1 \\ \hat{d}_2 \end{bmatrix}. \tag{5.12}$$

Thus we see

$$\begin{bmatrix} \hat{v}_1 \\ \hat{v}_2 \end{bmatrix} = \begin{bmatrix} \hat{M} \\ \hat{N} \end{bmatrix} \hat{q} = \begin{bmatrix} \hat{M} & 0 \\ \hat{N} & 0 \end{bmatrix} \begin{bmatrix} \hat{M} & -\hat{U} \\ -\hat{N} & \hat{V} \end{bmatrix}^{-1} \begin{bmatrix} \hat{d}_1 \\ \hat{d}_2 \end{bmatrix}.$$

The above transfer function inverse is in RH_∞ by assumption (5.11); therefore, we see that (a) holds.

Next we must show that (a) implies (5.11). To do this we use the Bezout identity.

$$\tilde{X}\hat{M} - \tilde{Y}\hat{N} = I .$$

Referring to the figure we obtain

$$\hat{M}\hat{q} = \hat{v}_1$$
$$\hat{N}\hat{q} = \hat{v}_2 .$$

Multiplying the Bezout identity by $\hat{q}$ and substituting we get

$$\hat{q} = \tilde{X}\hat{v}_1 - \tilde{Y}\hat{v}_2 .$$

Now by assumption the transfer functions from $\hat{d}_1$ and $\hat{d}_2$ to $\hat{v}_1$ and $\hat{v}_2$ are in RH_∞. Thus by the last equation the transfer function from the inputs $\hat{d}_1$ and $\hat{d}_2$ to $\hat{q}$ must be in RH_∞.

Similarly we can show that the transfer functions from the inputs to $\hat{p}$ are in RH_∞, by instead starting with a Bezout identity $\tilde{X}_K\hat{V} + \tilde{Y}_K\hat{U} = I$ for the controller. Recalling the relationship in (5.12) we see that (5.11) must be satisfied.

To show that (a) and (c) are equivalent we simply use the left coprime factorizations for G_{22} and $\hat{K}$, and follow the argument above.

∎

We can now prove the parametrization result of the chapter; this result is sometimes referred to as the *Youla parametrization*.

Theorem 5.12. *A controller K input–output stabilizes G_{22} in Figure 5.2 if and only if there exists $\hat{Q} \in RH_\infty$ such that*

$$\hat{K} = (\hat{Y} - \hat{M}\hat{Q})(\hat{X} - \hat{N}\hat{Q})^{-1} = (\tilde{X} - \hat{Q}\tilde{N})^{-1}(\tilde{Y} - \hat{Q}\tilde{M}) \qquad (5.13)$$

and the latter two inverses exist in the set of proper rational functions RP.

Proof. We begin by showing that the latter equality holds for any $\hat{Q} \in RH_\infty$ such that the inverses exist. Given such a $\hat{Q}$ we have by the doubly coprime factorization formula that

$$\begin{bmatrix} I & \hat{Q} \\ 0 & I \end{bmatrix} \begin{bmatrix} \tilde{X} & -\tilde{Y} \\ -\tilde{N} & \tilde{M} \end{bmatrix} \begin{bmatrix} \hat{M} & \hat{Y} \\ \hat{N} & \hat{X} \end{bmatrix} \begin{bmatrix} I & -\hat{Q} \\ 0 & I \end{bmatrix} = I ,$$

which yields

$$\begin{bmatrix} \tilde{X} - \hat{Q}\tilde{N} & -(\tilde{Y} - \hat{Q}\tilde{M}) \\ -\tilde{N} & \tilde{M} \end{bmatrix} \begin{bmatrix} \hat{M} & \hat{Y} - \hat{M}\hat{Q} \\ \hat{N} & \hat{X} - \hat{N}\hat{Q} \end{bmatrix} = I \qquad (5.14)$$

Taking this product we get finally

$$\begin{bmatrix} ? & (\tilde{X} - \hat{Q}\tilde{N})(\hat{Y} - \hat{M}\hat{Q}) - (\tilde{Y} - \hat{Q}\tilde{M})(\hat{X} - \hat{N}\hat{Q}) \\ ? & ? \end{bmatrix} = I .$$

Here "?" denotes irrelevant entries, and from the top right entry we see that the two quotients of the theorem must be equal if the appropriate inverses exist.

We now turn to showing that this parametrization is indeed stabilizing. So choose a $\hat{Q} \in RH_{\infty}$, where the above inverses exist, and define

$$\hat{U} = \hat{Y} - \hat{M}\hat{Q} \qquad \hat{V} = \hat{X} - \hat{N}\hat{Q}$$
$$\tilde{U} = \tilde{Y} - \hat{Q}\tilde{M} \qquad \tilde{V} = \tilde{X} - \hat{Q}\tilde{N} .$$

From (5.14) we see that

$$\begin{bmatrix} \tilde{V} & -\tilde{U} \\ -\tilde{N} & \tilde{M} \end{bmatrix} \begin{bmatrix} \hat{M} & \hat{U} \\ \hat{N} & \hat{V} \end{bmatrix} = I ,$$

which implies that $\hat{U}, \hat{V}$ and $\tilde{U}, \tilde{V}$ are right and left coprime factorizations of $\hat{K}$ respectively. Also it clearly says

$$\begin{bmatrix} \hat{M} & \hat{U} \\ \hat{N} & \hat{V} \end{bmatrix}^{-1} \in RH_{\infty} .$$

Therefore invoking Lemma 5.11 we see that $\hat{K}$ is stabilizing.

Finally, we show that every stabilizing controller is given by the parametrization in (5.13). Fix a controller $\hat{K}$ and let

$$\hat{K} =: \hat{U}\hat{V}^{-1}$$

be a right coprime factorization. Define

$$\hat{\Theta} := \tilde{M}\hat{V} - \tilde{N}\hat{U} ,$$

and observe

$$\begin{bmatrix} \tilde{X} & -\tilde{Y} \\ -\tilde{N} & \tilde{M} \end{bmatrix} \begin{bmatrix} \hat{M} & \hat{U} \\ \hat{N} & \hat{V} \end{bmatrix} = \begin{bmatrix} I & \tilde{X}\hat{U} - \tilde{Y}\hat{V} \\ 0 & \hat{\Theta} \end{bmatrix}$$

from the doubly coprime factorization of $\hat{G}_{22}$. Now both the matrix functions on the left-hand side have inverses in RH_{∞}: the first by the doubly coprime factorization; the second by Lemma 5.11. Immediately we see that

$$\hat{\Theta}^{-1} \in RH_{\infty} .$$

Set $\hat{Q} := -(\tilde{X}\hat{U} - \tilde{Y}\hat{V})\hat{\Theta}^{-1}$, which is in RH_{∞}, and left-multiply the above matrix equation by

$$\begin{bmatrix} \hat{M} & \hat{Y} \\ \hat{N} & \hat{X} \end{bmatrix}$$

to get

$$\begin{bmatrix} \hat{M} & \hat{U} \\ \hat{N} & \hat{V} \end{bmatrix} = \begin{bmatrix} \hat{M} & \hat{Y} \\ \hat{N} & \hat{X} \end{bmatrix} \begin{bmatrix} I & -\hat{Q}\hat{\Theta} \\ 0 & \hat{\Theta} \end{bmatrix} = \begin{bmatrix} \hat{M} & (\hat{Y} - \hat{M}\hat{Q})\hat{\Theta} \\ \hat{N} & (\hat{X} - \hat{N}\hat{Q})\hat{\Theta} \end{bmatrix} .$$

Just looking at the second block-column we see that

$$(\hat{X} - \hat{N}\hat{Q})^{-1} = \hat{\Theta}\hat{V}^{-1}$$

exists and is proper, and also that

$$\hat{K} = (\hat{Y} - \hat{M}\hat{Q})(\hat{X} - \hat{N}\hat{Q})^{-1}.$$

∎

In the above statement, we have described all stabilizing controllers for Figure 5.2 in terms of *parameter* $\hat{Q}(s)$. The only constraints on this $\hat{Q}$ are that it is RH_∞, and that the inverse

$$(\hat{X} - \hat{N}\hat{Q})^{-1}$$

exists and is proper. This latter condition can be expressed as the requirement that $\hat{X}(\infty) - \hat{N}(\infty)\hat{Q}(\infty)$ should be invertible, and ensures our family of controllers remains proper; see also Exercise 2.18(b).

5.3.3 *Closed-loop maps for the general system*

In this section we summarize and assemble the results we have developed, by returning to stabilization of the general setup of Figure 5.1. First recall our partition of the system G:

$$\hat{G} = \left[\begin{array}{c|cc} A & B_1 & B_2 \\ \hline C_1 & D_{11} & D_{12} \\ C_2 & D_{21} & D_{22} \end{array}\right] = \begin{bmatrix} \hat{G}_{11} & \hat{G}_{12} \\ \hat{G}_{21} & \hat{G}_{22} \end{bmatrix} .$$

Next take a doubly coprime factorization of $\hat{G}_{22}$

$$\begin{bmatrix} \tilde{X} & -\tilde{Y} \\ -\tilde{N} & \tilde{M} \end{bmatrix} \begin{bmatrix} \hat{M} & \hat{Y} \\ \hat{N} & \hat{X} \end{bmatrix} = I ,$$

where $\hat{G}_{22} = \hat{N}\hat{M}^{-1} = \tilde{M}^{-1}\tilde{N}$. With these definitions we have the following general result.

Theorem 5.13. *Suppose* (A, B_2, C_2) *is a stabilizable and detectable realization. Then all controllers that internally stabilize the system of Figure 5.1 are given by*

$$\hat{K} = (\hat{Y} - \hat{M}\hat{Q})(\hat{X} - \hat{N}\hat{Q})^{-1} = (\tilde{X} - \hat{Q}\tilde{N})^{-1}(\tilde{Y} - \hat{Q}\tilde{M}) ,$$

where $\hat{Q}(s) \in RH_\infty$ *and* $\hat{X}(\infty) - \hat{N}(\infty)\hat{Q}(\infty)$ *is invertible.*

Proof. This follows immediately from Corollary 5.9 and Theorem 5.12.

∎

Having parametrized all stabilizing controllers in terms of the RH_∞ parameter $\hat{Q}$, we now turn to how this parametrizes the closed-loop map from w to z. Define the three transfer functions

$$\hat{T}_1 = \hat{G}_{11} + \hat{G}_{12}\hat{Y}\tilde{M}\hat{G}_{21},$$
$$\hat{T}_2 = \hat{G}_{12}\hat{M},$$
$$\hat{T}_3 = \tilde{M}\hat{G}_{21} .$$

With these definitions in hand we have

Theorem 5.14.

(a) $\hat{T}_k \in RH_\infty$, for $1 \leq k \leq 3$.

(b) With K as in Theorem 5.13, the transfer function from w to z in Figure 5.1 is given by

$$\hat{T}_1 - \hat{T}_2\hat{Q}\hat{T}_3 \ .$$

Proof. This follows by routine algebra, and is left as an exercise. ∎

The expression in (b) above shows that the closed-loop transfer function from w to z is an affine function of the control parameter $\hat{Q}$. This is a property that can be exploited in many problems, and in particular is very important for optimization.

In this chapter we have defined stability for the general feedback configuration of Figure 5.1. We found conditions under which stabilization was possible, and gave a complete characterization of all stabilizing controllers. The next topic of the course is optimal controller synthesis, where our performance demands are more than just stability.

Exercises

5.1. Consider the region $\mathcal{R} = \left\{ s \in \mathbb{C} : \begin{bmatrix} s + s^* & \alpha(s - s^*) \\ \alpha(s^* - s) & s + s^* \end{bmatrix} < 0 \right\}$ in the complex plane. Here s^* denotes conjugation, and $\alpha > 0$.

   (a) Sketch the region $\mathcal{R}$.
   (b) For A, X in $\mathbb{C}^{n \times n}$ we consider the $2n \times 2n$ LMI

$$\begin{bmatrix} AX + XA^* & \alpha(AX - XA^*) \\ \alpha(XA^* - AX) & AX + XA^* \end{bmatrix} < 0$$

   that is obtained by formally replacing s by AX, s^* by XA^* in the definition of $\mathcal{R}$. Show that if there exists $X > 0$ satisfying this LMI, then the eigenvalues of A are in the region $\mathcal{R}$. It can be shown that the converse is also true.
   (c) Now we have a system $\dot{x} = Ax + Bu$, and we wish to design a state feedback $u = Fx$ such that the closed loop poles fall in $\mathcal{R}$. Derive an LMI problem to do this.

5.2. Complete the proof of Proposition 5.10 by showing that the given transfer functions satisfy Equation (5.9). To do this you must eliminate unobservable/uncontrollable states from the composition of the realizations. This is a little tedious, but significantly less so than working explicitly with matrices of rational functions.

5.3. In this problem we consider the system in the figure, where P is a stable plant, i.e., $\hat{P}(s) \in RH_\infty$.

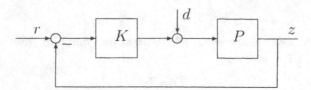

(a) Show that all stabilizing controllers are given by $\hat{K} = \hat{Q}(I - \hat{P}\hat{Q})^{-1}$, with $\hat{Q}(s) \in RH_\infty$ and such that the inverse is proper.

(b) Now consider the case of single input–output P and K. Parametrize all controllers that stabilize the plant and give perfect asymptotic tracking of a step applied in r, for $d = 0$. Is this always possible?

(c) Repeat part (b) for asymptotically rejecting a step applied in d.

5.4. Prove Theorem 5.14.

5.5. In the standard feedback picture of the figure we assume that H, Δ are both in RH_∞.

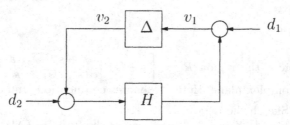

(a) Show that the system is input–output stable (i.e., the map from d to v is in RH_∞) if and only if $I - \Delta H$ is invertible in RH_∞.

(b) If $\|\Delta\|_\infty \|H\|_\infty < 1$, show the system is input–output stable.

5.6. Let $\hat{P}(s)$ be an element of RP. A right coprime factorization $\hat{P} = \hat{N}\hat{M}^{-1}$ is called *normalized* if $\begin{bmatrix} \hat{N}(s) \\ \hat{M}(s) \end{bmatrix}$ is inner (see Chapter 4). It is shown in the exercises of the next chapter that every $\hat{P}$ admits a normalized right coprime factorization (n.r.c.f.).

(a) Given a n.r.c.f. $\hat{P} = \hat{N}\hat{M}^{-1}$, we now consider the perturbed plant

$$\hat{P}_\Delta = (\hat{N} + \hat{\Delta}_N)(\hat{M} + \hat{\Delta}_M)^{-1},$$

where $\hat{\Delta}_N$, $\hat{\Delta}_M$ are in RH_∞ and $\left\| \begin{matrix} \hat{\Delta}_N \\ \hat{\Delta}_M \end{matrix} \right\|_\infty \leq \epsilon$.

Show that $\hat{P}_\Delta$ corresponds to the system inside the dashed box in the following figure.

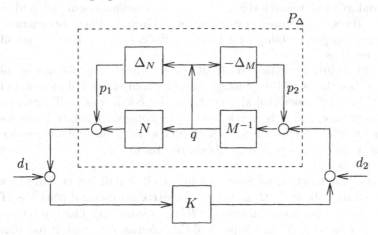

Further show that given a controller K, the above configuration can be redrawn into the equivalent figure below, where

$$\hat{H} = \hat{M}^{-1}(I - \hat{K}\hat{P})^{-1} \begin{bmatrix} \hat{K} & I \end{bmatrix}, \qquad p = \begin{bmatrix} p_1 \\ p_2 \end{bmatrix}, \qquad d = \begin{bmatrix} d_1 \\ d_2 \end{bmatrix}.$$

(b) Using Problem 5, show that if K stabilizes P and

$$\epsilon < \left\| \begin{bmatrix} \hat{P} \\ I \end{bmatrix} (I - \hat{K}\hat{P})^{-1} \begin{bmatrix} \hat{K} & I \end{bmatrix} \right\|_\infty^{-1}$$

then the controller will stabilize the perturbed plant.

(c) Based on these results, propose a method of *model reduction for unstable plants*. The method should be well behaved from the point of view of the stabilization question (explain why). Describe the algorithm in terms of concrete steps which can be calculated in state space.

Notes and references

The standard closed-loop formulation used here originates in [42], and owes much to the detailed presentation in [56]. These sources concentrate on pure input–output stability of a form equivalent to the internal stability considered here.

The LMI solution to the state-feedback stabilization problem goes back to [16], although the LMI terminology and its computational tools were not yet in place. Reference [16] also contains the conditions in Theorem 5.8, which were independently developed in the context of multi-dimensional systems by [96]. The present form with a constraint on controller order, is a special case of the optimal synthesis results in [58, 112], to be covered later in the book.

The transfer function parametrization of all stabilizing controllers was introduced in [180] and [92] in terms of matrix polynomial fractions. The parametrization given here in terms of RH_∞ is from [33]. This parametrization is not crucial in future chapters of this course. However, it has played an important part in the development of robust control theory, and continues to be the launching point for many synthesis approaches; see [20] for extensive uses of this tool, and [28] for its application to L_1 optimal control.

Coprime factorizations over RH_∞ were introduced in [164], to obtain a factorization theory suited to the study of stability. A more extensive presentation of this approach is given in [166]. To deal with the issue of computation, state space realizations for coprime factorizations were developed in [91], and the state space doubly coprime factorization formulae used here are from [109]. Coprime factorizations of this sort have numerous applications, and are used in many branches of robust control theory; see, for example, [60] and [65].

6

H_2 Optimal Control

In this chapter we begin our study of optimal synthesis and in particular
will derive controllers that optimize the H_2 performance criterion. We will
start by defining the synthesis problem to be solved, and will then provide
a number of motivating interpretations. Following this, we will develop
some new matrix tools for the task at hand, before proceeding to solve this
optimal control problem.

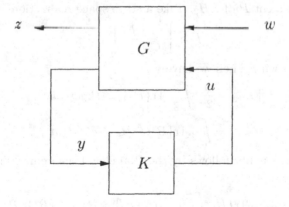

Figure 6.1. Synthesis arrangement

The performance criterion of the chapter is defined from the space RH_2
of matrix valued transfer functions. This space consists of rational, strictly
proper transfer functions, which have all their poles in the half-plane $\mathbb{C}^-$.

Thus it is the natural extension to matrix functions of the space RH_2 of vector valued functions introduced in Chapter 3.

The inner product on this space is defined as

$$\langle \hat{F}, \hat{P} \rangle_2 = \frac{1}{2\pi} \int_{-\infty}^{\infty} \text{Tr}\{\hat{F}^*(j\omega)\hat{P}(j\omega)\} d\omega$$

for two elements $\hat{F}$ and $\hat{P}$ in RH_2.

The synthesis problem we wish to address is concerned with the familiar feedback arrangement shown in Figure 6.1, where both G and K are state space systems. It is convenient to introduce the following notation for the transfer function from w to z in the diagram.

$$\underline{S}(\hat{G}, \hat{K}) := \hat{G}_{11} + \hat{G}_{12}\hat{K}(I - \hat{G}_{22}\hat{K})^{-1}\hat{G}_{21} \; = \; \text{transfer function } w \mapsto z .$$

This is sometimes called the *star product* between transfer functions $\hat{G}$ and $\hat{K}$, or equivalently their *linear fractional transformation*. $\underline{S}(\hat{G}, \hat{K})$ is well-defined and proper provided the interconnection is well-posed, which will be a basic requirement.

The optimal H_2 synthesis problem is

Given: nominal state space system G;

Find: a state space controller K, which internally stabilizes G and

$$\text{minimizes } \|\underline{S}(\hat{G}, \hat{K})\|_2 .$$

The major goal of the chapter is to construct, or *synthesize*, the solution K to this optimization problem. A motivation for this optimization will be given in the next section.

Let us first consider the pure *analysis* question of computing the H_2 norm of an element $\hat{P}$ of RH_2. Take a state space realization

$$\hat{P} = \left[\begin{array}{c|c} A & B \\ \hline C & 0 \end{array} \right] ,$$

where A is Hurwitz. Then we have

$$\|\hat{P}\|_2^2 = \frac{1}{2\pi} \int_{-\infty}^{\infty} \text{Tr}\{\hat{P}^*(j\omega)\hat{P}(j\omega)\} d\omega$$

$$= \int_0^{\infty} \text{Tr}\{B^* e^{A^* t} C^* C e^{At} B\} dt ,$$

where the last equality follows by the Plancherel theorem. Using linearity of the trace gives us

$$\|\hat{P}\|_2^2 = \text{Tr}\{B^* \int_0^{\infty} e^{A^* t} C^* C e^{At} dt B\} = \text{Tr} B^* Y_o B$$

with Y_o denoting the observability gramian. A similar argument shows that $\|\hat{P}\|_2^2 = \text{Tr} C X_c C^*$ where X_c is the controllability gramian of this realization. At the end of the chapter we will see that LMIs can also be used for this calculation.

6.1 Motivation for H_2 control

In this section we briefly discuss the motivation, from a control system perspective, for the design of a feedback control from the point of view of optimal H_2 performance.

First, a general comment on the desire to *minimize* a norm of the map from w to z; this comment also applies to H_∞ control, to be considered in the next chapter. Clearly, for this kind of objective to be meaningful the arrangement of Figure 6.1 must be such that z represents variables that must be "kept small." For example, in a tracking problem the configuration should be set up so that z contains the tracking *error*, rather than the tracking output. Also z usually contains some penalty on the control effort u, to make the problem meaningful. Although we will not discuss these modeling issues here, this philosophy underlies the objective of "making maps small" which we are adopting.

Generally speaking, the H_2 norm is an appropriate measure of performance whenever there is information about the spectral content of the driving variable w. The most common example of this is when w is stationary noise, which we can model for instance as a stationary random process. Frequently when given this interpretation H_2 optimal control is referred to as linear quadratic gaussian (LQG) control. This motivation is now briefly outlined, but no attempt will be made to present a self-contained treatment of stochastic concepts. This would take us far afield and is never required again in this course. Readers unfamiliar with these concepts can consult the references at the end of the chapter.

We consider a stationary stochastic process $w(t)$, and define its autocorrelation matrix by

$$R_w(\tau) = \mathcal{E}w(t+\tau)w^*(t),$$

where here $\mathcal{E}$ denotes expected value; the spectral density $\hat{S}_w(j\omega)$ is the Fourier transform of $R_w(\tau)$. While the specific signal trajectory $w(t)$ is a priori unknown, it is common that one has a good model of the spectral density. The signal variance or power is related to it by

$$\mathcal{E}|w(t)|^2 = \mathrm{Tr}R_w(0) = \frac{1}{2\pi}\int_{-\infty}^{\infty} \mathrm{Tr}\hat{S}_w(j\omega)d\omega.$$

If z and w are related by $z = Pw$ for a stable, linear time invariant system P, it can be shown that

$$\hat{S}_z(j\omega) = \hat{P}(j\omega)\hat{S}_w(j\omega)\hat{P}(j\omega)^*$$

and therefore

$$\mathcal{E}|z(t)|^2 = \frac{1}{2\pi} \int_{-\infty}^{\infty} \mathrm{Tr} \hat{P}(j\omega) \hat{S}_w(j\omega) \hat{P}(j\omega)^* d\omega$$

$$= \frac{1}{2\pi} \int_{-\infty}^{\infty} \mathrm{Tr} \hat{P}(j\omega)^* \hat{P}(j\omega) \hat{S}_w(j\omega) d\omega$$

The right-hand side of the previous equation looks exactly like $\|\hat{P}\|_2^2$, except for the weighting matrix $\hat{S}_w(j\omega)$. In fact the identity is exact for the case of *white noise*[1] $\hat{S}_w(j\omega) = I$ and therefore the H_2 norm can be interpreted as the output variance for white noise applied at the input. More generally, by writing $\hat{S}_w(j\omega) = \hat{W}(j\omega)\hat{W}(j\omega)^*$ and incorporating the "weight" $\hat{W}(j\omega)$ into the setup of Figure 6.1, the H_2 norm can be used to study the response to processes of any known spectral density.

A second example where we have information about the input spectrum is when the input signal $w(t)$ is known a priori. This arises, for example, in the case of tracking of fixed reference signals. Let us consider the special case where $w(t)$ is *scalar* valued, and in particular

$$w(t) = \delta(t) \ ,$$

the Dirac impulse function. Then directly from the definition we have that

$$\|z\|_2^2 = \int_0^{\infty} z^*(t) z(t) dt$$

$$= \frac{1}{2\pi} \int_{-\infty}^{\infty} \hat{z}^*(j\omega) \hat{z}(j\omega) d\omega$$

$$= \frac{1}{2\pi} \int_{-\infty}^{\infty} \hat{P}^*(j\omega) \hat{P}(j\omega) d\omega$$

$$= \|\hat{P}\|_2^2 \ .$$

Thus we see that when the system input is scalar valued, the H_2 norm of $\hat{P}$ gives us a direct measure of the system output energy. If instead of the Dirac delta function we wished to study the response to a different (transient) input, this signal can be generated as the impulse response of a known filter $\hat{W}(s)$, which once again can be incorporated into our H_2 optimization setup.

To extend this interpretation to vector valued inputs: suppose that the input to P has m spatial channels; that is, $w(t)$ takes values in $\mathbb{C}^m$. Then define

$$w_i(t) = \delta(t) \cdot e_i,$$

---

[1]Strictly speaking, this idealization escapes the theory of stationary random processes; a rigorous treatment is not in the scope of our course.

where e_i is the ith standard basis vector in $\mathbb{C}^m$. If $z_i = Pw_i$, then it follows that

$$\sum_{k=1}^{m} \|z_i\|_2^2 = \|\hat{P}\|_2^2$$

Details are left as an exercise. Thus $\|\hat{P}\|_2^2$ gives us the sum of the output energies, which can also be thought of as a measure of the average output energy over the impulsive inputs.

Finally, we present a third motivation for the H_2 norm which is of a different nature; here the input is an arbitrary L_2 signal and has otherwise no known characteristics, but we wish to measure the output signal in the L_∞ norm (over time). That is, our criterion is the induced norm from L_2 to L_∞,

$$\|P\|_{L_2 \to L_\infty} = \sup_{\|w\|_2 = 1} \|z\|_\infty$$

This interpretation applies to the case of a scalar *output* $z(t)$. We can write

$$|z(t)| = \left| \frac{1}{2\pi} \int_{-\infty}^{\infty} \hat{z}(j\omega)e^{j\omega t} d\omega \right|$$

$$= \left| \frac{1}{2\pi} \int_{-\infty}^{\infty} \hat{P}(j\omega)\hat{w}(j\omega)e^{j\omega t} d\omega \right| \leq \|\hat{P}\|_2 \|\hat{w}\|_2,$$

where we apply the Cauchy-Schwarz inequality. Hence $\|P\|_{L_2 \to L_\infty} \leq \|\hat{P}\|_2$, and equality is not difficult to show.

Here we have discussed three different views of the H_2 norm of a system transfer function. There is also a fourth which we leave until Chapter 10. Having motivated this norm we proceed to develop an important tool for tackling the optimal synthesis problem.

6.2 Riccati equation and Hamiltonian matrix

In this section we develop some matrix theory which will be required during our development of the H_2 optimal synthesis procedure.

A Hamiltonian matrix is a real matrix of the form

$$H = \begin{bmatrix} A & R \\ -Q & -A^* \end{bmatrix}$$

where Q and R are $n \times n$ symmetric matrices, and A is also in $\mathbb{R}^{n \times n}$. The Riccati equation associated with H is

$$A^*X + XA + XRX + Q = 0 .$$

The connection between the two is seen in the following similarity transformation of H which uses a solution to the Riccati equation

$$\begin{bmatrix} I & 0 \\ -X & I \end{bmatrix} \begin{bmatrix} A & R \\ -Q & -A^* \end{bmatrix} \begin{bmatrix} I & 0 \\ X & I \end{bmatrix} = \begin{bmatrix} A + RX & R \\ 0 & -A^* - XR \end{bmatrix}. \qquad (6.1)$$

Since the transformed representation of H is a zero in its lower left corner, we see that X provides us with the natural coordinate system for an invariant subspace of H. Specifically Im $\begin{bmatrix} I \\ X \end{bmatrix}$ is an invariant subspace of H. This observation, and the fact that $A + RX$ is a representation for H on that subspace is the key to this section. Hamiltonian matrices have the following elementary property.

Proposition 6.1. *The matrices H and $-H$ are similar (i.e., there exists T such that $H = -T^{-1}HT$).*

Proof. Let

$$J = \begin{bmatrix} 0 & -I \\ I & 0 \end{bmatrix},$$

and observe that $J^{-1}HJ = -H^*$. So H is similar to $(-H)^*$, the *transpose* of the real matrix $-H$. Now any matrix is similar to its transpose; see the Chapter 1 exercises. Therefore, H is similar to $-H$.   ∎

This result implies that if λ is an eigenvalue of H, $-\lambda$ is an eigenvalue of the same multiplicity; in particular, their corresponding invariant subspaces are of the same dimension. Note that since H is real its eigenvalues are also mirrored across the real axis in $\mathbb{C}$; so eigenvalues will be mirrored over both axes in the complex plane.

If H has no purely imaginary eigenvalues, it must have an equal number in the left and right half-planes of $\mathbb{C}$. Further, the similarity of H and $-H$ implies (see Chapter 1) that the invariant subspaces corresponding to stable and unstable eigenvalues have both dimension n in this case. This leads to the following definition:

Definition 6.2. *Suppose that H is a Hamiltonian matrix. Then H is said to be in the domain of the Riccati operator if there exist square, $n \times n$ matrices H_- and X such that*

$$\begin{bmatrix} I \\ X \end{bmatrix} H_- = H \begin{bmatrix} I \\ X \end{bmatrix}, \qquad (6.2)$$

where H_- is Hurwitz. In this case we define the function $\mathrm{Ric}(H) = X$.

The identity (6.2) implies that

$$\mathrm{Im} \begin{bmatrix} I \\ X \end{bmatrix}$$

is an invariant subspace of H, of dimension n. Since H_- is Hurwitz this subspace corresponds to all eigenvalues of H in $\mathbb{C}^-$. From the previous discussion this implies that H can have no purely imaginary eigenvalues; however, (6.2) is slightly stronger in that it sets a particular structure on the subspace (the first n coordinates are independent). Also it should be clear that X is necessarily unique, and thus the notation $\mathrm{Ric}(H)$ is unambiguous.

For Hamiltonian matrices of the above type we prove the following crucial properties.

Theorem 6.3. *Suppose H is in the domain of the Riccati operator and $X = \mathrm{Ric}(H)$. Then*

(a) X is a symmetric matrix;

(b) X satisfies the Riccati equation;

(c) $A + RX$ is Hurwitz.

Proof. We prove each of the properties in turn, starting with (a): again let

$$J = \begin{bmatrix} 0 & -I \\ I & 0 \end{bmatrix},$$

and left-multiply property (6.2) by J and then $\begin{bmatrix} I & X^* \end{bmatrix}$ to get

$$\begin{bmatrix} I \\ X \end{bmatrix}^* J \begin{bmatrix} I \\ X \end{bmatrix} H_- = \begin{bmatrix} I \\ X \end{bmatrix}^* JH \begin{bmatrix} I \\ X \end{bmatrix}.$$

Now JH is symmetric and therefore the left-hand side above must be symmetric, meaning

$$\begin{bmatrix} I \\ X \end{bmatrix}^* J \begin{bmatrix} I \\ X \end{bmatrix} H_- = -H_-^* \begin{bmatrix} I \\ X \end{bmatrix}^* J \begin{bmatrix} I \\ X \end{bmatrix}.$$

From this we have that

$$(X^* - X)H_- + H_-^*(X^* - X) = 0.$$

Recall that H_- is Hurwitz, and therefore this latter Lyapunov equation has a unique solution; that is, $X^* - X = 0$.

We now show property (b): left-multiply (6.2) by $\begin{bmatrix} -X & I \end{bmatrix}$ to arrive at

$$\begin{bmatrix} -X & I \end{bmatrix} H \begin{bmatrix} I \\ X \end{bmatrix} = \begin{bmatrix} -X & I \end{bmatrix} \begin{bmatrix} I \\ X \end{bmatrix} H_- = 0.$$

Now expand the left-hand side, using the definition of H, and get the Riccati equation.

Finally we demonstrate that property (c) holds. Again we use the relationship (6.2); this time left-multiply by $\begin{bmatrix} I & 0 \end{bmatrix}$ and see that

$$\begin{bmatrix} I & 0 \end{bmatrix} H \begin{bmatrix} I \\ X \end{bmatrix} = \begin{bmatrix} I & 0 \end{bmatrix} \begin{bmatrix} I \\ X \end{bmatrix} H_- = H_- .$$

The left-hand side is equal to $A + RX$.

■

Recalling the identity (6.1) for any solution to the Riccati equation, and the fact that any X satisfying (6.2) must be unique, we immediately have the following:

Proposition 6.4. *There is at most one solution to the Riccati equation such that $A + RX$ is Hurwitz.*

To complete our work on Hamiltonian matrices and the Riccati equation we consider a result which ties it directly to systems theory.

Theorem 6.5. *Suppose that H is a Hamiltonian matrix and*

(a) H has no purely imaginary eigenvalues;

(b) R is either positive or negative semidefinite;

(c) (A, R) is a stabilizable matrix pair.

Then H is in the domain of the Riccati operator.

Proof. Since H has no imaginary axis eigenvalue we see that by Proposition 6.1 it has n eigenvalues in each of the open half-planes, where $A \in \mathbb{R}^{n \times n}$. Thus there exist matrices X_1, X_2 and a Hurwitz matrix H_- in $\mathbb{R}^{n \times n}$ such that

$$H \begin{bmatrix} X_1 \\ X_2 \end{bmatrix} = \begin{bmatrix} X_1 \\ X_2 \end{bmatrix} H_-, \tag{6.3}$$

where the rank of $\begin{bmatrix} X_1 \\ X_2 \end{bmatrix}$ is n. It is sufficient to show that X_1 is nonsingular; then multiplication by X_1^{-1} turns (6.3) to the form (6.2) with $X = X_2 X_1^{-1}$. Equivalently, we must show that $\operatorname{Ker} X_1 = 0$. We accomplish this in two steps.

Step A: Show that $X_1^* X_2$ is symmetric.

Left-multiply the last equation by

$$\begin{bmatrix} X_1 \\ X_2 \end{bmatrix}^* J \quad \text{where} \quad J = \begin{bmatrix} 0 & -I \\ I & 0 \end{bmatrix}$$

to arrive at

$$\begin{bmatrix} X_1 \\ X_2 \end{bmatrix}^* J H \begin{bmatrix} X_1 \\ X_2 \end{bmatrix} = \begin{bmatrix} X_1 \\ X_2 \end{bmatrix}^* J \begin{bmatrix} X_1 \\ X_2 \end{bmatrix} H_- .$$

As noted earlier JH is symmetric, and so from the right-hand side above we get

$$(-X_1^* X_2 + X_2^* X_1)H_- = -H_-^*(-X_1^* X_2 + X_2^* X_1) .$$

Moving everything to one side of the equality we get a Lyapunov equation; since H_-^* is Hurwitz it has the unique solution of zero. That is

$$-X_1^* X_2 + X_2^* X_1 = 0 .$$

Step B: Show that $\operatorname{Ker} X_1 = 0$.

First we demonstrate that $\operatorname{Ker} X_1$ is an invariant subspace of H_-. Left multiply (6.3) by $\begin{bmatrix} I & 0 \end{bmatrix}$ to get this is

$$AX_1 + RX_2 = X_1 H_- . \tag{6.4}$$

Now for any $x \in \operatorname{Ker} X_1$ we have

$$x^* X_2^*(AX_1 + RX_2)x = x^* X_2^* X_1 H_- x = x^* X_1^* X_2 H_- x = 0 ,$$

where we have used the result of Step A. We have apparently

$$x^* X_2^* RX_2 x = 0 ,$$

and since R is semidefinite that $RX_2 x = 0$. Finally multiply (6.4) by x to get

$$RX_2 x = X_1 H_- x = 0 ,$$

which says that $\operatorname{Ker} X_1$ is invariant to H_-.

We now use this latter fact to show that $\operatorname{Ker} X_1 = 0$. We know that if this kernel is nonzero, then there exists a nonzero element $x \in \operatorname{Ker} X_1$ such that

$$H_- x = \lambda x ,$$

where the real part of λ is negative. Thus it is enough to show that any x satisfying this equation must be zero, which we now do.

Left multiply (6.3) by $[0 \ I]$ to get

$$-QX_1 - A^* X_2 = X_2 H_- ,$$

which therefore implies

$$(A^* + \lambda I)X_2 x = 0 .$$

Also from above $RX_2 x = 0$ and so

$$x^* X_2^*[A + \lambda^* I \quad R] = 0 .$$

Now $x^* X_2^* = 0$ since (A, R) is stabilizable, and so

$$\begin{bmatrix} X_1 \\ X_2 \end{bmatrix} x = 0 .$$

Since the left-hand matrix has full rank we see that x is necessarily zero. ∎

To end we have the following exercise.

Corollary 6.6. *Suppose (A, B, C) is a stabilizable and detectable matrix triple, and*

$$H = \begin{bmatrix} A & -BB^* \\ -C^*C & -A^* \end{bmatrix}.$$

Then H is in the domain of the Riccati operator.

We have developed all Riccati results that are required to pursue the optimal H_2 synthesis.

6.3   Synthesis

Our solution to the optimal synthesis problem will be strongly based on the concept of an *inner* function, introduced in §4.6.3. In fact the connection between the H_2 problem and Riccati equation techniques will be based on a key lemma about inner functions.

We recall that a transfer function $\hat{U}(s) \in H_\infty$ is called inner when it satisfies

$$\hat{U}^\sim(j\omega)\hat{U}(j\omega) = I \quad \text{for all } \omega \in \mathbb{R}.$$

Multiplication by an inner function defines an isometry on H_2. This was indicated before; however, the property also extends to the matrix space RH_2 which we are considering in the present chapter. To see this, suppose $\hat{U}$ is inner and that $\hat{G}_1$ and $\hat{G}_2$ are in RH_2, then

$$\begin{aligned}
\langle \hat{U}\hat{G}_1, \hat{U}\hat{G}_2 \rangle_2 &= \frac{1}{2\pi} \int_{-\infty}^{\infty} \text{Tr}(\hat{U}(j\omega)\hat{G}_1(j\omega))^*\hat{U}(j\omega)\hat{G}_2(j\omega)d\omega \\
&= \frac{1}{2\pi} \int_{-\infty}^{\infty} \text{Tr}\,\hat{G}_1^*(j\omega)\hat{G}_2(j\omega)d\omega \\
&= \langle \hat{G}_1, \hat{G}_2 \rangle_2.
\end{aligned}$$

Namely, inner products are unchanged when such a function is inserted.

We now relate inner functions with Riccati equations.

Lemma 6.7. *Consider the state space matrices A, B, C, D, where the realization (A, B, C) is stabilizable and detectable and $D^*[C \ D] = [0 \ I]$. For*

$$H = \begin{bmatrix} A & -BB^* \\ -C^*C & -A^* \end{bmatrix},$$

define $X = \text{Ric}(H)$ and $F = -B^*X$. Then

$$\hat{U}(s) = \left[\begin{array}{c|c} A + BF & B \\ \hline C + DF & D \end{array} \right] \tag{6.5}$$

is inner.

Proof. Since X is a stabilizing solution, we have that $A - BB^*X = A + BF$ is Hurwitz, therefore $\hat{U}(s) \in RH_\infty$. Now from the Riccati equation

$$A^*X + XA + C^*C - XBB^*X = 0,$$

the definition of F, and the hypothesis it is verified routinely that

$$(A + BF)^*X + X(A + BF) + (C + DF)^*(C + DF) = 0.$$

Therefore X is the observability gramian corresponding to the realization (6.5). Also, the gramian X satisfies

$$(C + DF)^*D + XB = F^* + XB = 0.$$

Therefore we are in a position to apply Lemma 4.16, and conclude that $\hat{U}(s)$ is inner. ∎

As a final preliminary, we have the following matrix version of our earlier vector valued result, which followed from the Paley-Wiener theorem; this extension is a straightforward exercise.

Proposition 6.8. *If $\hat{G}_1$ and $\hat{G}_2$ are matrix valued functions in RH_2 and $RH_2^\perp$, respectively, then*

$$\langle \hat{G}_1, \hat{G}_2 \rangle_{L_2} = 0 .$$

Observe that if $\hat{Q} \in RH_2$ then the function $\hat{Q}^\sim \in RH_2^\perp$.

We are now in a position to state the main result, which explicitly gives the optimal solution to the H_2 synthesis problem. We spend the remainder of the section proving the result using a sequence of lemmas. The technique used exploits the properties of the Riccati equation, and in particular its stabilizing solution.

Theorem 6.9. *Suppose G is a state space system with realization*

$$\hat{G}(s) = \left[\begin{array}{c|cc} A & B_1 & B_2 \\ \hline C_1 & 0 & D_{12} \\ C_2 & D_{21} & 0 \end{array} \right],$$

where

(a) (A, B_1, C_1) is stabilizable and detectable;

(b) (A, B_2, C_2) is stabilizable and detectable;

(c) $D_{12}^* \begin{bmatrix} C_1 & D_{12} \end{bmatrix} = \begin{bmatrix} 0 & I \end{bmatrix}$;

(d) $D_{21} \begin{bmatrix} B_1^* & D_{21}^* \end{bmatrix} = \begin{bmatrix} 0 & I \end{bmatrix}$.

Then the optimal stabilizing controller to the H_2 synthesis problem is given by

$$\hat{K}_2(s) = \left[\begin{array}{c|c} A + B_2 F + L C_2 & -L \\ \hline F & 0 \end{array} \right],$$

with

$$L = -Y C_2^*, \quad F = -B_2^* X,$$

where $X = \mathrm{Ric}(H_c)$, $Y = \mathrm{Ric}(H_f)$, *and the Hamiltonian matrices are*

$$H_c = \left[\begin{array}{cc} A & -B_2 B_2^* \\ -C_1^* C_1 & -A^* \end{array} \right], \quad H_f = \left[\begin{array}{cc} A^* & -C_2^* C_2 \\ -B_1 B_1^* & -A \end{array} \right].$$

Furthermore the performance achieved by K_2 is

$$\|\hat{G}_c B_1\|_2^2 + \|F \hat{G}_f\|_2^2,$$

where

$$\hat{G}_c = \left[\begin{array}{c|c} A + B_2 F & I \\ \hline C_1 + D_{12} F & 0 \end{array} \right] \quad and \quad \hat{G}_f = \left[\begin{array}{c|c} A + L C_2 & B_1 + L D_{21} \\ \hline I & 0 \end{array} \right].$$

There are a number of conditions imposed on the state space matrices of $\hat{G}(s)$ which we now comment on. The first two are that D_{11} and D_{22} both be zero. That $D_{11} = 0$ is a reasonable assumption, which ensures the RH_2 space requirement that the closed-loop transfer is strictly proper provided $\hat{K}$ is strictly proper. That $D_{22} = 0$ is done purely out of convenience and can be removed.

Conditions (a) and (b) guarantee that both H_c and H_f are in the domain of the Riccati operator; (a) is a technical condition that can be relaxed at the expense of a more complicated proof. Clearly, however the condition in (b) is necessary for a stabilizing controller to exist.

The final two conditions are also made out of convenience. Condition (c) can be interpreted as saying that the plant output $C_1 x$ and the weight $D_{12} u$ on the control effort are orthogonal, and all the control channels affect the system performances directly. Condition (d) states that the system disturbances that drive the dynamics are orthogonal to measurement noise, and furthermore that the noise channels affect all the system measurements. The relaxing of these assumptions results in more complicated controller expressions and derivations, but the main method now pursued remains largely unchanged in such a general scenario.

We can now embark on our quest to prove Theorem 6.9. The next result is absolutely central to our technique of proof. Note that the hypothesis are a subset of those in Theorem 6.9.

Lemma 6.10. *Suppose G is a system with state space realization*

$$\hat{G}(s) = \left[\begin{array}{c|cc} A & B_1 & B_2 \\ \hline C_1 & 0 & D_{12} \\ C_2 & D_{21} & 0 \end{array} \right]$$

satisfying

(a) (A, B_1, C_1) *is stabilizable and detectable;*

(b) (A, B_2, C_2) *is stabilizable and detectable;*

(c) $D_{12}^* \begin{bmatrix} C_1 & D_{12} \end{bmatrix} = \begin{bmatrix} 0 & I \end{bmatrix}$.

If a controller K internally stabilizes G, then the closed-loop performance satisfies

$$\|\underline{S}(\hat{G}, \hat{K})\|_2^2 = \|\hat{G}_c B_1\|_2^2 + \|\underline{S}(\hat{G}_{tmp}, \hat{K})\|_2^2 , \tag{6.6}$$

with

$$\hat{G}_c = \left[\begin{array}{c|c} A + B_2 F & I \\ \hline C_1 + D_{12}F & 0 \end{array} \right] \quad and \quad \hat{G}_{tmp} = \left[\begin{array}{c|cc} A & B_1 & B_2 \\ \hline -F & 0 & I \\ C_2 & D_{21} & 0 \end{array} \right],$$

where $F = -B_2^* X$, $X = \mathrm{Ric}(H_c)$, and

$$H_c = \begin{bmatrix} A & -B_2 B_2^* \\ -C_1^* C_1 & -A^* \end{bmatrix}.$$

Proof. We first remark that by hypothesis (a) and (b), and invoking Corollary 6.6, H_c is in the domain of the Riccati operator. This means X and F are well-defined, and further that $A + B_2 F$ is Hurwitz, and so $\hat{G}_c \in RH_2$.

To understand the role of $\hat{G}_{tmp}$ it is useful to first write the closed-loop state equations for $\underline{S}(\hat{G}, \hat{K})$. These are the plant equations

$$\dot{x}(t) = Ax(t) + B_1 w(t) + B_2 u(t) \tag{6.7}$$
$$z(t) = C_1 x(t) + D_{12} u(t) \tag{6.8}$$
$$y(t) = C_2 x(t) + D_{21} w(t) , \tag{6.9}$$

together with the controller equations

$$\dot{x}_K(t) = A_K x_K(t) + B_K y(t),$$
$$u(t) = C_K x_K(t) + D_K y(t).$$

Now notice that if we were to represent instead $\underline{S}(\hat{G}_{tmp}, \hat{K})$, all equations would be the same except that (6.8) would be replaced by a new output equation

$$v(t) = -Fx(t) + u(t),$$

where the output $v = \underline{S}(G_{tmp}, K)w$.

A first consequence of the above observation is that K must internally stabilize G_{tmp}, since the closed-loop state equations are the same. Also it is easily verified that $\underline{S}(\hat{G}_{tmp}, \hat{K})$ is strictly proper if and only if $\underline{S}(\hat{G}, \hat{K})$ is; we thus focus on this case where $\underline{S}(\hat{G}_{tmp}, \hat{K}) \in RH_2$, since otherwise both sides of (6.6) are infinite.

The auxiliary variable v will aid us in the computation of $\|\underline{S}(\hat{G}, \hat{K})\|_2$. For this purpose, substitute $u(t) = v(t) + Fx(t)$ into (6.7)–(6.8) to get

$$\dot{x}(t) = (A + B_2 F)x(t) + B_1 w(t) + B_2 v(t)$$
$$z(t) = (C_1 + D_{12} F)x(t) + D_{12} v(t) .$$

Converting to transfer functions we have

$$\hat{z}(s) = \hat{G}_c(s) B_1 \hat{w}(s) + \hat{U}(s) \hat{v}(s) ,$$

with $\hat{G}_c(s)$ defined above and

$$\hat{U}(s) = \left[\begin{array}{c|c} A + B_2 F & B_2 \\ \hline C_1 + D_{12} F & D_{12} \end{array} \right] .$$

Also since $\hat{v}(s) = \underline{S}(\hat{G}_{tmp}, \hat{K}) \, \hat{w}(s)$ we conclude that

$$\underline{S}(\hat{G}, \hat{K}) = \hat{G}_c B_1 + \hat{U}\underline{S}(\hat{G}_{tmp}, \hat{K}) .$$

Next let us take the norm of $\underline{S}(\hat{G}, \hat{K})$ making use of this new expression:

$$\|\underline{S}(\hat{G}, \hat{K})\|_2^2 = \|\hat{G}_c B_1\|_2^2 + \|\hat{U}\underline{S}(\hat{G}_{tmp}, \hat{K})\|_2^2$$
$$+ 2\mathrm{Re}\left\langle \hat{U}\underline{S}(\hat{G}_{tmp}, \hat{K}), \, \hat{G}_c B_1 \right\rangle_2 .$$

The key to the proof is that

(i) $\hat{U}$ is inner; this was proved in Lemma 6.7;

(ii) $\hat{U}^\sim \hat{G}_c B_1$ is in $RH_2^\perp$; this fact follows by similar state space manipulations and is left as an exercise.

The conclusion now follows since $\|\underline{S}(\hat{G}_{tmp}, \hat{K})\|_2 = \|\hat{U}\underline{S}(\hat{G}_{tmp}, \hat{K})\|_2$, and

$$\left\langle \hat{U}\underline{S}(\hat{G}_{tmp}, \hat{K}), \hat{G}_c B_1 \right\rangle_2 = \left\langle \underline{S}(\hat{G}_{tmp}, \hat{K}), \hat{U}^\sim \hat{G}_c B_1 \right\rangle_2 = 0 ,$$

where the orthogonality is clear because $\underline{S}(\hat{G}_{tmp}, \hat{K}) \in RH_2$ as noted above.   ■

The major point of this lemma is that it puts a lower bound on the achievable performance, which is $\|\hat{G}_c B_1\|_2^2$. Now K stabilizes G of the theorem if and only if G_{tmp} is stabilized, and thus minimizing the closed-loop performance is equivalent to minimizing $\|\underline{S}(\hat{G}_{tmp}, \hat{K})\|_2$ by choosing K.

An additional remark is that in the special case of *state feedback*; that is, when $C_2 = I$, $D_{21} = 0$, this second term can be made zero by the static control law $K = F$, as follows from the direct substitution into $\underline{S}(\hat{G}_{tmp}, \hat{K})$. Alternatively, revisiting the proof we note that $u = Fx$ gives $v = 0$, and the second term of the cost would not appear. Therefore Lemma 6.10 provides a solution to the H_2 problem in the state feedback case.

In the general case, the auxiliary variable v with its associated cost reflects the price paid by not having the state available for measurement. This additional cost can now be optimized as well. Before addressing this, we state a so-called duality result, which relates a feedback loop to another one involving the transpose transfer functions. Its proof is an exercise involving only the basic properties of matrix transpose and trace.

Lemma 6.11. *Suppose G has a realization (A, B, C, D), controller K has a realization (A_K, B_K, C_K, D_K), and $\gamma > 0$. Then the following are equivalent.*

(i) K internally stabilizes G and $\|\underline{S}(\hat{G}, \hat{K})\|_2 = \gamma$;

(ii) K' internally stabilizes G' and $\|\underline{S}(\hat{G}', \hat{K}')\|_2 = \gamma$.

Here

$$\hat{G}' = \left[\begin{array}{c|c} A^* & C^* \\ \hline B^* & D^* \end{array}\right], \quad and \quad \hat{K}' = \left[\begin{array}{c|c} A_K^* & C_K^* \\ \hline B_K^* & D_K^* \end{array}\right].$$

The previous three results combine to give a further breakdown of the closed-loop transfer function of G and K.

Lemma 6.12. *Suppose the hypothesis of Theorem 6.9 hold, and let G_{tmp} be as defined in Lemma 6.10. If K internally stabilizes G_{tmp}, then*

$$\|\underline{S}(\hat{G}_{tmp}, \hat{K})\|_2^2 = \|F\hat{G}_f\|_2^2 + \|\underline{S}(\hat{E}, \hat{K})\|_2^2,$$

where

$$\hat{E}(s) = \left[\begin{array}{c|cc} A & -L & B_2 \\ \hline -F & 0 & I \\ C_2 & I & 0 \end{array}\right].$$

Proof. We only outline the argument, leaving algebraic details as an exercise.

Start by applying Lemma 6.11 to show that

$$\|\underline{S}(\hat{G}_{tmp}, \hat{K})\|_2 = \|\underline{S}(\hat{G}'_{tmp}, \hat{K}')\|_2,$$

with $\hat{K}'$ stabilizing $\hat{G}'_{tmp}$. Now we can set G to G'_{tmp} and K to K' in Lemma 6.10; that the relevant conditions are satisfied follows from the hypothesis of Theorem 6.9. This gives

$$\|\underline{S}(\hat{G}'_{tmp}, \hat{K}')\|_2^2 = \|\hat{G}'_f F^*\|_2^2 + \|\underline{S}(\hat{E}', \hat{K}')\|_2^2.$$

Finally applying Lemma 6.11 to the right-hand side we get

$$\|\underline{S}(\hat{G}'_{tmp}, \hat{K}')\|_2^2 = \|F\hat{G}_f\|_2^2 + \|\underline{S}(\hat{E}, \hat{K})\|_2^2.$$

∎

We can now prove the optimal synthesis theorem.

Proof of Theorem 6.9. Observe that controller K_2 of the theorem statement is an observer based controller of the standard form; also by

Theorem 6.3, Corollary 6.6, and our assumption, we see $A + B_2F$ and $A + LC_2$ are Hurwitz. Therefore K_2 internally stabilizes G.

Invoke Lemma 6.10 and Lemma 6.12 to get

$$\|\underline{S}(\hat{G}, K_2)\|_2^2 = \|\hat{G}_c B_1\|_2^2 + \|F\hat{G}_f\|_2^2 + \|\underline{S}(\hat{E}, \hat{K}_2)\|_2^2 .$$

Since the first two terms on the right do not depend on $\hat{K}_2$ at all, it is sufficient to show that $\|\underline{S}(\hat{E}, \hat{K}_2)\|_2 = 0$. This is easily shown by writing out the state space equations for the closed-loop system $\underline{S}(\hat{E}, \hat{K}_2)$. ∎

At this point we have completely proved our optimal synthesis result. Notice from the last proof that regardless of the controller K we have

$$\|\underline{S}(\hat{G}, \hat{K})\|_2^2 = \|\hat{G}_c B_1\|_2^2 + \|F\hat{G}_f\|_2^2 + \|\underline{S}(\hat{E}, \hat{K})\|_2^2 ,$$

providing that K internally stabilizes G, or equivalently E.

An alternative approach to the H_2 synthesis problem we just solved involves using the controller parameterization of Chapter 5 in terms of the RH_∞ parameter $\hat{Q}$. Then our synthesis problem becomes

$$\text{minimize} \quad \|\hat{T}_1 - \hat{T}_2 \hat{Q} \hat{T}_3\|_2 \quad \text{with } \hat{Q} \in RH_\infty,$$

a so-called *model matching* problem. If we use the stabilizing feedback gains F and L of this section, in the state space coprime factorization of Chapter 5, we can show that the optimal solution to this model matching problem occurs when $\hat{Q} = 0$. We leave this as a possible exercise.

6.4   State feedback H_2 synthesis via LMIs

We have just described a method of constructing the optimal solution to the H_2 synthesis problem, which uses the Riccati equation as a central tool. In this section we investigate obtaining a solution via LMI machinery instead, and focus our attention on the simplified case of state feedback. Our goal is to catch a quick glimpse of how LMI techniques can be used to solve this type of synthesis problem.

Our first result characterizes the RH_2 norm and internal stability in terms of linear matrix inequalities.

Proposition 6.13. *Suppose P is a state space system with realization (A, B, C). Then*

$$A \text{ is Hurwitz and } \|\hat{P}\|_2 < 1$$

if and only if there exists a symmetric matrix $X > 0$ such that $\text{Tr} CXC^* < 1$ and

$$AX + XA^* + BB^* < 0.$$

Proof. To show the "only if" part: by hypothesis the system controllability gramian X_c satisfies

$$\text{Tr} CX_c C^* < 1$$

and A is Hurwitz. Now the expression

$$X = \int_0^\infty e^{At}(BB^* + \epsilon I)e^{A^*t}dt,$$

is continuous in ϵ, and equal to X_c when $\epsilon = 0$. Therefore, for some $\epsilon > 0$, we have $\text{Tr} CXC^* < 1$. This matrix X clearly satisfies the Lyapunov equation

$$AX + XA^* + BB^* + \epsilon I = 0,$$

which means that the inequality of the claim is met.

The "if" direction follows the same idea. By assumption there exists X which is a positive definite solution to

$$AX + XA^* + BB^* < 0,$$

and satisfies $\text{Tr} CXC^* < 1$. Thus A is necessarily Hurwitz, and X is a generalized controllability gramian; so we know from Chapter 4 that $X \geq X_c$, where X_c is the controllability gramian. This implies that $\text{Tr} CXC^* \geq \text{Tr} CX_c C^*$ and thus $\|\hat{P}\|_2 < 1$.   ∎

We are now ready to formulate the state feedback synthesis problem. The plant G has the following form

$$\hat{G}(s) := \left[\begin{array}{c|cc} A & B_1 & B_2 \\ \hline C_1 & 0 & D_{12} \\ I & 0 & 0 \end{array} \right],$$

where $A \in \mathbb{R}^{n \times n}, B_1 \in \mathbb{R}^{n \times p}, B_2 \in \mathbb{R}^{n \times m}, C_1 \in \mathbb{R}^{q \times n}$. The controller is taken to be simply a static feedback gain $\hat{K}(s) = F$. While more generally we could use a dynamic controller, we have already seen in the previous section that the static restriction entails no loss of performance in the state feedback case. Our first synthesis result is based on the same change of matrix variables which was encountered in the study of stabilization.

Proposition 6.14. *There exists a feedback gain $\hat{K}(s) = F$ that internally stabilizes G and satisfies*

$$\|\underline{S}(\hat{G}, \hat{K})\|_2 < 1$$

if and only if there exist a rectangular matrix $Z \in \mathbb{R}^{m \times n}$ such that

$$F = ZX^{-1},$$

where $X > 0$ and satisfies the inequalities

$$[A \quad B_2] \begin{bmatrix} X \\ Z \end{bmatrix} + [X \quad Z^*] \begin{bmatrix} A^* \\ B_2^* \end{bmatrix} + B_1 B_1^* < 0, \tag{6.10}$$

$$\text{Tr}(C_1 X + D_{12}Z)X^{-1}(C_1 X + D_{12}Z)^* < 1. \tag{6.11}$$

Proof. Simply write out the state space equations corresponding to the interconnection shown in Figure 6.1.

$$\dot{x}(t) = (A + B_2 F)x(t) + B_1 w(t)$$
$$z(t) = (C_1 + D_{12} F)x(t).$$

Now invoking Proposition 6.13 we see that our problem can be solved if and only if there exist F and $X > 0$ satisfying

$$(A + B_2 F)X + X(A + B_2 F)^* + B_1 B_1^* < 0$$
$$\operatorname{Tr}(C_1 + D_{12}F)X(C_1 + D_{12}F)^* < 1$$

Now introducing the change of variables $FX = Z$ these conditions are equivalent to (6.10) and (6.11). ∎

While we have transformed the H_2 synthesis problem to a set of matrix inequalities, this is not yet an LMI problem since (6.11) is not convex in X, Z. Now we recall our discussion from Chapter 1, where it was mentioned that often problems which do not appear to be convex can be transformed into LMI problems by a *Schur complement* operation. This is indeed the case here, although it will also be necessary to introduce an additional variable, a so-called slack variable. We explain these techniques while obtaining our main result.

Theorem 6.15. *There exists a feedback gain $\hat{K}(s) = F$ that internally stabilizes G and satisfies*

$$\|\underline{S}(\hat{G}, \hat{K})\|_2 < 1$$

if and only if there exist symmetric matrices $X \in \mathbb{R}^{n \times n}$, $W \in \mathbb{R}^{q \times q}$, and a rectangular matrix $Z \in \mathbb{R}^{m \times n}$ such that

$$F = ZX^{-1},$$

and the inequalities

$$\begin{bmatrix} A & B_2 \end{bmatrix} \begin{bmatrix} X \\ Z \end{bmatrix} + \begin{bmatrix} X & Z^* \end{bmatrix} \begin{bmatrix} A^* \\ B_2^* \end{bmatrix} + B_1 B_1^* < 0, \qquad (6.12)$$

$$\begin{bmatrix} X & (C_1 X + D_{12}Z)^* \\ (C_1 X + D_{12}Z) & W \end{bmatrix} > 0, \qquad (6.13)$$

$$\operatorname{Tr}(W) < 1 \qquad (6.14)$$

are satisfied.

Proof. It suffices to show that conditions (6.13) and (6.14) are equivalent to (6.11) and $X > 0$.

Suppose (6.11) holds; since the trace is monotonic under matrix inequalities, then we can always find a matrix W satisfying $\operatorname{Tr}(W) < 1$ and

$$(C_1 X + D_{12}Z)X^{-1}(C_1 X + D_{12}Z)^* < W.$$

Indeed it suffices to perturb the matrix on the left by a small positive matrix. Now from the Schur complement formula of Theorem 1.10, the above inequality and $X > 0$ are equivalent to (6.13).

The previous steps can be reversed to obtain the converse implication.   ∎

Thus we have reduced the static state feedback H_2 synthesis problem to a set of convex conditions in the three variables W, X and Z. Given a feasible point, the control gain F can be obtained by an algebraic operation.

The above derivations exhibit a common feature of tackling problems via LMIs: there is always an element of "art" involved in finding the appropriate transformation or change of variables that would render a problem convex, and success is never guaranteed. In the next chapter we will present additional tools to aid us in this process. The references contain a more extensive set of such tools, in particular how to tackle the general (dynamic output feedback) H_2 synthesis problem via LMI techniques.

We are now ready to turn our attention to a new performance criterion; it will be studied purely from an LMI point of view.

Exercises

6.1. Verify that for a multi-input system P, the norm $\|\hat{P}\|_2^2$ is the sum of output energies corresponding to impulses $\delta(t)e_i$ applied in each input channel.

6.2. Prove Corollary 6.6.

6.3. Normalized coprime factorizations. This exercise complements Exercise 5.6. We recall that a right coprime factorization $\hat{P} = \hat{N}\hat{M}^{-1}$ of a proper $\hat{P}(s)$ is called normalized when $\begin{bmatrix} \hat{N}(s) \\ \hat{M}(s) \end{bmatrix}$ is inner.

(a) Consider $\hat{P} = \left[\begin{array}{c|c} A & B \\ \hline C & 0 \end{array} \right]$, and assume the realization is minimal. Let $F = -B^*X$, where

$$X = \text{Ric} \begin{bmatrix} A & -BB^* \\ -C^*C & -A^* \end{bmatrix}.$$

Show that

$$\begin{bmatrix} \hat{N}(s) \\ \hat{M}(s) \end{bmatrix} = \left[\begin{array}{c|c} A+BF & B \\ \hline C & 0 \\ F & I \end{array} \right]$$

defines a normalized right coprime factorization for $\hat{P}$.

(b) Extend the result for any $\hat{P}$ (not necessarily strictly proper).

6.4. In the proof of Lemma 6.10, verify the claim (ii) that $\hat{U}^\sim \hat{G}_c B_1$ is in $RH_2^\perp$.

6.5. Prove the dualization result of Lemma 6.11.

6.6. Fill in the details in the proof of Lemma 6.12. In particular verify that Lemma 6.10 can be applied to $\hat{G}'_{tmp}$ and K', and that the resulting $(\hat{G}'_{tmp})_{tmp} = \hat{E}'$.

6.7. Show that the performance of the optimal controller in Theorem 6.9 has the following expression in terms of the Riccati matrices.

$$\|\hat{G}_c B_1\|_2^2 + \|F\hat{G}_f\|_2^2 = \operatorname{Tr}(B_1^* X B_1) + \operatorname{Tr}(FYF^*).$$

6.8. Let X_s be denote the stabilizing solution of the Riccati equation

$$A^* X + X A + C^* C - X B B^* X = 0, \tag{6.15}$$

where we assume that (A, B, C) is controllable and observable.

(a) For any other symmetric solution X to (6.15) show that

$$(A - BB^* X_s)^*(X_s - X) + (X_s - X)(A - BB^* X_s) +$$
$$(X_s - X)BB^*(X_s - X) = 0.$$

Use this to prove $X \leq X_s$; i.e., X_s is the maximizing solution to (6.15).

(b) Now introduce the LMI

$$\begin{bmatrix} A^* X + X A + C^* C & X B \\ B^* X & I \end{bmatrix} \geq 0.$$

Show that if X satisfies the LMI, then also $X \leq X_s$.

(c) Show that there exists an *anti-stabilizing* solution X_a to the Riccati equation (6.15), that is a solution such that all the eigenvalues of the matrix $A - BB^* X_a$ are on the right half-plane $\operatorname{Re}(s) > 0$.

(d) Prove that $X_a \leq X \leq X_s$ for any solution to (6.15), or the LMI in (b).

(e) Show that the inequality $X_a < X_s$ is strict, and that these are the only two solutions to (6.15) with this property. *Hint:* relate $X_s - X_a$ to the controllability gramian of the pair $(A - BB^* X_s, B)$.

6.9. Here we investigate some uniqueness properties of H_2 optimal control.

(a) Show by counterexample that the optimal controller constructed in this chapter need not be the only controller that achieves optimum performance.

(b) As already noted, the H_2 optimal control synthesis can be posed as the model matching problem

$$\text{minimize } \|\hat{T}_1 - \hat{T}_2\hat{Q}\hat{T}_3\|_2, \text{ with } \hat{Q} \in RH_\infty.$$

This is simply done by invoking the Youla parametrization. Show that the optimal closed loop transfer function, if one exists, is unique. *Hint:* recall that RH_2 is an inner product space, and realize that the above model matching problem can be viewed as finding the nearest point in a subspace of RH_2 to $\hat{T}_1$.

Notes and references

Optimal H_2 control has a long history, and is more commonly known as linear quadratic gaussian (LQG) control, because of its stochastic interpretation outlined at the beginning of this chapter. It gets its roots from the papers [81, 84] on Kalman filters. A more extensive historical treatment of this subject, including extensive references, as well as more details on the stochastic interpretation of this problem, can be found in [4, 93]; see also [23]. The solution given here is based on the famous paper [38], where in addition to the optimal controller derived here, all possible controllers satisfying a particular H_2 performance level are parametrized. The state feedback solution given here in terms of LMIs is from [52].

The main analytical tool used in this chapter is the Riccati equation, which has a substantial research literature associated with it. We have just provided a glimpse for our purposes; however, see the recent monograph [94], and also the classic survey article [171] for additional details and references on these equations. For work on computational algorithms see for instance [35].

LMI solutions to many control synthesis problems have appeared in recent years. For a general synthesis method applicable in particular to output-feedback H_2 synthesis, see [146].

7
H_∞ Synthesis

In this chapter we consider optimal synthesis with respect to the H_∞ norm introduced in Chapter 3. Again we are concerned with the feedback arrangement of Figure 6.1 where we have two state space systems G and K, each having their familiar role.

We will pursue the answer to the following question: does there exist a state space controller K such that

- the closed-loop system is internally stable;
- the closed-loop performance satisfies

$$\|\underline{S}(\hat{G}, \hat{K})\|_\infty < 1?$$

Thus we only plan to consider the problem of making the closed-loop norm less than one; for short, we say the closed loop is *contractive* in the sense of H_∞. It is clear, however, that determining whether there exists a stabilizing controller so that $\|\underline{S}(\hat{G}, \hat{K})\|_\infty < \gamma$, for some constant γ, can be achieved by rescaling the γ dependent problem to arrive at the contractive version given above. Furthermore, by searching over γ, our approach will allow us to get as close to the minimal H_∞ norm as we desire, but in contrast to our work on H_2 optimal control, we will not seek a controller that exactly optimizes the H_∞ norm.

There are many approaches for solving the H_∞ control problem. Probably the most celebrated solution is in terms of Riccati equations of a similar style to the H_2 solution of Chapter 6. Here we will present a solution based entirely on linear matrix inequalities, which has the main advantage that it can be obtained with relatively straightforward matrix tools, and without any restrictions on the problem data. In fact Riccati equations and LMIs are intimately related, an issue we will explain when proving the Kalman–Yakubovich–Popov lemma concerning the *analysis* of the H_∞ norm of a system, which will be key to the subsequent synthesis solution.

Before getting into the details of the problem, we make a few comments about the motivation for this optimization.

As discussed in Chapter 3, the H_∞ norm is the L_2-induced norm of a causal, stable, linear-time invariant system. More precisely, given a causal linear time invariant operator $G : L_2(-\infty, \infty) \rightarrow L_2(-\infty, \infty)$, the corresponding operator in the isomorphic space $\hat{L}_2(j\mathbb{R})$ is a multiplication operator $M_{\hat{G}}$ for a certain $\hat{G}(s) \in H_\infty$, and

$$\|G\|_{L_2 \rightarrow L_2} = \|M_{\hat{G}}\|_{\hat{L}_2 \rightarrow \hat{L}_2} = \|\hat{G}\|_\infty.$$

What is the motivation for minimizing such an induced norm? If we refer back to the philosophy of "making error signals small" discussed in Chapter 6, we are minimizing the maximum "gain" of the system in the energy or L_2 sense. Equivalently, the excitation w is considered to be an arbitrary L_2 signal and we wish to minimize its *worst-case* effect on the energy of z. This may be an appropriate criterion if, as opposed to the situation of Chapter 6, we know little about the spectral characteristics of w. We will discuss, in more detail, alternatives and tradeoffs for noise modeling in Chapter 10.

There is, however, a more important reason than noise rejection that motivates an induced norm criterion; as seen in §3.2.1, a contractive operator Q has the property that the invertibility of $I - Q$ is ensured; this so-called *small gain* property will be key to ensuring stability of certain feedback systems, in particular when some of the components are not precisely specified. This reason has made H_∞ control a central subject in control theory; further discussion of this application will be pursued in the next two chapters.

7.1   Two important matrix inequalities

The entire synthesis approach of the chapter revolves around the two technical results presented here. The first of these is a result purely about matrices; the second is an important systems theory result and is frequently called the Kalman–Yakubovich–Popov lemma, or KYP lemma for short.

We begin by stating the following which the reader can prove as an exercise.

Lemma 7.1. *Suppose $\bar{P}$ and $\bar{Q}$ are matrices satisfying* $\text{Ker}\,\bar{P} = 0$ *and* $\text{Ker}\,\bar{Q} = 0$. *Then for every matrix Y there exists a solution J to*

$$\bar{P}^* J \bar{Q} = Y.$$

The above lemma is used to prove the next one which is one of the two major technical results of this section.

Lemma 7.2. *Suppose*

(a) P, Q and H are matrices and that H is symmetric;

(b) The matrices N_P and N_Q are full rank matrices satisfying $\operatorname{Im} N_P = \operatorname{Ker} P$ and $\operatorname{Im} N_Q = \operatorname{Ker} Q$.

Then there exists a matrix J such that

$$H + P^* J Q + Q^* J^* P < 0 , \qquad (7.1)$$

if and only if the inequalities

$$N_P^* H N_P < 0 \quad and \quad N_Q^* H N_Q < 0$$

both hold.

Remark 7.3. *Observe that when either of the kernels of P and Q are zero, the result does not apply as stated. However, it is readily seen from Lemma 7.1 that if both of the kernels are zero, then there is always a solution J. If, for example, only $\operatorname{Ker} P = 0$, then $N_Q^* H N_Q < 0$ is a necessary and sufficient condition for a solution to (7.1) to exist, as follows by simplified version of the following proof.*

In short, the convention is that whenever one of the kernels is trivial, the corresponding condition on H is dropped. In the sequel we will encounter other results of a similar form, and will not make explicit allusion to these degenerate cases; however, the same convention applies to all these results.

Proof. We will show the equivalence of the conditions directly by construction. To begin define V_1 to be a matrix such that

$$\operatorname{Im} V_1 = \operatorname{Ker} P \cap \operatorname{Ker} Q,$$

and V_2 and V_3 such that

$$\operatorname{Im} \begin{bmatrix} V_1 & V_2 \end{bmatrix} = \operatorname{Ker} P \quad and \quad \operatorname{Im} \begin{bmatrix} V_1 & V_3 \end{bmatrix} = \operatorname{Ker} Q.$$

Without loss of generality we assume that V_1, V_2, and V_3 have full column rank and define V_4 so that

$$V = \begin{bmatrix} V_1 & V_2 & V_3 & V_4 \end{bmatrix}$$

is square and nonsingular. Therefore the LMI in (7.1) above holds, if and only if

$$V^* H V + V^* P^* J Q V + V^* Q^* J^* P V < 0 \quad \text{does.} \qquad (7.2)$$

Now PV and QV are simply the matrices P and Q on the domain basis defined by V; therefore they have the form

$$PV = \begin{bmatrix} 0 & 0 & P_1 & P_2 \end{bmatrix} \quad and \quad QV = \begin{bmatrix} 0 & Q_1 & 0 & Q_2 \end{bmatrix};$$

we also define the block components

$$V^*HV =: \begin{bmatrix} H_{11} & H_{12} & H_{13} & H_{14} \\ H_{12}^* & H_{22} & H_{23} & H_{24} \\ H_{13}^* & H_{23}^* & H_{33} & H_{34} \\ H_{14}^* & H_{24}^* & H_{34}^* & H_{44} \end{bmatrix}.$$

Further define the variable Y by

$$Y = \begin{bmatrix} Y_{11} & Y_{12} \\ Y_{21} & Y_{22} \end{bmatrix} = \begin{bmatrix} P_1^* \\ P_2^* \end{bmatrix} J \begin{bmatrix} Q_1 & Q_2 \end{bmatrix}.$$

From their definitions $\mathrm{Ker} \begin{bmatrix} P_1 & P_2 \end{bmatrix} = 0$ and $\mathrm{Ker} \begin{bmatrix} Q_1 & Q_2 \end{bmatrix} = 0$, and so by Lemma 7.1 we see that Y is freely assignable by choosing an appropriate matrix J.

Writing out inequality (7.2) using the above definitions, we get

$$\begin{bmatrix} H_{11} & H_{12} & H_{13} & H_{14} \\ H_{12}^* & H_{22} & H_{23} + Y_{11}^* & H_{24} + Y_{21}^* \\ H_{13}^* & H_{23}^* + Y_{11} & H_{33} & H_{34} + Y_{12} \\ H_{14}^* & H_{24}^* + Y_{21} & H_{34}^* + Y_{12}^* & H_{44} + Y_{22} + Y_{22}^* \end{bmatrix} < 0.$$

Apply the Schur complement formula to the upper 3×3 block, and we see the above holds if and only if the two following inequalities are met.

$$\bar{H} := \begin{bmatrix} H_{11} & H_{12} & H_{13} \\ H_{12}^* & H_{22} & H_{23} + Y_{11}^* \\ H_{13}^* & H_{23}^* + Y_{11} & H_{33} \end{bmatrix} < 0$$

$$\text{and} \quad H_{44} + Y_{22} + Y_{22}^* - \begin{bmatrix} H_{14} \\ H_{24} + Y_{21}^* \\ H_{34} + Y_{12} \end{bmatrix}^* \bar{H}^{-1} \begin{bmatrix} H_{14} \\ H_{24} + Y_{21}^* \\ H_{34} + Y_{12} \end{bmatrix} < 0$$

As already noted above Y is freely assignable and so we see that provided the first inequality can be achieved by choosing Y_{11}, the second can always be met by appropriate choice of Y_{12}, Y_{21} and Y_{22}. That is the above two inequalities can be achieved if and only if $\bar{H} < 0$ holds for some Y_{11}. Now applying a Schur complement on $\bar{H}$ with respect to H_{11}, we obtain

$$\begin{bmatrix} H_{11} & 0 & 0 \\ 0 & H_{22} - H_{12}^* H_{11}^{-1} H_{12} & Y_{11}^* + X^* \\ 0 & Y_{11} + X & H_{33} - H_{13}^* H_{11}^{-1} H_{13} \end{bmatrix} < 0,$$

where $X = H_{23}^* - H_{13}^* H_{11}^{-1} H_{12}$. Now since Y_{11} is freely assignable we see readily that the last condition can be satisfied if and only if the diagonal entries of the left-hand matrix are all negative definite. Using the Schur complement result twice, these three conditions can be converted to the equivalent conditions

$$\begin{bmatrix} H_{11} & H_{12} \\ H_{12}^* & H_{22} \end{bmatrix} < 0 \quad \text{and} \quad \begin{bmatrix} H_{11} & H_{13} \\ H_{13}^* & H_{33} \end{bmatrix} < 0.$$

By the choice of our basis we see that these hold if and only if $N_P^* H N_P < 0$ and $N_Q^* H N_Q < 0$ are both met.   ∎

Having proved this matrix result we move on to our second result, the KYP lemma.

7.1.1   The Kalman–Yakubovich–Popov Lemma

There are many versions of this result, which establishes the equivalence between a frequency domain inequality and a state space condition in terms of either a Riccati equation or an LMI. The version given below turns an H_∞ norm condition into an LMI. Being able to do this is very helpful for attaining our goal of controller synthesis; however, it is equally important simply as a finite dimensional analysis test for transfer functions.

Lemma 7.4. *Suppose* $\hat{M}(s) = \left[\begin{array}{c|c} A & B \\ \hline C & D \end{array}\right]$. *Then the following are equivalent conditions.*

(i)   The matrix A is Hurwitz and

$$\|\hat{M}\|_\infty < 1.$$

(ii)   There exists a matrix $X > 0$ such that

$$\begin{bmatrix} C^* \\ D^* \end{bmatrix} [C \quad D] + \begin{bmatrix} A^*X + XA & XB \\ B^*X & -I \end{bmatrix} < 0 . \tag{7.3}$$

The condition in (ii) is clearly an LMI and gives us a very convenient way to evaluate the H_∞ norm of a transfer function. In the proof below we see that (i) follows from (ii) in a reasonably straightforward way, by showing the direct connection between the above LMI and the state space equations that describe $\hat{M}$. Proving the converse is considerably harder; fortunately we will be able to exploit the Riccati equation techniques which were introduced in Chapter 6. An alternative proof, which employs only matrix arguments, will be given later in the course; see Proposition 8.28.

Proof. We begin by showing (ii) implies (i). The top left block in (7.3) states that $A^*X + XA + C^*C < 0$. Since $X > 0$, we see that A must be Hurwitz.

It remains to show contractiveness, which we do by employing a system-theoretic argument based on the state equations for M. Using the strict inequality (7.3), choose $0 < \epsilon < 1$ such that

$$\begin{bmatrix} C^* \\ D^* \end{bmatrix} [C \quad D] + \begin{bmatrix} A^*X + XA & XB \\ B^*X & -(1-\epsilon)I \end{bmatrix} < 0 \tag{7.4}$$

holds. Let $w \in L_2[0, \infty)$ and realize that in order to show that M is contractive, it is sufficient to show that $\|z\| \le (1 - \epsilon)\|w\|$, where $z := Mw$.

The state space equations relating w and z are

$$\dot{x}(t) = Ax(t) + Bw(t), \quad x(0) = 0,$$
$$z(t) = Cx(t) + Dw(t).$$

Now multiplying inequality (7.4) on the left by $\begin{bmatrix} x^*(t) & w^*(t) \end{bmatrix}$ and on the right by the adjoint we have

$$|z(t)|^2 + 2\operatorname{Re}\{x^*(t)X(Ax(t) + Bw(t))\} - (1-\epsilon)|w(t)|^2 \leq 0$$

By introducing the storage function $V : \mathbb{C}^n \to \mathbb{R}$, defined by $V(x(t)) = x^*(t)Xx(t)$, we arrive at the so-called dissipation inequality

$$\dot{V} + |z(t)|^2 \leq (1-\epsilon)|w(t)|^2.$$

Integrating on an interval $[0, T]$, recalling that $x(0) = 0$, gives

$$x(T)^*Xx(T) + \int_0^T |z(t)|^2 dt \leq (1-\epsilon)\int_0^T |w(t)|^2 dt.$$

Now let $T \to \infty$; since $w \in L_2$ and A is Hurwitz, it follows from Exercise 3.13 that $x(T)$ converges to zero.[1] Therefore we find

$$\|z\|^2 \leq (1-\epsilon)\|w\|^2,$$

which completes this direction of the proof.

We now tackle the direction (i) implies (ii). To simplify the expressions we will write the derivation in the special case $D = 0$, but an analogous argument applies to the general case (see the exercises). Starting from

$$\hat{M}(s) = \left[\begin{array}{c|c} A & B \\ \hline C & 0 \end{array}\right],$$

and recalling the definition of $\hat{M}^\sim(s)$ from Chapter 6, we derive the state space representation

$$I - \hat{M}^\sim(s)\hat{M}(s) = \left[\begin{array}{cc|c} A & 0 & -B \\ -C^*C & -A^* & 0 \\ \hline 0 & B^* & I \end{array}\right].$$

It is easy to verify that

$$[I - \hat{M}^\sim(s)\hat{M}(s)]^{-1} = \left[\begin{array}{cc|c} A & BB^* & B \\ -C^*C & -A^* & 0 \\ \hline 0 & B^* & I \end{array}\right]. \tag{7.5}$$

Since $\|\hat{M}\|_\infty < 1$ by hypothesis, we conclude that $[I - \hat{M}^\sim(s)\hat{M}(s)]^{-1}$ has no poles on the imaginary axis. Furthermore we now show, using the PBH

---

[1] Alternatively, we can use here the fact that $X > 0$ to eliminate the term in $x(T)$.

7.1. Two important matrix inequalities 223

test, that the realization (7.5) has no unobservable eigenvalues that are purely imaginary. Suppose that

$$\begin{bmatrix} j\omega_0 I - A & -BB^* \\ C^*C & j\omega_0 I + A^* \\ 0 & B^* \end{bmatrix} \begin{bmatrix} x_1 \\ x_2 \end{bmatrix} = 0,$$

for some vectors x_1 and x_2. Then we have the following chain of implications:

$B^*x_2 = 0$ implies $(j\omega_0 I - A)x_1 = 0$;

therefore $x_1 = 0$ since A is Hurwitz;

this means $(j\omega_0 I + A^*)x_2 = 0$;

which implies $x_2 = 0$ again because A is Hurwitz.

We conclude that (7.5) has no unobservable eigenvalues on the imaginary axis; an analogous argument shows the absence of uncontrollable eigenvalues. This means that the matrix

$$H = \begin{bmatrix} A & BB^* \\ -C^*C & -A^* \end{bmatrix}$$

has no purely imaginary eigenvalues. Referring to Theorem 6.5, notice that $BB^* \geq 0$ and (A, BB^*) is stabilizable since A is Hurwitz. Hence H is in the domain of the Riccati operator, and we can define $X_0 = \text{Ric}(H)$ satisfying

$$A^*X_0 + X_0 A + C^*C + X_0 BB^* X_0 = 0 \tag{7.6}$$

and $A + BB^*X_0$ Hurwitz. Also note that (7.6) implies $A^*X_0 + X_0 A \leq 0$; therefore from our work on Lyapunov equations we see that

$$X_0 \geq 0$$

since A is Hurwitz. To obtain the LMI characterization of (ii) we must slightly strengthen the previous relationships. For this purpose define $\bar{X}$ to be the solution of the Lyapunov equation

$$(A + BB^*X_0)^*\bar{X} + \bar{X}(A + BB^*X_0) = -I. \tag{7.7}$$

Since $(A + BB^*X_0)$ is Hurwitz we have $\bar{X} > 0$. Now let $X = X_0 + \epsilon\bar{X} > 0$, which is positive definite for all $\epsilon > 0$. Using (7.6) and (7.7) we have

$$A^*X + XA + C^*C + XBB^*X = -\epsilon I + \epsilon^2 \bar{X}BB^*\bar{X}.$$

Choose $\epsilon > 0$ sufficiently small so that this equation is negative definite. Hence we have found $X > 0$ satisfying the strict *Riccati inequality*

$$A^*X + XA + C^*C + XBB^*X < 0.$$

Now applying a Schur complement operation, this inequality is equivalent to

$$\begin{bmatrix} A^*X + XA + C^*C & XB \\ B^*X & -I \end{bmatrix} < 0,$$

which is (7.3) for the special case $D = 0$.   ∎

The preceding proof illustrates some of the deepest relationships of linear systems theory. We have seen that frequency domain inequalities are associated with dissipativity of storage functions in the time domain, and also the connection between LMIs (linked to dissipativity) and Riccati equations (which arise in quadratic optimization).

In fact this latter connection extends as well to problems of H_∞ synthesis, where both Riccati equation and LMI approaches can be used. In this course we will pursue the LMI solution. Surprisingly the two results of this section are all we require, together with basic matrix algebra, to solve this control problem.

7.2   Synthesis

We start with the state space realizations that describe the systems G and K:

$$
\hat{G}(s) = \left[\begin{array}{c|cc} A & B_1 & B_2 \\ \hline C_1 & D_{11} & D_{12} \\ C_2 & D_{21} & 0 \end{array}\right], \quad \hat{K}(s) = \left[\begin{array}{c|c} A_K & B_K \\ \hline C_K & D_K \end{array}\right].
$$

Notice that we have assumed $D_{22} = 0$. Removing this assumption leads to more complicated formulas, but the technique is identical. We make no other assumptions about the state space systems. The state dimensions of the nominal system and controller will be important: $A \in \mathbb{R}^{n \times n}$, $A_K \in \mathbb{R}^{n_K \times n_K}$.

Our first step is to combine these two state space realizations into one which describes the map from w to z. We obtain

$$
\underline{S}(\hat{G}, \hat{K}) = \left[\begin{array}{c|c} A_{cl} & B_{cl} \\ \hline C_{cl} & D_{cl} \end{array}\right]
$$

$$
= \left[\begin{array}{cc|c} A + B_2 D_K C_2 & B_2 C_K & B_1 + B_2 D_K D_{21} \\ B_K C_2 & A_K & B_K D_{21} \\ \hline C_1 + D_{12} D_K C_2 & D_{12} C_K & D_{11} + D_{12} D_K D_{21} \end{array}\right].
$$

Now define the matrix

$$
J = \begin{bmatrix} A_K & B_K \\ C_K & D_K \end{bmatrix},
$$

which collects the representation for K into one matrix. We can parametrize the closed-loop relation in terms of the controller realization as follows. First make the following definitions.

$$\bar{A} = \begin{bmatrix} A & 0 \\ 0 & 0 \end{bmatrix} \qquad\qquad \bar{B} = \begin{bmatrix} B_1 \\ 0 \end{bmatrix} \qquad (7.8)$$

$$\bar{C} = [C_1 \quad 0] \qquad\qquad \underline{C} = \begin{bmatrix} 0 & I \\ C_2 & 0 \end{bmatrix}$$

$$\underline{B} = \begin{bmatrix} 0 & B_2 \\ I & 0 \end{bmatrix} \qquad\qquad \underline{D}_{12} = [0 \quad D_{12}]$$

$$\underline{D}_{21} = \begin{bmatrix} 0 \\ D_{21} \end{bmatrix},$$

which are entirely in terms of the state space matrices for G. Then we have

$$\begin{aligned} A_{cl} &= \bar{A} + \underline{B}J\underline{C} & B_{cl} &= \bar{B} + \underline{B}J\underline{D}_{21} \\ C_{cl} &= \bar{C} + \underline{D}_{12}J\underline{C} & D_{cl} &= D_{11} + \underline{D}_{12}J\underline{D}_{21}. \end{aligned} \qquad (7.9)$$

The crucial point here is that the parametrization of the closed-loop state space matrices is *affine* in the controller matrix J.

Now we are looking for a controller K such that the closed loop is contractive and internally stable. The following form of the KYP Lemma will help us.

Corollary 7.5. *Suppose* $\hat{M}_{cl}(s) = \left[\begin{array}{c|c} A_{cl} & B_{cl} \\ \hline C_{cl} & D_{cl} \end{array}\right]$ *Then the following are equivalent conditions.*

(a) The matrix A_{cl} is Hurwitz and $\|\hat{M}_{cl}\|_\infty < 1$;

(b) There exists a symmetric positive definite matrix X_{cl} such that

$$\begin{bmatrix} A_{cl}^* X_{cl} + X_{cl} A_{cl} & X_{cl} B_{cl} & C_{cl}^* \\ B_{cl}^* X_{cl} & -I & D_{cl}^* \\ C_{cl} & D_{cl} & -I \end{bmatrix} < 0.$$

This result is readily proved from Lemma 7.4 by applying the Schur complement formula. Notice that the matrix inequality in (b) is affine in X_{cl} and J individually, but it is not jointly affine in both variables. The main task now is to obtain a characterization where we do have a convex problem.

Now define the matrices

$$P_{X_{cl}} = [\underline{B}^* X_{cl} \quad 0 \quad \underline{D}_{12}^*]$$
$$Q = [\underline{C} \quad \underline{D}_{21} \quad 0]$$

and further

$$H_{X_{cl}} = \begin{bmatrix} \bar{A}^* X_{cl} + X_{cl}\bar{A} & X_{cl}\bar{B} & \bar{C}^* \\ \bar{B}^* X_{cl} & -I & D_{11}^* \\ \bar{C} & D_{11} & -I \end{bmatrix}.$$

It follows that the inequality in (b) above is exactly

$$H_{X_{cl}} + Q^* J^* P_{X_{cl}} + P_{X_{cl}}^* J Q < 0 .$$

Lemma 7.6. *Given the above definitions there exists a controller synthesis K if and only if there exists a symmetric matrix $X_{cl} > 0$ such that*

$$N_{P_{X_{cl}}}^* H_{X_{cl}} N_{P_{X_{cl}}} < 0 \quad and \quad N_Q^* H_{X_{cl}} N_Q < 0 ,$$

where $N_{P_{X_{cl}}}$ and N_Q are as defined in Lemma 7.2.

Proof. From the discussion above we see that a controller K exists if and only if there exists $X_{cl} > 0$ satisfying

$$H_{X_{cl}} + Q^* J^* P_{X_{cl}} + P_{X_{cl}}^* J Q < 0 .$$

Now invoke Lemma 7.2.

∎

This lemma says that a controller exists if and only if the two matrix inequalities can be satisfied. Each of the inequalities is given in terms of the state space matrices of G and the variable X_{cl}. However we must realize that since X_{cl} appears in both $H_{X_{cl}}$ and $N_{P_{X_{cl}}}$, that these are not LMI conditions. Converting to an LMI formulation is our next goal, and will require a number of steps. Given a matrix $X_{cl} > 0$, define the related matrix

$$T_{X_{cl}} = \begin{bmatrix} \bar{A} X_{cl}^{-1} + X_{cl}^{-1} \bar{A}^* & \bar{B} & X_{cl}^{-1} \bar{C}^* \\ \bar{B}^* & -I & D_{11}^* \\ \bar{C} X_{cl}^{-1} & D_{11} & -I \end{bmatrix}, \qquad (7.10)$$

and the matrix

$$P = \begin{bmatrix} \underline{B}^* & 0 & \underline{D}_{12}^* \end{bmatrix}, \qquad (7.11)$$

which only depends on the state space realization of G. The next lemma converts one of the two matrix inequalities of the lemma, involving $H_{X_{cl}}$, to one in terms of $T_{X_{cl}}$.

Lemma 7.7. *Suppose $X_{cl} > 0$. Then*

$$N_{P_{X_{cl}}}^* H_{X_{cl}} N_{P_{X_{cl}}} < 0 \quad if and only if \quad N_P^* T_{X_{cl}} N_P < 0.$$

Proof. Start by observing that

$$P_{X_{cl}} = PS ,$$

where

$$S = \begin{bmatrix} X_{cl} & 0 & 0 \\ 0 & I & 0 \\ 0 & 0 & I \end{bmatrix} .$$

Therefore we have

$$\mathrm{Ker}\, P_{X_{cl}} = S^{-1} \mathrm{Ker}\, P .$$

Then using the definitions of $N_{P_{X_{cl}}}$ and N_P we can set

$$N_{P_{X_{cl}}} = S^{-1} N_P .$$

Finally we have that $N_{P_{X_{cl}}}^* H_{X_{cl}} N_{P_{X_{cl}}} < 0$ if and only if

$$N_P^* (S^{-1})^* H_{X_{cl}} S^{-1} N_P < 0,$$

and it is routine to verify $(S^{-1})^* H_{X_{cl}} S^{-1} = T_{X_{cl}}$.   ∎

Combining the last two lemmas we see that there exists a controller of state dimension n_K if and only if there exists a symmetric matrix $X_{cl} > 0$ such that

$$N_P^* T_{X_{cl}} N_P < 0 \quad \text{and} \quad N_Q^* H_{X_{cl}} N_Q < 0 . \tag{7.12}$$

The first of these inequalities is an LMI in the matrix variable X_{cl}^{-1}, whereas the second is an LMI in terms of X_{cl}. However, the system of both inequalities is not an LMI. Our intent is to convert these seemingly non-convex conditions into an LMI condition.

Recall that X_{cl} is a real and symmetric $(n+n_K) \times (n+n_K)$ matrix; here n and n_K are state dimensions of G and K. Let us now define the matrices X and Y which are $n \times n$ submatrices of X_{cl} and X_{cl}^{-1}, by

$$X_{cl} =: \begin{bmatrix} X & X_2 \\ X_2^* & X_3 \end{bmatrix} \quad \text{and} \quad X_{cl}^{-1} =: \begin{bmatrix} Y & Y_2 \\ Y_2^* & Y_3 \end{bmatrix} . \tag{7.13}$$

We now show that the two inequality conditions listed in (7.12), only constrain the submatrices X and Y.

Lemma 7.8. *Suppose X_{cl} is a positive definite $(n+n_K) \times (n+n_K)$ matrix and X and Y are $n \times n$ matrices defined as in (7.13). Then*

$$N_P^* T_{X_{cl}} N_P < 0 \quad \text{and} \quad N_Q^* H_{X_{cl}} N_Q < 0$$

if and only if the following two matrix inequalities are satisfied:

(a)

$$\begin{bmatrix} N_o & 0 \\ 0 & I \end{bmatrix}^* \begin{bmatrix} A^*X + XA & XB_1 & C_1^* \\ B_1^*X & -I & D_{11}^* \\ C_1 & D_{11} & -I \end{bmatrix} \begin{bmatrix} N_o & 0 \\ 0 & I \end{bmatrix} < 0 ;$$

(b)

$$\begin{bmatrix} N_c & 0 \\ 0 & I \end{bmatrix}^* \begin{bmatrix} AY + YA^* & YC_1^* & B_1 \\ C_1Y & -I & D_{11} \\ B_1^* & D_{11}^* & -I \end{bmatrix} \begin{bmatrix} N_c & 0 \\ 0 & I \end{bmatrix} < 0 ,$$

where N_o and N_c are full rank matrices whose images satisfy

$$\text{Im} N_o = \text{Ker} \begin{bmatrix} C_2 & D_{21} \end{bmatrix}$$
$$\text{Im} N_c = \text{Ker} \begin{bmatrix} B_2^* & D_{12}^* \end{bmatrix} .$$

Proof. The proof amounts to writing out the definitions and removing redundant constraints. Let us show that $N_P^* T_{X_{cl}} N_P < 0$ is equivalent to the LMI in (b).

From the definitions of $T_{X_{cl}}$ in (7.10), and $\bar{A}$, $\bar{B}$, and $\bar{C}$ in (7.8) we get

$$
T_{X_{cl}} = \begin{bmatrix} AY + YA^* & AY_2 & B_1 & YC_1^* \\ Y_2^* A^* & 0 & 0 & Y_2^* C_1^* \\ B_1^* & 0 & -I & D_{11}^* \\ C_1 Y & C_1 Y_2 & D_{11} & -I \end{bmatrix}.
$$

Also recalling the definition of P in (7.11), and substituting for $\underline{B}$ and $\underline{D}_{12}$ from (7.8) yields

$$
P = \begin{bmatrix} 0 & I & 0 & 0 \\ B_2^* & 0 & 0 & D_{12}^* \end{bmatrix}.
$$

Thus the kernel of P is the image of

$$
N_P = \begin{bmatrix} V_1 & 0 \\ 0 & 0 \\ 0 & I \\ V_2 & 0 \end{bmatrix},
$$

where

$$
\begin{bmatrix} V_1 \\ V_2 \end{bmatrix} = N_c
$$

spans the kernel of $\begin{bmatrix} B_2^* & D_{12}^* \end{bmatrix}$ as defined above. Notice that the second block row of N_P is exactly zero, and therefore the second block-row and block-column of $T_{X_{cl}}$, as explained above, do not enter into the constraint $N_P^* T_{X_{cl}} N_P < 0$. Namely, this inequality is

$$
\begin{bmatrix} V_1 & 0 \\ 0 & I \\ V_2 & 0 \end{bmatrix}^* \begin{bmatrix} AY + YA^* & B_1 & YC_1^* \\ B_1^* & -I & D_{11}^* \\ C_1 Y & D_{11} & -I \end{bmatrix} \begin{bmatrix} V_1 & 0 \\ 0 & I \\ V_2 & 0 \end{bmatrix} < 0.
$$

By applying the permutation

$$
\begin{bmatrix} V_1 & 0 \\ 0 & I \\ V_2 & 0 \end{bmatrix} = \begin{bmatrix} I & 0 & 0 \\ 0 & 0 & I \\ 0 & I & 0 \end{bmatrix} \begin{bmatrix} N_c & 0 \\ 0 & I \end{bmatrix}
$$

we arrive at (b).

Using a nearly identical argument, we can readily show that condition $N_Q^* H_{X_{cl}} N_Q < 0$ is equivalent to LMI (a) in the theorem statement. ∎

What we have shown is that a controller synthesis exists if and only if there exists an $(n + n_K) \times (n + n_K)$ matrix X_{cl} that satisfies conditions (a) and (b) of the last lemma. These latter two conditions only involve X and Y, which are submatrices of X_{cl} and X_{cl}^{-1}, respectively. Our next result tells us under what conditions, given arbitrary matrices X and Y, it is possible to find a positive definite matrix X_{cl} that satisfies (7.13).

Lemma 7.9. *Suppose X and Y are symmetric, positive definite matrices in $\mathbb{R}^{n\times n}$; and n_K is a positive integer. Then there exist matrices $X_2, Y_2 \in \mathbb{R}^{n\times n_K}$ and symmetric matrices $X_3, Y_3 \in \mathbb{R}^{n_K\times n_K}$, satisfying*

$$\begin{bmatrix} X & X_2 \\ X_2^* & X_3 \end{bmatrix} > 0 \quad and \quad \begin{bmatrix} X & X_2 \\ X_2^* & X_3 \end{bmatrix}^{-1} = \begin{bmatrix} Y & Y_2 \\ Y_2^* & Y_3 \end{bmatrix}$$

if and only if

$$\begin{bmatrix} X & I \\ I & Y \end{bmatrix} \geq 0 \quad and \quad \mathrm{rank}\begin{bmatrix} X & I \\ I & Y \end{bmatrix} \leq n + n_K . \tag{7.14}$$

Proof. First we prove that the first two conditions imply the second two. From

$$\begin{bmatrix} X & X_2 \\ X_2^* & X_3 \end{bmatrix} \begin{bmatrix} Y & Y_2 \\ Y_2^* & Y_3 \end{bmatrix} = I \tag{7.15}$$

it is routine to verify that

$$0 \leq \begin{bmatrix} I & 0 \\ Y & Y_2 \end{bmatrix} \begin{bmatrix} X & X_2 \\ X_2^* & X_3 \end{bmatrix} \begin{bmatrix} I & Y \\ 0 & Y_2^* \end{bmatrix} = \begin{bmatrix} X & I \\ I & Y \end{bmatrix}.$$

Since the above product includes a $(n + n_K) \times (n + n_K)$ matrix, we have

$$\mathrm{rank}\begin{bmatrix} X & I \\ I & Y \end{bmatrix} \leq n + n_k.$$

To prove "if" we start with the assumption that (7.14) holds. From the Schur complement relationship

$$\begin{bmatrix} X & I \\ I & Y \end{bmatrix} = \begin{bmatrix} I & Y^{-1} \\ 0 & I \end{bmatrix} \begin{bmatrix} X - Y^{-1} & 0 \\ 0 & Y \end{bmatrix} \begin{bmatrix} I & 0 \\ Y^{-1} & I \end{bmatrix}$$

we conclude that

$$X - Y^{-1} \geq 0 \quad \text{and rank } (X - Y^{-1}) \leq n_K .$$

These conditions ensure that there exists a matrix $X_2 \in \mathbb{R}^{n\times n_K}$ so that

$$X - Y^{-1} = X_2 X_2^* \geq 0,$$

which in particular yields $X - X_2 X_2^* > 0$. From this and the Schur complement argument we see that

$$\begin{bmatrix} X & X_2 \\ X_2^* & I \end{bmatrix} > 0 .$$

Also

$$\begin{bmatrix} X & X_2 \\ X_2^* & I \end{bmatrix}^{-1} = \begin{bmatrix} Y & -YX_2 \\ -X_2^*Y & X_2^*YX_2 + I \end{bmatrix}$$

and so we set $X_3 = I$. ∎

The lemma states that a matrix X_{cl} in $\mathbb{R}^{(n+n_K)\times(n+n_K)}$, satisfying (7.13), can be constructed from X and Y exactly when the LMI and rank conditions in (7.14) are satisfied. The rank condition is not in general an LMI, but notice that

$$\operatorname{rank} \begin{bmatrix} X & I \\ I & Y \end{bmatrix} \le 2n .$$

Therefore we see that if $n_K \ge n$ in the lemma, the rank condition becomes vacuous and we are left with only the LMI condition. We can now prove the synthesis theorem.

Theorem 7.10. *A synthesis exists for the H_∞ problem, if and only if there exist symmetric matrices $X > 0$ and $Y > 0$ such that*

(a)

$$\begin{bmatrix} N_o & 0 \\ 0 & I \end{bmatrix}^* \begin{bmatrix} A^*X + XA & XB_1 & C_1^* \\ B_1^*X & -I & D_{11}^* \\ C_1 & D_{11} & -I \end{bmatrix} \begin{bmatrix} N_o & 0 \\ 0 & I \end{bmatrix} < 0 ;$$

(b)

$$\begin{bmatrix} N_c & 0 \\ 0 & I \end{bmatrix}^* \begin{bmatrix} AY + YA^* & YC_1^* & B_1 \\ C_1Y & -I & D_{11} \\ B_1^* & D_{11}^* & -I \end{bmatrix} \begin{bmatrix} N_c & 0 \\ 0 & I \end{bmatrix} < 0 ;$$

(c)

$$\begin{bmatrix} X & I \\ I & Y \end{bmatrix} \ge 0,$$

where N_o and N_c are full rank matrices whose images satisfy

$$\operatorname{Im} N_o = \operatorname{Ker} \begin{bmatrix} C_2 & D_{21} \end{bmatrix}$$
$$\operatorname{Im} N_c = \operatorname{Ker} \begin{bmatrix} B_2^* & D_{12}^* \end{bmatrix} .$$

We note that if any of the above kernels is equal to the trivial subspace, then the result must be modified by simply eliminating the corresponding LMI ((a) or (b)) from our conditions. See Remark 7.3.

Proof. Suppose a controller exists; then by Lemma 7.8 a controller exists if and only if the inequalities

$$N_P^* T_{X_{cl}} N_P < 0 \quad \text{and} \quad N_Q^* H_{X_{cl}} N_Q < 0$$

hold for some symmetric, positive definite matrix X_{cl} in $\mathbb{R}^{(n+n_K)\times(n+n_K)}$. By Lemma 7.8 these LMIs being satisfied imply that (a) and (b) are met. Also invoking Lemma 7.9 we see that (c) is satisfied.

Showing that (a-c) imply the existence of a synthesis is essentially the reverse process. We choose $n_K \ge n$, in this way the rank condition in Lemma 7.9 is automatically satisfied, and thus there exists an X_{cl} in

$\mathbb{R}^{(n+n_K) \times (n+n_K)}$ which satisfies (7.13). The proof is now completed by using X_{cl} and (a-b) together with Lemma 7.8. ∎

This theorem gives us exact conditions under which a solution exists to our H_∞ synthesis problem. Notice that in the sufficiency direction we required that $n_k \geq n$, but clearly it suffices to choose $n_k = n$. In other words a synthesis exists if and only if one exists with state dimension $n_K = n$.

What if we want controllers of order n_K less than n? Then clearly from the above proof we have the following characterization.

Corollary 7.11. *A synthesis of order n_k exists for the H_∞ problem, if and only if there exist symmetric matrices $X > 0$ and $Y > 0$ satisfying (a), (b), and (c) in Theorem 7.10 plus the additional constraint*

$$\text{rank} \begin{bmatrix} X & I \\ I & Y \end{bmatrix} \leq n + n_K .$$

Unfortunately this constraint is not convex when $n_K < n$, so this says that in general the reduced order H_∞ problem is computationally much harder than the full order problem. Nevertheless, the above explicit condition can be exploited in certain situations.

7.3 Controller reconstruction

The results of the last section provide us with an explicit way to determine whether a synthesis exists which solves the H_∞ problem. Implicit in our development is a method to construct controllers when the conditions of Theorem 7.10 are met. We now outline this procedure, which simply retraces our steps so far.

Suppose X and Y have been found satisfying Theorem 7.10; then by Lemma 7.9 there exists a matrix $X_{cl} \in \mathbb{R}^{n \times n_K}$ satisfying

$$X_{cl} = \begin{bmatrix} X & ? \\ ? & ? \end{bmatrix} \quad \text{and} \quad X_{cl}^{-1} = \begin{bmatrix} Y & ? \\ ? & ? \end{bmatrix} .$$

From the proof of the lemma we can construct X_{cl} by finding a matrix $X_2 \in \mathbb{R}^{n \times n_K}$ such that $X - Y^{-1} = X_2 X_2^*$. Then

$$X_{cl} = \begin{bmatrix} X & X_2^* \\ X_2 & I \end{bmatrix}$$

has the properties desired above. As seen before, the order n_K need be no larger than n, and in general can be chosen to be the rank of $X - Y^{-1}$.

Next by Lemma 7.2 we know that there exists a solution to

$$H_{X_{cl}} + Q^* J^* P_{X_{cl}} + P_{X_{cl}}^* J Q < 0 ,$$

and that any such solution J provides the state space realization for a feasible controller K. The solution of this LMI can be accomplished using standard techniques, and there is clearly an open set of solutions J.

Exercises

7.1. Prove Lemma 7.1.

7.2. Consider the state space system

$$\dot{x} = Ax + Bw, \qquad x(0) = x_0,$$
$$z = Cx,$$

and suppose that $\left\| \left[\begin{array}{c|c} A & B \\ \hline C & 0 \end{array} \right] \right\|_\infty < 1$. Let X be the stabilizing solution of the Riccati equation

$$A^*X + XA + C^*C + XBB^*X = 0.$$

(a) Show that the trajectories of the system satisfy the identity

$$|z(t)|^2 - |w(t)|^2 = -|w(t) - B^*Xx(t)|^2 - \frac{d}{dt}\left(x(t)^*Xx(t) \right).$$

(b) Show that

$$\sup_{w \in L_2[0,\infty)} \left(\|z\|^2 - \|w\|^2 \right) = x_0^*Xx_0,$$

and find the signal $w(t)$ which achieves that optimum.

7.3. Generalization of the KYP Lemma. Let A be a Hurwitz matrix, and let $\Psi = \begin{bmatrix} Q & S \\ S^* & R \end{bmatrix}$ be a symmetric matrix with $R > 0$. We define

$$\hat{\psi}(j\omega) = \begin{bmatrix} (j\omega I - A)^{-1}B \\ I \end{bmatrix}^* \Psi \begin{bmatrix} (j\omega I - A)^{-1}B \\ I \end{bmatrix}$$

Show that the following are equivalent:

(i) $\hat{\psi}(j\omega) \geq \epsilon I$ for all $\omega \in \mathbb{R}$ and some $\epsilon > 0$.

(ii) The Hamiltonian matrix

$$H = \begin{bmatrix} A - BR^{-1}S^* & BR^{-1}B^* \\ Q - SR^{-1}S^* & -(A - BR^{-1}S^*)^* \end{bmatrix}$$

is in the domain of the Riccati operator.

(iii) The LMI

$$\begin{bmatrix} A^*X + XA & XB \\ B^*X & 0 \end{bmatrix} + \Psi > 0$$

admits a symmetric solution X.

(iv) There exists a quadratic storage function $V(x) = x^* \bar{X} x$ such
the dissipation inequality

$$
\dot{V} \le \begin{bmatrix} x(t) \\ w(t) \end{bmatrix}^* \begin{bmatrix} Q & S \\ S^* & R - \epsilon I \end{bmatrix} \begin{bmatrix} x(t) \\ w(t) \end{bmatrix}
$$

is satisfied over any solutions to the equation $\dot{x} = Ax + Bw$.

Hint: The method of proof from §7.1.1 can be replicated here.

7.4. **Spectral factorization.** This exercise is a continuation of the previous
one on the KYP Lemma. We take the same definitions for ψ, H, etc.,
and assume the above equivalent conditions are satisfied. Now set

$$
\hat{M}(s) = \left[\begin{array}{c|c} A & B \\ \hline R^{-\frac{1}{2}}(S^* - B^* X_0) & R^{\frac{1}{2}} \end{array} \right],
$$

where $X_0 = \mathrm{Ric}(H)$. Show that $\hat{M}(s) \in RH_\infty$, $\hat{M}(s)^{-1} \in RH_\infty$, and
the factorization

$$
\hat{\psi}(j\omega) = \hat{M}(j\omega)^* \hat{M}(j\omega)
$$

holds for every $\omega \in \mathbb{R}$.

7.5. **Mixed H_2/H_∞ control.**

(a) We are given a stable system with the inputs partitioned in two
channels w_1, w_2 and a common output z:

$$
\hat{P} = [\hat{P}_1 \quad \hat{P}_2] = \left[\begin{array}{c|cc} A & B_1 & B_2 \\ \hline C & D & 0 \end{array} \right].
$$

Suppose there exists $X > 0$ satisfying

$$
\begin{bmatrix} A^*X + XA & XB_1 & C^* \\ B_1^*X & -I & D^* \\ C & D & -I \end{bmatrix} < 0, \tag{7.16}
$$

$$
\mathrm{Tr}(B_2^* X B_2) < \gamma^2. \tag{7.17}
$$

Show that P satisfies the specifications $\|\hat{P}_1\|_{H_\infty} < 1$ and
$\|\hat{P}_2\|_{H_2} < \gamma$. Is the converse true?

(b) We now wish to use part (a) for state-feedback synthesis. In
other words, given an open-loop system

$$
\dot{x} = A_0 x + B_1 w_1 + B_2 w_2 + B_0 u,
$$
$$
z = C_0 x + D w_1 + D_0 u,
$$

we want to find a state feedback $u = Fx$ such that the closed
loop satisfies (7.16)-(7.17). Substitute the closed-loop matrices
into (7.16); does this give an LMI problem for synthesis?

(c) Now modify (7.16) to an LMI in the variable X^{-1}, and show
how to replace (7.17) by *two* convex conditions in X^{-1} and an
appropriately chosen slack variable Z.

(d) Use part (c) to obtain a convex method for mixed H_∞/H_2 state feedback synthesis.

7.6. Show that in a *state feedback* H_∞ synthesis problem (i.e. when $y = x$ is the measurement), the controller can be taken to be static without any loss of performance.

7.7. As a special case of reduced order H_∞ synthesis, prove Theorem 4.20 on model reduction.

7.8. Prove Theorem 5.8 involving stabilization. *Hint:* reproduce the steps of the H_∞ synthesis proof, using a Lyapunov inequality in place of the KYP Lemma.

7.9. Connections to Riccati solutions for the H_∞ problem. Let

$$\hat{G}(s) = \left[\begin{array}{c|cc} A & B_1 & B_2 \\ \hline C_1 & 0 & D_{12} \\ C_2 & D_{21} & 0 \end{array}\right]$$

satisfy the normalization conditions

$$D_{12}^* \begin{bmatrix} C_1 & D_{12} \end{bmatrix} = \begin{bmatrix} 0 & I \end{bmatrix} \quad \text{and} \quad D_{21} \begin{bmatrix} B_1^* & D_{21}^* \end{bmatrix} = \begin{bmatrix} 0 & I \end{bmatrix}.$$

Notice that these (and $D_{11} = 0$) are part of the conditions we imposed in our solution to the H_2-optimal control in the previous chapter.

(a) Show that the H_∞ synthesis is equivalent to the feasibility of the LMIs $X > 0$, $Y > 0$ and

$$\begin{bmatrix} A^*X + XA + C_1^*C_1 - C_2^*C_2 & XB_1 \\ B_1^*X & -I \end{bmatrix} < 0,$$

$$\begin{bmatrix} AY + YA^* + B_1B_1^* - B_2B_2^* & YC_1^* \\ C_1Y & -I \end{bmatrix} < 0,$$

$$\begin{bmatrix} X & I \\ I & Y \end{bmatrix} \geq 0.$$

(b) Now denote $P = Y^{-1}$, $Q = X^{-1}$. Convert the above conditions to the following:

$$A^*P + PA + C_1^*C_1 + P(B_1B_1^* - B_2B_2^*)P < 0,$$
$$AQ + QA^* + B_1B_1^* + Q(C_1^*C_1 - C_2^*C_2)Q < 0,$$
$$\rho(PQ) \leq 1.$$

These are two *Riccati inequalities* plus a spectral radius coupling condition. Formally analogous conditions involving the corresponding Riccati *equations* can be obtained when the plant satisfies some additional technical assumptions. For details consult the references.

Notes and references

The H_∞ control problem was formulated in [184], and was motivated by the necessity for a control framework that could systematically incorporate errors in the plant model. At the time this had been a goal of control research for a number of years, and H_2 control seemed poorly suited for this task [36]. The main observation of [184] was that these requirements could be met by working in a Banach algebra such as H_∞, but not H_2 which lacks this structure. We will revisit this question in subsequent chapters.

The formulation of the H_∞ problem precipitated an enormous research effort into its solution. The initial activity was based on the parametrization of stabilizing controllers discussed in Chapter 5, which reduced the problem to approximation in analytic function space. This problem was solved in the multivariable case by a combination of function theory and state space methods, notably Riccati equations. For an extensive account of this approach to the H_∞ problem, see [56]. Recent extensions to infinite dimensional systems can be found in [54].

Ultimately, these efforts led to a solution to the H_∞ problem in terms of two Riccati equations and based entirely on state space methods [38]; see the books [67, 188] for an extensive presentation of this approach and historical references. This solution has close ties to the theory of differential games (see [10]). For extensions of the Riccati equation method to distributed parameter systems, see [162].

One of the drawbacks of the Riccati equation theory was that it required unnecessary rank conditions on the plant; the ensuing research in removing such restrictions [26, 154] led to the use of Riccati inequalities [142, 145] which pointed in the direction of LMIs. Complete LMI solutions to the problem with no unnecessary system requirements appeared in [58, 112]; these papers form the basis for this chapter, particularly the presentation in [58].

The LMI solution has, however, other advantages beyond this regularity question. In the first place, the solution method puts H_∞ synthesis on a common footing with other performance specifications, namely those which involve a closed loop Lyapunov matrix. This has led to applications to multiobjective control design; see [146] and references therein.

Also, the LMI solution has led to powerful generalizations: in [29] a more general version is solved where spatial constraints can be specified; [46] solves the time varying and periodic problems; finally, the extension of this solution to multi-dimensional systems forms the basis of linear parameter-varying control, a important method for gain-scheduling design [112]. The latter two generalizations are considered in Chapter 11.

Reference [24] provides methods and results showing the equivalence of time-varying and time invariant controllers in a norm-based context.

8
Uncertain Systems

In the last three chapters we have developed synthesis techniques for feedback systems where the plant model was completely specified, in the sense that given any input there is a uniquely determined output. Also our plant models were linear, time invariant, and finite dimensional. We now return our focus to *analysis*, but move beyond our previous restriction of having complete system knowledge to the consideration of *uncertain* systems.

In a narrow sense, uncertainty arises when some aspect of the system model is not completely known at the time of analysis and design. A typical example here is the value of a parameter which may vary according to operating conditions. More generally, such unpredictability is present because of fine-scale effects that cannot be accurately characterized. However as discussed in the introduction to this course, we will use the term *uncertainty* in a broader sense to include also the result of deliberate under-modeling, when this occurs to avoid very detailed and complex models.

To illustrate this latter point, let us briefly and informally examine some of the issues involved with modeling a complex system. In the figure below we have a conceptual illustration of a complex input–output system. As depicted it has an input $\bar{w}$ and an output $\bar{z}$. A complete description of such a system is given by the set of input–output pairs

$$\mathcal{P} = \{(\bar{w}, \bar{z}) : \bar{z} \text{ is the output given input } \bar{w}\}.$$

Here the complex system can be nonlinear or infinite dimensional, but is assumed to be deterministic.

Now suppose we attempt to model this complex system with a mapping G. Then similar to the complex system, we can describe this model by an

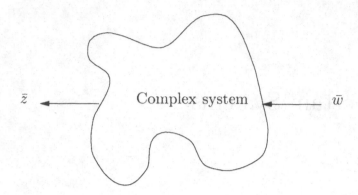

input–output set

$$\mathcal{R}_G = \{(w, z) : z = Gw\} \ .$$

Our intuitive notion of a good model is, given an input–output pair $(\bar{w}, \bar{z})$ in the set of system behaviors $\mathcal{P}$ and the corresponding model pair $(\bar{w}, z) \in \mathcal{R}_G$ that

z is a good approximation to $\bar{z}$.

If, however, G is much simpler than the system it is intended to model, this is clearly an unreasonable expectation for all possible inputs $\bar{w}$.

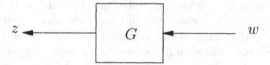

Now consider instead a set $\mathcal{R}$ of input–output pairs generated by a *family* $\mathcal{G}$ which is a specified set of input–output maps.

$$\mathcal{R} = \{(w, z) : \text{ there exists } G \in \mathcal{G} \text{ such that } z = Gw\}. \qquad (8.1)$$

In words this set is just the union of the sets $\mathcal{R}_G$ when G ranges over the set $\mathcal{G}$. Such a set $\mathcal{R}$ is a *relation* with respect to input–output pairs, but may no longer specify a function; namely, there could be many pairs $(w, z_0), (w, z_1), \ldots, (w, z_n)$, with distinct z_k, associated with a given input w. Notice this can never be the case in set $\mathcal{R}_G$. When would $\mathcal{R}$ provide a good model for the system given by $\mathcal{P}$? Descriptively, the basic requirement would be that

most elements of $\mathcal{P}$ are close to some element of $\mathcal{R}$.

In this way $\mathcal{R}$ would approximately contain or cover the complex model (strict covering $\mathcal{P} \subset \mathcal{R}$ is also desirable, but may be more difficult to ascertain). Now the key observation is that this requirement can indeed be satisfied even when the description of $\mathcal{G}$ is much simpler than the complex system; for example, each G can be a linear mapping, yet the system to be modeled could be significantly nonlinear.

Example:

Consider the static input–output system specified by the set

$$\mathcal{P} = \{(w, z) : z(t) = f(w(t)), \text{ for each } t\},$$

where f is a highly complex nonlinear function, but is known to satisfy the so-called sector bound condition

$$|f(w(t))| \leq |w(t)|, \text{ for each } t.$$

Suppose we focus on signals in L_2. Then clearly the relation

$$\mathcal{R} = \{(w, z) : \|z\| \leq \|w\|\}$$

will satisfy $\mathcal{R} \supset \mathcal{P}$ and thus cover the complex behavior. Now it is easily shown (see Lemma 8.4 below) that $\mathcal{R}$ has the form (8.1) where $\mathcal{G}$ is the unit ball of *linear* operators on L_2. Also, recall that in Chapter 0 we discussed some additional related examples. □

What is the price paid for this containment procedure? Mainly, that the relation $\mathcal{R}$ often describes a much larger behavior than $\mathcal{P}$; in particular it is usually *not* true that "most elements of $\mathcal{R}$ are close to some element of $\mathcal{P}$." In the above example, pairs $(w, z) \in \mathcal{R}$ may, for instance, have the two signals supported in disjoint time intervals, a situation which is not present in $\mathcal{P}$.

In other words, this modeling technique decreases *complexity* by introducing *uncertainty* into the system description. This kind of tradeoff is implicitly present in most of engineering modeling. What is special about the robust control theory that follows is that the uncertainty is made explicit in terms of a set of input–output models $\mathcal{G}$. For the remainder of the chapter we study an important way to describe such sets, and provide methods to analyze such descriptions. In the next chapter we will explicitly relate our current work to robust stability and robust performance of complex feedback systems, and will discuss synthesis under these conditions.

8.1 Uncertainty modeling and well-connectedness

In this section we introduce precisely the type of uncertain systems we will be studying. The idea is to parametrize the set $\mathcal{G}$ from the above discussion using perturbations around a nominal mapping. Then by varying the nature of these perturbations we can alter the characteristics of the set $\mathcal{G}$ and ultimately the set of input–output pairs $\mathcal{R}$. We will seldom need to directly refer to the sets $\mathcal{G}$ and $\mathcal{R}$; however, it is important to keep the above motivation for our system models in mind as they will help us to chart our course. The basic setup we now use is shown in Figure 8.1.

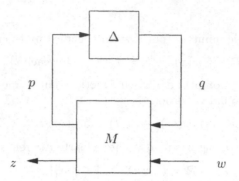

Figure 8.1. Uncertain system model

Here M and Δ are bounded operators on $L_2[0, \infty)$, and w the system input is in $L_2[0, \infty)$. The picture represents the map $w \mapsto z$, which is formally defined by the equations

$$q = \Delta p$$
$$\begin{bmatrix} p \\ z \end{bmatrix} = \begin{bmatrix} M_{11} & M_{12} \\ M_{21} & M_{22} \end{bmatrix} \begin{bmatrix} q \\ w \end{bmatrix},$$

where M is compatibly partitioned with respect to the inputs and outputs. More precisely, M is an operator between Cartesian products of L_2 spaces, $M : L_2^{m_1} \times L_2^{m_2} \to L_2^{p_1} \times L_2^{p_2}$, and each M_{ik} is in $\mathcal{L}(L_2)$ of appropriate dimensions. The notation above allows us to work with these operators formally as if they were matrices.

Example:

If $\begin{bmatrix} I & 0 \\ Q & I \end{bmatrix} : L_2^{m_1} \times L_2^{m_2} \to L_2^{m_1} \times L_2^{m_2}$ is a partitioned operator, it is nonsingular and its inverse is given by

$$\begin{bmatrix} I & 0 \\ Q & I \end{bmatrix}^{-1} = \begin{bmatrix} I & 0 \\ -Q & I \end{bmatrix}.$$

This can be directly verified by composing the above operators. □

Our goal will be to understand the possible maps $w \mapsto z$ when all that is known about Δ is that it resides in a pre-specified subset $\mathbf{\Delta}$ of the bounded linear operators on L_2. We will consider two fundamental types of sets in this chapter which are specified by spatial structure, norm bounds and their dynamical characteristics.

Before embarking on this specific investigation let us quickly look at the algebraic form of maps that can be obtained using the above arrangement.

Observe that if $(I - M_{11}\Delta)$ is nonsingular, then

$$w \mapsto z = M_{22} + M_{21}\Delta(I - M_{11}\Delta)^{-1}M_{12} =: \bar{S}(M, \Delta),$$

where this expression defines the notation $\bar{S}(M, \Delta)$, the *upper* star product. Recall that we defined the lower star product $\underline{S}(\cdot, \cdot)$ earlier when studying synthesis.

Examples:

We now present two examples of the most common types of uncertain models. Suppose that

$$M = \begin{bmatrix} 0 & I \\ I & G \end{bmatrix}.$$

Then we see

$$\bar{S}(M, \Delta) = G + \Delta.$$

That is the sets of operators generated by $\bar{S}(M, \Delta)$, when $\Delta \in \mathbf{\Delta}$ is simply that of an *additive perturbation* to an operator G.

Similarly let

$$M = \begin{bmatrix} 0 & G \\ I & G \end{bmatrix}$$

and get that

$$\bar{S}(M, \Delta) = G + \Delta G = (I + \Delta)G,$$

which is a *multiplicative perturbation* to the operator G.

□

Two examples have been presented above, which show that the setup of Figure 8.1 can be used to capture two simple sets. It turns out that this arrangement has a surprisingly rich number of possibilities, which are prescribed by choosing the form of the operator M and the uncertainty or perturbation set $\mathbf{\Delta}$. One reason for this is that this modeling technique is ideally suited to handle system interconnection (e.g. cascade or feedback), as discussed in more detail below. The exercises at the end of the chapter will also highlight some of these possibilities. First, though, we lay the basis for a rigorous treatment of these representations by means of the following definition.

Definition 8.1. *Given an operator M and a set $\mathbf{\Delta}$, the uncertain system in Figure 8.1 is robustly well-connected if*

$$I - M_{11}\Delta \text{ is nonsingular for all } \Delta \in \mathbf{\Delta}.$$

We will also use the terminology $(M_{11}, \mathbf{\Delta})$ is robustly well-connected to refer to this property.

When the uncertain system is robustly well-connected, we are assured that the map $w \mapsto z$ of Figure 8.1 is bounded and given by $\bar{S}(M, \Delta)$, for every $\Delta \in \mathbf{\Delta}$. At first sight, we may get the impression that robust well-connectedness is just a technicality. It turns out, however, that it is the central question for robustness analysis, and many fundamental system properties reduce to robust well-connectedness; we will, in fact, dedicate this chapter to answering this question in a variety of cases. Subsequently, in Chapter 9 these techniques will be applied to some important problems of robust feedback design.

A simple first case of evaluation of robust well-connectedness is when the set $\mathbf{\Delta}$ is the unit ball in operator space. In this case of so-called *unstructured uncertainty*, the analysis can be reduced to a small gain property. The term unstructured uncertainty is used because we are not imposing any structure on the perturbations considered except that they are contractive.

Theorem 8.2. *Let Q be an operator and $\mathbf{\Delta} = \{\Delta \in \mathcal{L}(L_2) : \|\Delta\| \leq 1\}$. Then $I - Q\Delta$ is nonsingular for all $\Delta \in \mathbf{\Delta}$ if and only if*

$$\|Q\| < 1.$$

Proof. The "if" direction: observe that for any $\Delta \in \mathbf{\Delta}$, we have the norm inequalities

$$\|Q\Delta\| \leq \|Q\|\|\Delta\| < 1.$$

Then it follows from the Small gain theorem in Chapter 3 that $I - Q\Delta$ is invertible.

For the "only if" direction, we must show that if $\|Q\| \geq 1$, then there exists $\Delta \in \mathbf{\Delta}$ with $I - Q\Delta$ singular. From our work in Chapter 3, we obtain the spectral radius condition

$$\rho(QQ^*) = \|Q\|^2 = \|Q^*\|^2 \geq 1,$$

where QQ^* has only non-negative spectral elements, so $I\lambda - QQ^*$ is singular with $\lambda = \|Q\|^2$.

Dividing by λ we see that $I - QQ^*\lambda^{-1}$ is singular, and thus we set $\Delta = \lambda^{-1}Q^*$, which is contractive since $\|\Delta\| = \|Q^*\|^{-1} \leq 1$. ∎

The preceding result reduces the analysis of whether the system of Figure 8.1 is robustly well-connected under unstructured uncertainty, to evaluating the norm of M_{11}. If for instance M_{11} is a finite dimensional LTI operator, a computational evaluation follows from the KYP lemma of Chapter 7. In the sequel we will refer to the norm inequality in Theorem 8.2 as the small gain *test* or *condition*.

We remark that there is nothing special about perturbations of size one. If the uncertainty set were specified as $\{\Delta \in \mathcal{L}(L_2) : \|\Delta\| \leq \beta\}$, the corresponding test would be $\|M_{11}\| < 1/\beta$; now, since a normalizing constant can always be included in the description of M, we will assume from now on that this normalization has already been performed and our uncertainty balls are of unit size.

Therefore the analysis of well-connectedness is simple for unstructured uncertainty. There are, however, usually important reasons to impose additional *structure* on the perturbation set $\mathbf{\Delta}$, in addition to a norm bound. We explain two such reasons here; the first is illustrated by the following example.

Example:

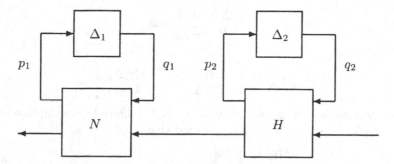

Figure 8.2. Cascade of uncertain systems

Figure 8.2 depicts the cascade interconnection of two uncertain systems of our standard form; i.e., a composition of the uncertain operators. It is a routine exercise, left for the reader, to show that the interconnection is equivalent to a system of the form of Figure 8.1, for an appropriately chosen M and

$$ p = \begin{bmatrix} p_1 \\ p_2 \end{bmatrix}, \quad q = \begin{bmatrix} q_1 \\ q_2 \end{bmatrix}, \quad \Delta = \begin{bmatrix} \Delta_1 & 0 \\ 0 & \Delta_2 \end{bmatrix}. $$

Therefore we find that the uncertainty set $\mathbf{\Delta}$ for the composite system will have, by construction, a block diagonal structure.   □

Thus we see that system *interconnection* generates structure; while components can be modeled by unstructured balls as described in previous examples, a spatially structured uncertainty set can be used to reflect the complexity of their interconnection. We remark that other forms of interconnection (e.g., feedback) can also be accommodated in this setting.

We now move to a second source of uncertainty structure; this arises when in addition to robust well-connectedness, one wishes to study the norm of the resulting operator set $\bar{S}(M, \mathbf{\Delta})$. That is, assuming $(M_{11}, \mathbf{\Delta})$ is well-connected, does

$$ \|\bar{S}(M, \Delta)\| < 1 $$

hold for all $\Delta \in \mathbf{\Delta}$. We have the following result, which states that the latter contractiveness problem can be recast as a well-connectedness question.

Proposition 8.3. *Define the perturbation set*

$$\boldsymbol{\Delta_p} = \left\{ \Delta = \begin{bmatrix} \Delta_u & 0 \\ 0 & \Delta_p \end{bmatrix} : \Delta_u \in \boldsymbol{\Delta}, \quad and \ \Delta_p \in \mathcal{L}(L_2) \ with \ \|\Delta_p\| \le 1 \right\}.$$

Then $(M_{11}, \boldsymbol{\Delta})$ *is robustly well-connected and*

$$\|\bar{S}(M, \Delta_u)\| < 1, \quad for \ all \ \Delta_u \in \boldsymbol{\Delta},$$

if and only if $I - M\Delta$ *is invertible for all* $\Delta \in \boldsymbol{\Delta_p}$.

Proof. We first demonstrate "only if." For any

$$\Delta = \begin{bmatrix} \Delta_u & 0 \\ 0 & \Delta_p \end{bmatrix} \in \boldsymbol{\Delta_p},$$

we have

$$I - M\Delta = \begin{bmatrix} I - M_{11}\Delta_u & -M_{12}\Delta_p \\ -M_{21}\Delta_u & I - M_{22}\Delta_p \end{bmatrix}.$$

Since by definition of $\boldsymbol{\Delta_p}$ we know $\Delta_u \in \boldsymbol{\Delta}$, it follows by assumption that $(I - M_{11}\Delta_u)^{-1}$ exists. It is easily verified that

$$\begin{bmatrix} I - M_{11}\Delta_u & -M_{12}\Delta_p \\ -M_{21}\Delta_u & I - M_{22}\Delta_p \end{bmatrix} = \begin{bmatrix} I & 0 \\ -M_{21}\Delta_u(I - M_{11}\Delta_u)^{-1} & I \end{bmatrix} \cdot \quad (8.2)$$
$$\begin{bmatrix} I - M_{11}\Delta_u & -M_{12}\Delta_p \\ 0 & I - \bar{S}(M, \Delta_u)\Delta_p \end{bmatrix}$$

Since the first operator on the right of the equality is nonsingular, $(I - M\Delta)^{-1}$ exists if $I - \bar{S}(M, \Delta_u)\Delta_p$ is nonsingular. This follows from Theorem 8.2, setting $Q = \bar{S}(M, \Delta_u)$ which by assumption is contractive.

The "if" direction: suppose $\Delta_u \in \boldsymbol{\Delta}$, then clearly

$$\Delta := \begin{bmatrix} \Delta_u & 0 \\ 0 & 0 \end{bmatrix} \in \boldsymbol{\Delta_p}$$

and

$$I - M\Delta = \begin{bmatrix} I - M_{11}\Delta_u & 0 \\ -M_{21}\Delta_u & I \end{bmatrix}$$

is nonsingular by assumption. Therefore $(I - M_{11}\Delta_u)^{-1}$ exists for all Δ_u in the set $\boldsymbol{\Delta}$.

Now looking at (8.2) for any fixed $\Delta_u \in \boldsymbol{\Delta}$, and all Δ_p satisfying $\|\Delta_p\| \le 1$, we have that

$$I - \bar{S}(M, \Delta_u)\Delta_p \text{ is nonsingular }.$$

Once again, Theorem 8.2 implies that we must have $\|\bar{S}(M, \Delta_u)\| < 1$ for all $\Delta_u \in \boldsymbol{\Delta}$. ∎

Thus we have identified two reasons for the introduction of spatial, block-diagonal structure to our perturbation set $\boldsymbol{\Delta}$. The first is that if a system is formed by the interconnection of subsystems, then this structure arises immediately owing to perturbations which may be present in the subsystems. Our second motivation is because this will enable us to evaluate the induced norm of the resulting set of mappings. The next section is devoted to studying the implications of this structure in the analysis of well-connectedness. Later on in this chapter we will explore additional structural constraints which can be imposed on our uncertainty set.

8.2 Arbitrary block-structured uncertainty

This section is devoted to developing an analysis test for robust well-connectedness when the only restriction imposed on the uncertainty is that it have a block-diagonal spatial structure. In particular, the main question is whether one can generalize the small gain test which was valid under unstructured perturbations.

We start by defining the block-diagonal uncertainty set $\boldsymbol{\Delta}_a$, which is a spatially structured subset of the unit ball in $\mathcal{L}(L_2^m)$.

$$\boldsymbol{\Delta}_a = \{\operatorname{diag}(\Delta_1, \ldots, \Delta_d) : \Delta_k \in \mathcal{L}(L_2^{m_k}) \text{ and } \|\Delta_k\| \leq 1 \text{ holds.}\},$$

where the spatial dimensions m_k are fixed and satisfy $m = m_1 + \cdots + m_d$. The above notation means that every perturbation Δ in this set $\boldsymbol{\Delta}_a$ is of the form

$$\Delta = \begin{bmatrix} \Delta_1 & 0 & \cdots & 0 \\ 0 & \ddots & \ddots & \vdots \\ \vdots & \ddots & \ddots & 0 \\ 0 & \cdots & 0 & \Delta_d \end{bmatrix},$$

where each Δ_k can be any contractive linear operator on the space $L_2^{m_k}$. This set constrains the spatial structure of its members, but allows any other dynamical characteristics.

In line with our discussion at the beginning of the chapter, it is useful to study the relation formed by the input–output pairs which are parametrized by $\boldsymbol{\Delta}_a$. Referring to Figure 8.1, consider the set

$$\mathcal{R}_a = \{(p, q) \in L_2 : \|p_k\| \geq \|q_k\|, \text{ where } k = 1, \ldots, d\}$$

Here p_k, q_k denote the parts of the vectors p, q which correspond to the spatial structure $\boldsymbol{\Delta}_a$. We claim that

$$\mathcal{R}_a \text{ is equal to the set } \{(p, q) \in L_2 : q = \Delta p, \text{ for some } \Delta \in \boldsymbol{\Delta}_a\}; \quad (8.3)$$

this follows from the following lemma applied to each block.

Lemma 8.4. *Suppose p and q are elements of $L_2[0, \infty)$. The following are equivalent.*

(i) $\|p\| \geq \|q\|$.

(ii) There exists an operator $\Delta \in \mathcal{L}(L_2)$ with $\|\Delta\| \leq 1$, such that $\Delta p = q$.

Proof. It is clear that (ii) implies (i), so we focus on the other direction. Assume that p is nonzero; otherwise, the result is immediate. Define Δ by

$$\Delta u := q \frac{\langle p, u \rangle}{\|p\|^2}.$$

Now $\Delta p = q$ as required, and the norm bound on Δ follows by an application of the Cauchy–Schwartz inequality.   ∎

This lemma completes the example in the introduction of this chapter, where a relation in terms of a single norm constraint was used to cover a complex nonlinearity. Here the relation $\mathcal{R}_a$ allows us to impose a finite number of such constraints.

A word on terminology: because the operators in $\boldsymbol{\Delta}_a$ are linear and in general not time invariant, this structure is sometimes called linear time-varying (LTV) uncertainty. We have used the more general term arbitrary block-structured uncertainty since

(i) from a modeling perspective, this uncertainty set is often not motivated by time variation. As explained above, we could well be modeling a nonlinearity;

(ii) from a mathematical perspective, the *relation* $\mathcal{R}_a$ is indeed *time invariant*; that is, if $(p, q) \in \mathcal{R}_a$, then $(S_\tau p, S_\tau q) \in \mathcal{R}_a$, where S_τ is the shift on $L_2[0, \infty)$. While a linear *parametrization* of $\mathcal{R}_a$ involves time-varying operators, we will see below that the time invariance of $\mathcal{R}_a$ is indeed the central property required for our analysis.

We now proceed to the study of robust well-connectedness over the set $\boldsymbol{\Delta}_a$. Throughout we will assume that the nominal system M is an LTI operator. Our objective is to study under what conditions $(M_{11}, \boldsymbol{\Delta}_a)$ is robustly well-connected. Recall that by this we mean

$$I - M_{11}\Delta \quad \text{is invertible, for all } \Delta \in \boldsymbol{\Delta}_a.$$

For convenience, we will drop the subindex from M_{11} in the following discussion and simply write M.

8.2.1   *A scaled small gain test and its sufficiency*

We begin with the simple observation that since $\boldsymbol{\Delta}_a$ is a subset of the unit ball of $\mathcal{L}(L_2)$, then clearly the small gain condition

$$\|M\| < 1$$

for unstructured uncertainty must be a sufficient test for robust well-connectedness over $\boldsymbol{\Delta}_a$. However, the above norm condition is in general conservative since the block-diagonal restriction did not come into play. In other words the condition $\|M\| \geq 1$ does not necessarily imply that $I - M\Delta$ is singular for some perturbation in $\mathbf{\Delta}_a$.

A basic tool to reduce the conservatism of the contractiveness condition above is to introduce operators which commute with the perturbations. We define the *commutant* of the uncertainty set $\mathbf{\Delta}_a$ by

$$\mathbf{\Theta}_a = \{\Theta \in \mathcal{L}(L_2) : \Theta \text{ is invertible and } \Theta\Delta = \Delta\Theta \text{ for all } \Delta \in \mathbf{\Delta}_a\}.$$

Notice that the inverse Θ^{-1} of any $\Theta \in \mathbf{\Theta}_a$ automatically has the same commuting property, and we can write the identity

$$\Theta(I - M\Delta)\Theta^{-1} = I - \Theta M \Theta^{-1}\Delta, \text{ for all } \Delta \in \mathbf{\Delta}_a.$$

Thus $I - M\Delta$ is singular if and only if $I - \Theta M \Theta^{-1}\Delta$ is singular. This means that if we can find an element Θ of the commutant satisfying

$$\|\Theta M \Theta^{-1}\| < 1,$$

then we can guarantee that $(M, \mathbf{\Delta}_a)$ is robustly well-connected simply by invoking the small gain condition. This motivates us to explicitly describe these commuting operators, which are often called *scalings*.

You will show in the exercises at the end of the chapter, that $\Theta \in \mathbf{\Theta}_a$ if and only if Θ is of the form

$$\Theta = \begin{bmatrix} \theta_1 I & 0 & \cdots & 0 \\ 0 & \ddots & \ddots & \\ \vdots & \ddots & \ddots & 0 \\ 0 & \cdots & 0 & \theta_d I \end{bmatrix}, \tag{8.4}$$

for some constant nonzero scalars θ_k in $\mathbb{C}$. Clearly since such an operator Θ is memoryless it can also be thought of as a matrix.

With this description we recapitulate our previous discussion as follows.

Proposition 8.5. *Suppose M is a bounded operator. If there exists Θ in $\mathbf{\Theta}_a$ such that*

$$\|\Theta M \Theta^{-1}\| < 1,$$

then $(M, \mathbf{\Delta}_a)$ is robustly well-connected.

It will also be convenient to consider the set of *positive* matrices of the same spatial structure; we define the set

$$\mathbf{P\Theta}_a = \{\Theta \in \mathbb{R}^{m \times m} : \Theta = \operatorname{diag}(\theta_1 I_{m_1}, \ldots, \theta_d I_{m_d}), \text{ where } \theta_k > 0\}. \tag{8.5}$$

Observe that $\mathbf{P\Theta}_a$ is a *convex cone* of matrices, as defined in Chapter 1. Also notice that:

- If $\Theta \in \mathbf{\Theta}_a$, then $\Theta^*\Theta \in \mathbf{P\Theta}_a$.

- Θ is in $P\Theta_a$ if and only if its square root $\Theta^{\frac{1}{2}}$ is in $P\Theta_a$.

These properties allow us to claim that in the above proposition it suffices to restrict attention to the positive elements of the commutant. We now state this equivalence more explicitly, together with two other restatements of the above test.

Proposition 8.6. *Suppose M is a time invariant bounded operator, with transfer function $\hat{M}(s) = C(Is - A)^{-1}B + D$, and A is Hurwitz of order n. Then the following are equivalent:*

(i) *there exists Θ in Θ_a such that $\|\Theta M\Theta^{-1}\| < 1$;*

(ii) *there exists Θ in $P\Theta_a$ such that $\|\Theta^{\frac{1}{2}} M\Theta^{-\frac{1}{2}}\| < 1$;*

(iii) *there exists Θ in $P\Theta_a$ such that*

$$M^*\Theta M - \Theta < 0; \tag{8.6}$$

(iv) *there exists Θ in $P\Theta_a$ and a symmetric $n \times n$ matrix $X > 0$ such that*

$$\begin{bmatrix} C^* \\ D^* \end{bmatrix} \Theta \begin{bmatrix} C & D \end{bmatrix} + \begin{bmatrix} A^*X + XA & XB \\ B^*X & -\Theta \end{bmatrix} < 0. \tag{8.7}$$

Here we have added the additional assumption that $\hat{M}$ is in RH_∞ for condition (iv); however, the first three conditions are equivalent for any operator M. The reformulation given in (iii) expresses our norm condition in terms of a convex operator inequality; we recall that (8.6) means the operator is negative definite, that is there exists $\epsilon > 0$ such that

$$\langle v, (M^*\Theta M - \Theta)v \rangle \leq -\epsilon \|v\|^2 \quad \text{for all } v \in L_2.$$

The last reformulation says that (i) can be reduced to an equivalent LMI feasibility condition.

Proof. We first show that (i) implies (iii). For this purpose we write

$$\|\Theta M\Theta^{-1}v\|^2 \leq (1 - \eta)\|v\|^2$$

for any $v \in L_2$ and some $\eta > 0$; equivalently, setting $u = \Theta^{-1}v$ we find that

$$\|\Theta Mu\|^2 \leq (1 - \eta)\|\Theta u\|^2$$

for any $u \in L_2$. Now by rewriting the above inequality we obtain

$$\langle u, (M^*\Theta^*\Theta M - \Theta^*\Theta)u \rangle \leq -\eta\|\Theta u\|^2 \leq -\frac{\eta}{\|\Theta^{-1}\|^2}\|u\|^2,$$

which implies (iii) since $\Theta^*\Theta \in P\Theta_a$.

The above argument can be essentially reversed to show that (iii) implies (ii); also, since $\Theta \in P\Theta_a$ implies $\Theta^{\frac{1}{2}} \in P\Theta_a \subset \Theta_a$, it is clear that (ii) implies (i).

This shows the equivalence of the first three conditions, with no reference to finite dimensionality or time invariance of the operator M. We now impose this restriction and show that (ii) implies (iv). Since M is LTI and finite dimensional, and Θ is constant, the norm condition in (i) amounts to the H_∞ norm condition

$$\|\Theta^{\frac{1}{2}}\hat{M}\Theta^{-\frac{1}{2}}\|_\infty = \left\|\left[\begin{array}{c|c} A & B\Theta^{-\frac{1}{2}} \\ \hline \Theta^{\frac{1}{2}}C & \Theta^{\frac{1}{2}}D\Theta^{-\frac{1}{2}} \end{array}\right]\right\|_\infty < 1.$$

Since A is Hurwitz, we can apply the KYP Lemma and conclude that there exists a symmetric matrix $X > 0$ such that

$$\begin{bmatrix} C^* \\ \Theta^{-\frac{1}{2}}D^* \end{bmatrix} \Theta \begin{bmatrix} C & D\Theta^{-\frac{1}{2}} \end{bmatrix} + \begin{bmatrix} A^*X + XA & XB\Theta^{-\frac{1}{2}} \\ \Theta^{-\frac{1}{2}}B^*X & -I \end{bmatrix} < 0.$$

Now left- and right-multiply this inequality by the symmetric matrix $\mathrm{diag}(I, \Theta^{\frac{1}{2}})$ to get

$$\begin{bmatrix} C^* \\ D^* \end{bmatrix} \Theta \begin{bmatrix} C & D \end{bmatrix} + \begin{bmatrix} A^*X + XA & XB \\ B^*X & -\Theta \end{bmatrix} < 0.$$

This is precisely the LMI in (8.7). Clearly the above steps can be reversed to establish that (iv) implies (ii). ∎

The preceding results provide a tractable means to establish robust well-connectedness over $\boldsymbol{\Delta}_a$. Indeed the existence of an appropriate scaling matrix $\Theta \subset \boldsymbol{P\Theta}_a$ can be determined from (8.7) by LMI computation. Since the identity matrix is a member of $\boldsymbol{P\Theta}_a$, there always exists a Θ in $\boldsymbol{P\Theta}_a$ such that $\|\Theta^{\frac{1}{2}}M\Theta^{-\frac{1}{2}}\| \leq \|M\|$. Therefore this scaled small gain test provides a less conservative test for robust well-connectedness than the previous condition $\|M\| < 1$. Our next task is to determine whether this scaling procedure has eliminated all conservatism from the small gain test. The main result of this section is that this is indeed the case: the scaled small gain condition is necessary as well as sufficient.

8.2.2 Necessity of the scaled small gain test

In this section we will concentrate our efforts on showing that the scaled small gain condition must hold if $(M, \boldsymbol{\Delta}_a)$ is a well-connected uncertain system. In order to achieve this goal we will employ some of the ideas and results from convex analysis introduced in §1.2.2. Our strategy will be to translate our problem to one in terms of the separation of two specific sets in $\mathbb{R}^d$. We will be able to show that well-connectedness implies that these sets are strictly separated, and that the scaled small gain condition is equivalent to the existence of a hyperplane that strictly separates these sets. We will then obtain our necessity result by showing that the strict separation of the sets implies the stronger condition that there exists a strictly separating hyperplane between them.

It will be advantageous to again look at the relation $\mathcal{R}_a$ introduced in the preliminary discussion of this section and defined in (8.3), rather than its parametrization by $\mathbf{\Delta}_a$. Also we will wish to describe our relation $\mathcal{R}_a$ in terms of *quadratic inequalities* of the form $\|p_k\|^2 - \|q_k\|^2 \geq 0$. Out of convenience we rewrite these as

$$\|E_k p\|^2 - \|E_k q\|^2 \geq 0, \qquad k = 1 \ldots d, \tag{8.8}$$

where the $m_k \times m$ matrix

$$E_k = \begin{bmatrix} 0 & \cdots & 0 & I & 0 & \cdots & 0 \end{bmatrix}$$

selects the kth block of the corresponding vector. The spatial dimensions of E_k are such that any element $\Theta = \mathrm{diag}(\theta_1 I, \ldots \theta_d I)$ in the set $\boldsymbol{P\Theta}_a$ can be written as

$$\Theta = \theta_1 E_1^* E_1 + \cdots \theta_d E_d^* E_d. \tag{8.9}$$

Inequalities such as the one in (8.8) are called integral quadratic constraints (IQCs); the name derives from the fact that they are quadratic in p and q, and involve the energy integral. We will have more to say about IQCs in Chapter 10.

We wish to study the interconnection of $\mathcal{R}_a$ to the nominal system M; for this purpose, we impose the equation $p = Mq$ on the expression (8.8) and define the *quadratic form* $\phi_k : L_2 \to \mathbb{R}$ by

$$\phi_k(q) := \|E_k M q\|^2 - \|E_k q\|^2 = \langle q, (M^* E_k^* E_k M - E_k^* E_k) q \rangle \tag{8.10}$$

With this new notation, we make the following observation.

Proposition 8.7. *If* $(M, \mathbf{\Delta}_a)$ *is robustly well-connected, then there cannot exist a nonzero* $q \in L_2$ *such that the following inequalities hold:*

$$\phi_k(q) \geq 0, \quad \textit{for each } k = 1, \ldots, d. \tag{8.11}$$

This proposition follows from the reasoning: if such a q existed, we would be assured that $(Mq, q) \in \mathcal{R}_a$. Thus by the earlier discussion, see Lemma 8.4, there would exist a $\Delta \in \mathbf{\Delta}_a$ such that

$$\Delta M q = q,$$

implying that the operator $(I - \Delta M)$ would be singular. And therefore, invoking Lemma 3.12, the operator $I - M\Delta$ would also be singular. Thus we have related robust well-connectedness to the infeasibility of the set of inequalities in (8.11).

We now wish to relate the quadratic forms ϕ_k to the scaled small gain condition. In fact this connection follows readily from version (iii) in Proposition 8.6 by expressing the elements of $\boldsymbol{P\Theta}_a$ as in (8.9), which leads to the identity

$$\langle q, (M^* \Theta M - \Theta) q \rangle = \sum_{k=1}^{d} \theta_k \langle q, (M^* E_k^* E_k M - E_k^* E_k) q \rangle = \sum_{k=1}^{d} \theta_k \phi_k(q).$$

The following result follows immediately by considering condition (iii) of Proposition 8.6; the reader can furnish a proof.

Proposition 8.8. *The equivalent conditions in Proposition 8.6 hold if and only if there exist scalars $\theta_k > 0$ and $\epsilon > 0$ such that*

$$\theta_1\phi_1(q) + \cdots + \theta_d\phi_d(q) \leq -\epsilon\|q\|^2, \text{ for all } q \in L_2. \tag{8.12}$$

Furthermore, if such a solution exists for (8.12), then there does not exist a nonzero q in L_2 satisfying (8.11).

We have thus related both robust well-connectedness and the scaled small gain test, to two different conditions involving the quadratic forms ϕ_k. The former condition states the infeasibility of the set of constraints (8.11); the latter condition is expressed as a single quadratic condition in (8.12) with *multipliers* θ_k, and implies the former. In the field of optimization a step of this sort, using multipliers to handle a set of quadratic constraints, is sometimes called the S-procedure. The S-procedure is termed *lossless* when there is no conservatism involved, namely, when the conditions are equivalent. In what follows we will show this is the case here for us.

To demonstrate this we first express our findings in geometric terms so that we can bring convex analysis to bear on the problem. To this end introduce the following subsets of $\mathbb{R}^d$:

$$\Pi = \{(r_1, \ldots, r_d) \in \mathbb{R}^d : \ r_k \geq 0, \text{ for each } k = 1, \ldots, d\};$$

$$\nabla = \{(\phi_1(q), \ldots, \phi_d(q)) \in \mathbb{R}^d : \ q \in L_2, \text{ with } \|q\| = 1\}.$$

Here Π is the positive *orthant* in $\mathbb{R}^d$, and ∇ describes the range of the quadratic forms ϕ_k as q varies in the unit sphere of L_2. For brevity we will write $\phi(q) := (\phi_1(q), \ldots, \phi_d(q))$ and $\theta = (\theta_1, \ldots \theta_d)$. Therefore with the standard inner product in $\mathbb{R}^d$, we have

$$\langle \theta, \phi(q) \rangle = \theta_1\phi_1(q) + \cdots + \theta_d\phi_d(q).$$

Now we are ready to interpret our conditions geometrically. From the discussion and Proposition 8.7 we see that robust well-connectedness implies that the closed set Π and the set ∇ are disjoint, that is $\Pi \cap \nabla = \emptyset$. Now ∇ is a bounded set but it is not necessarily closed. Therefore we cannot invoke Proposition 1.3 to conclude these sets are strictly separated when they are disjoint. Nonetheless this turns out to be true.

Proposition 8.9. *Suppose that (M, Δ_a) is robustly well-connected. Then the sets Π and ∇ are strictly separated, namely,*

$$D(\Pi, \nabla) := \inf_{r \in \Pi, y \in \nabla} |r - y| > 0.$$

This result states that not only are the sets Π and ∇ disjoint, but they are separated by a positive distance. Its proof is technically quite involved and adds little of conceptual significance to our discussion. For this reason the proof is relegated to Appendix B.

Turning now to the condition in (8.12), we see that it specifies the existence of a vector θ with positive entries and an $\epsilon > 0$, such that

$$\langle \theta, y \rangle \leq -\epsilon, \text{ for all } y \in \nabla.$$

In other words, the set ∇ is constrained to lie strictly inside the negative half-space defined by the hyperplane $\langle \theta, y \rangle = 0$. Notice also that since the entries θ_k of θ are positive, the set Π will automatically lie in the opposite half-space. We have the following result.

Proposition 8.10. *The conditions of Proposition 8.6 are satisfied if and only if there is a hyperplane strictly separating the sets ∇ and Π.*

Proof. From the definition of a strictly separating hyperplane, we need to show that the conditions in Proposition 8.6 are equivalent to the existence of a vector $\theta \in \mathbb{R}^d$, and scalars α and β, such that

$$\langle \theta, y \rangle \leq \alpha < \beta \leq \langle \theta, r \rangle \text{ holds, for all } y \in \nabla, r \in \Pi. \tag{8.13}$$

The "only if" direction follows from the above comments and Proposition 8.8 by choosing $\alpha = -\epsilon$ and $\beta = 0$.

To establish "if," we first invoke Proposition 8.8 to see that it is sufficient to prove that (8.12) is satisfied. Suppose that (8.13) holds for some θ, α and β. First note, from the structure of Π that we must have $\theta_k \geq 0$, for each k; otherwise we could make the corresponding r_k arbitrarily large and violate the bound $\beta \leq \langle \theta, r \rangle$. Hence, we see that $0 \leq \langle \theta, r \rangle$, for all r in Π; therefore without loss of generality we set $\beta = 0$. Observe this implies that $\alpha < 0$.

Finally we show there exists an $\epsilon > 0$ such that each of the entries θ_k can be perturbed to be strictly positive, satisfying a bound of the form $\langle \theta, y \rangle \leq -\epsilon < 0$, for all $y \in \nabla$. To see that this is always possible simply recall that each θ_k is non-negative, $\alpha < 0$ and ∇ is bounded. This establishes inequality (8.12). ■

Figure 8.3 contains an illustration of the preceding conditions for the case where $d = 2$. In part (a) of the figure we see the pair of sets ∇ and Π where they have a positive distance between them, as implied by robust well-connectedness according to Proposition 8.9. However, the depiction in (b) of Figure 8.3 imposes a strict *hyperplane* separation between the two sets, which is equivalent to the scaled small gain condition by the proposition just proved.

Consequently, if we wish to show that the conditions of Proposition 8.6 are *necessary* for robust well-connectedness, we need to prove that strict separation of the sets ∇ and Π *automatically* implies the existence of a strictly separating hyperplane. This leads us to think of *convex* separation theory, reviewed in §1.2.2. The closed positive orthant Π is clearly a convex set. As for the set ∇, we have the following result. Its proof will rely strongly

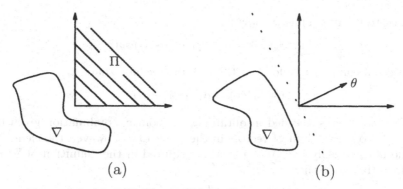

Figure 8.3. (a) $D(\Pi, \nabla) > 0$; (b) Π, ∇ separated by a hyperplane

on the *time invariance* of the operator M since it implies the forms ϕ_k also enjoy such a property.

Lemma 8.11. *The closure $\bar{\nabla}$ is convex.*

Proof. We introduce the notation

$$T_k := M^* E_k^* E_k M - E_k^* E_k,$$

which lets us write $\phi_k(q) = \langle T_k q, q \rangle$ from (8.10).

For each $k = 1, \ldots, d$, the operator T_k is self-adjoint and satisfies $S_\tau^* T_K S_\tau = T_k$, where S_τ is the usual time shift operator for $\tau \geq 0$. This follows from the time invariance of M and the fact that $S_\tau^* S_\tau = I$.

Choose two elements $y = \phi(q)$ and $\tilde{y} = \phi(\tilde{q})$ from the set ∇. By definition, $\|q\| = \|\tilde{q}\| = 1$. We wish to demonstrate any point on the line segment that joins them is an element of $\bar{\nabla}$. That is, for any $\alpha \in [0,1]$ we have $\alpha y + (1 - \alpha)\tilde{y} \in \bar{\nabla}$. Given such an α let

$$q_\tau := \sqrt{\alpha}\, q + \sqrt{1 - \alpha}\, S_\tau\, \tilde{q}.$$

Our first goal is to examine the behavior of $\phi(q_\tau)$ as τ tends to infinity. It is convenient to do this by considering each component $\phi_k(q_\tau)$. So for any given k we compute

$$\phi_k(q_\tau) = \langle T_k q_\tau, q_\tau \rangle$$
$$= \alpha \langle T_k q, q \rangle + (1 - \alpha)\langle T_k S_\tau \tilde{q}, S_\tau \tilde{q} \rangle + 2\sqrt{\alpha(1 - \alpha)}\mathrm{Re}\langle T_k q, S_\tau \tilde{q} \rangle. \tag{8.14}$$

We observe that the second term on the right-hand side satisfies

$$\langle T_k S_\tau \tilde{q}, S_\tau \tilde{q} \rangle = \langle S_\tau^* T_k S_\tau \tilde{q}, \tilde{q} \rangle = \langle T_k \tilde{q}, \tilde{q} \rangle = \phi_k(\tilde{q}).$$

With this observation we can rewrite (8.14) as

$$\phi_k(q_\tau) = \alpha\phi_k(q) + (1 - \alpha)\phi_k(\tilde{q}) + 2\sqrt{\alpha(1 - \alpha)}\mathrm{Re}\langle T_k q, S_\tau \tilde{q} \rangle.$$

Now we let $\tau \to \infty$. Clearly the inner product on the right converges to zero since the first element is fixed, and the support of the second is being shifted to infinity. Thus we have

$$\lim_{\tau \to \infty} \phi_k(q_\tau) = \alpha \phi_k(q) + (1 - \alpha)\phi_k(\tilde{q}).$$

Collect all components, for $k = 1, \ldots, d$, to conclude that

$$\lim_{\tau \to \infty} \phi(q_\tau) = \alpha y + (1 - \alpha)\tilde{y}.$$

Thus we have succeeded in obtaining the convex combination given by $\alpha y + (1 - \alpha)\tilde{y}$ as a limit of vectors in the range of ϕ; however, the elements q_τ do not generally have unit norm, as required in the definition of ∇; to address this, note that

$$\lim_{\tau \to \infty} \|q_\tau\|^2 = \alpha \|q\|^2 + (1 - \alpha)\|\tilde{q}\|^2 = 1.$$

This follows by the same argument as the above, replacing T_k by the identity operator. Thus the q_τ have asymptotically unit norm, which implies that

$$\lim_{\tau \to \infty} \phi \left(\frac{q_\tau}{\|q_\tau\|} \right) = \alpha y + (1 - \alpha)\tilde{y}.$$

Now by definition the elements on the left are in ∇, for every τ, so we conclude that

$$\alpha y + (1 - \alpha)\tilde{y} \in \bar{\nabla}.$$

Finally, by continuity the same will hold if we choose the original y and $\tilde{y}$ from $\bar{\nabla}$ rather than ∇; this establishes convexity.   ∎

We can now assemble all the results of this section to show that the scaled small gain condition is indeed necessary for well-connectedness of (M, Δ_a). This result is a direct consequence of the following.

- Suppose the uncertain system (M, Δ_a) is robustly well-connected, then by Proposition 8.9 we conclude that sets ∇ and Π are strictly separated.

- Since ∇ and Π are strictly separated, so must be their closures $\bar{\nabla}$ and $\bar{\Pi}$; now the positive orthant Π is convex, an so is $\bar{\nabla}$ by Lemma 8.11; therefore the conditions of Theorem 1.5 are satisfied, and there exists a strictly separating hyperplane; that is, there exists $\theta \in \mathbb{R}^d$ such that (8.13) is satisfied.

- Now by Proposition 8.10, any and all of the equivalent scaled small gain conditions of Proposition 8.6 must hold.

This argument clearly shows that well-connectedness implies the scaled small gain condition, and we already know the converse statement is true from Proposition 8.5. Thus we have obtained a necessary and sufficient condition for robustness analysis over Δ_a, which is summarized now as the main result of this section.

Theorem 8.12. *Suppose M is a time invariant bounded operator on $L_2[0, \infty)$. The uncertain system $(M, \boldsymbol{\Delta}_a)$ is robustly well-connected if and only if*

$$\inf_{\Theta \in \boldsymbol{\Theta}_a} \|\Theta M \Theta^{-1}\| < 1. \qquad (8.15)$$

The theorem states that if the operator M is time invariant, then the infimum condition is an exact test for well-connectedness. The importance of this result is immediately apparent from Proposition 8.6. Part (iii) shows that the above test is equivalent to a convex condition over the positive scaling set $\boldsymbol{P\Theta}_a$. Also if M is a state space system, then well-connectedness of $(M, \boldsymbol{\Delta}_a)$ reduces to an LMI feasibility problem by part (iv) of the proposition. It is the fact that this condition can be checked with surety which makes the above theorem very attractive.

Having provided computable conditions for well-connectedness with respect to the uncertainty set $\boldsymbol{\Delta}_a$ of spatially structured, but otherwise arbitrary perturbations, we are now ready to move on to consider uncertainty sets which are further constrained in their dynamical characteristics.

8.3   The structured singular value

After having studied some robustness problems at some level of detail in the preceding section, we will now extend some of the lessons learned to robustness analysis of more general uncertainty models. In §8.4 these general tools will be applied to another special case of robust well-connectedness, structured LTI uncertainty.

So far our uncertain systems have been characterized by a nominal component M, which we have taken to be LTI, and an uncertain component. For the latter, we have alternated between a description in terms of a *relation* $\mathcal{R}$ between signals, and a parameterization of the relation by a *set of operators* $\boldsymbol{\Delta}$. While these are interchangeable notions for the analysis we have performed, some aspects of the problem are best illuminated with each version.

Specifically, we first studied the *unstructured* uncertainty set where $\boldsymbol{\Delta}$ was just the unit ball of operators; in this case the analysis of well-connectedness reduced to small gain condition on M. We also studied spatially *structured* uncertainty, where the analysis of well-connectedness reduced to a *scaled* small gain condition (8.15). In both cases there is the common theme that robust well-connectedness depends only on a gain condition involving the nominal system M. We now explore how this notion of small gain can be generalized to more complex uncertainty structures.

The structures we are interested in consist of imposing some additional property $\mathcal{P}(\Delta)$ to the *unit ball* of operators; that is

$$\boldsymbol{\Delta} = \{\Delta \in \mathcal{L}(L_2) : \|\Delta\| \leq 1 \text{ and the property } \mathcal{P}(\Delta) \text{ holds.}\}$$

Examples of constraints imposed by $\mathcal{P}(\Delta)$ are block-diagonal structures as in $\boldsymbol{\Delta}_a$, or dynamic restrictions on some operator blocks (e.g., specifying that they are LTI, or static, or memoryless, etc.). We will assume that $\mathcal{P}(\Delta)$ imposes no norm restrictions, and in fact will further assume that if Δ satisfies $\mathcal{P}(\Delta)$, then so does $\gamma\Delta$ for any $\gamma > 0$. Namely, the set

$$C\boldsymbol{\Delta} := \{\Delta \in \mathcal{L}(L_2) : \text{the property } \mathcal{P}(\Delta) \text{ is satisfied}\}$$

is a *cone* (the cone generated by $\boldsymbol{\Delta}$). These assumptions are implicit in what follows.

Definition 8.13. *The structured singular value of an operator M with respect to a set $\boldsymbol{\Delta}$ which satisfies the above assumptions, is*

$$\mu(M, \boldsymbol{\Delta}) := \frac{1}{\inf\{\|\Delta\| : \Delta \in C\boldsymbol{\Delta} \text{ and } I - M\Delta \text{ is singular}\}} \tag{8.16}$$

when the infimum is defined. Otherwise $\mu(M, \boldsymbol{\Delta})$ is defined to be zero.

The infimum above is undefined only in situations were there are no perturbations that satisfy the singularity condition. This occurs, for instance, if $M = 0$. If perturbations exist which make $I - M\Delta$ singular, the infimum will be strictly positive and therefore $\mu(M, \boldsymbol{\Delta})$ will be finite. We remark that the terminology *structured singular value* originates with the matrix case, which was the first to be considered historically, and is discussed in the next section.

To start our investigation we have the following result, which provides an upper bound on the value of the structured singular value. It amounts to a restatement of the small gain result.

Proposition 8.14. $\mu(M, \boldsymbol{\Delta}) \leq \|M\|$, *with equality in the unstructured case* $\boldsymbol{\Delta} = \{\Delta \in \mathcal{L}(L_2) : \|\Delta\| \leq 1\}$.

Proof. We work with the non-trivial case $\mu(M, \boldsymbol{\Delta}) \neq 0$; in particular, $M \neq 0$ and the set in (8.16) is nonempty.

If $\|\Delta\| < \|M\|^{-1}$, we know that $I - \Delta M$ is nonsingular by small gain. Therefore the infimum in (8.16) is no less than $\|M\|^{-1}$, and thus the first statement follows by inversion.

In the unstructured uncertainty case, $C\boldsymbol{\Delta}$ is the entire space $\mathcal{L}(L_2)$; by scaling in Theorem 8.2, we can always construct $\Delta \in \mathcal{L}(L_2)$, $\|\Delta\| = \|M\|^{-1}$, and $I - \Delta M$ singular. This Δ achieves the infimum and proves the desired equality. ∎

So we see that the structured singular value reduces to the norm if $\boldsymbol{\Delta}$ is unstructured; more precisely $\mu(M, \mathcal{BL}(L_2)) = \|M\|$, where $\mathcal{BL}(L_2)$ denotes the unit ball of $\mathcal{L}(L_2)$. Furthermore if the uncertainty is arbitrary block-structured, we have the following version of Theorem 8.12.

Proposition 8.15. *Let M be an LTI, bounded operator on L_2. Then*

$$\mu(M, \boldsymbol{\Delta}_a) = \inf_{\Theta \in \Theta_a} \|\Theta M \Theta^{-1}\|.$$

This proposition follows from Theorem 8.12. We leave the details as an exercise.

Returning to the general case, it is clear by now that the structured singular value is closely related to robust well-connectedness. In fact one can write the following chain of statements:

$\mu(M, \mathbf{\Delta}) < 1$ if and only if $\inf\{\|\Delta\| : \Delta \in C\mathbf{\Delta}, I - M\Delta \text{ singular}\} > 1$;
    only if $I - M\Delta$ nonsingular, for all $\Delta \in \mathbf{\Delta}$;
    if and only if $(M, \mathbf{\Delta})$ is robustly well-connected.

The first and last equivalences are by definition. In the intermediate step, we did not make an "if" statement because the infimum might be unity, not achieved by any Δ. In other words, in general one can only say that robust well-connectedness implies $\mu(M, \mathbf{\Delta}) \leq 1$. This difficulty does not appear in the two cases considered up to now where the infima are indeed achieved. The issue is a little more delicate in the case of LTI uncertainty, to be considered in the next section. Nevertheless it is customary to regard the condition $\mu(M, \mathbf{\Delta}) < 1$ as interchangeable with robust well-connectedness; this is our desired generalization of a small gain test.

The following example gives more insight into the broad scope of problems covered by the structured singular value, as well as its mathematical structure.

Example:

Let the uncertainty set $\mathbf{\Delta} = \{\delta I : \delta \in \mathbb{C}, |\delta| \leq 1\}$. Then we have the following chain of equalities.

$$\mu(M, \mathbf{\Delta}) = (\inf\{|\delta| : \delta \in \mathbb{C}, I - M\delta \text{ is singular}\})^{-1}$$
$$= \sup\{|\delta^{-1}| : \delta^{-1} \in \mathbb{C} \text{ and } I\delta^{-1} - M \text{ is singular}\}$$
$$= \sup\{|\lambda| : \lambda \in \text{spec}(M)\}$$
$$= \rho(M),$$

where $\rho(\cdot)$ is the operator spectral radius.   □

So we see that the spectral radius also fits into this very general notion of system gain. Notice in particular that while it easy to show that the structured singular value satisfies the scaling property

$$\mu(\alpha M, \mathbf{\Delta}) = |\alpha|\mu(M, \mathbf{\Delta}), \text{ for all } \alpha \in \mathbb{C},$$

in general it does *not* define a norm. For instance the spectral radius is not a norm.

We have thus introduced a system property, the structured singular value, which is directly linked to the analysis of robust well-connectedness under very general uncertainty structures. At this level of generality, however, our definition has not yet accomplished much, since it may not be easy to evaluate μ. What we would like is a characterization, as the one obtained in Proposition 8.15 for μ with respect to $\mathbf{\Delta}_a$, that directly lends itself to computation. Such a strong result, however, will not always be available, which should not be surprising given the effort which was required to obtain this special case. Still we will find it valuable to again adopt our previous approach based on scaled norms for the general situation.

Analogously to §8.2.1, we introduce a set of *scalings* which commute with perturbations in our uncertainty structure. Define the nonsingular commutant

$$\mathbf{\Theta_\Delta} := \{\Theta \in \mathcal{L}(L_2) : \Theta \text{ is invertible and } \Theta\Delta = \Delta\Theta \text{ for all } \Delta \in \mathbf{\Delta}\}. \tag{8.17}$$

This set does not in general have the structure (8.4) of constant, block-diagonal matrices, but is nevertheless a well-defined entity. The following can then be easily established with the methods of §8.2.1 and Proposition 8.14. Note that this result is true for any bounded operator M; that is, M need not be a state space system or even time invariant.

Proposition 8.16. *Let $\mathbf{\Theta_\Delta}$ be the commutant of a general uncertainty set $\mathbf{\Delta}$, as defined in (8.17). Then*

$$\mu(M, \mathbf{\Delta}) \leq \inf_{\Theta \in \mathbf{\Theta_\Delta}} \|\Theta M \Theta^{-1}\|.$$

Thus we have obtained an upper bound for the structured singular value based on commuting scalings; equivalently, the existence of Θ in the appropriate set $\mathbf{\Theta_\Delta}$ satisfying

$$\|\Theta M \Theta^{-1}\| < 1 \tag{8.18}$$

is always *sufficient* for robust well-connectedness over $\mathbf{\Delta}$. What is not true in general is the necessity of the above condition. In fact only in very special cases is the bound of Proposition 8.16 known to be an equality; one of these is of course when M is time invariant and $\mathbf{\Delta} = \mathbf{\Delta}_a$. Thus the scaling method provides in general a conservative test of well-connectedness.

It is also possible in this general setting to introduce a set of positive scaling operators, by

$$\mathbf{P\Theta_\Delta} = \{\Theta \in \mathcal{L}(L_2) : \Theta = \tilde{\Theta}^* \tilde{\Theta} \text{ for some } \tilde{\Theta} \in \mathbf{\Theta_\Delta}\}.$$

With this definition, it follows that (8.18) is feasible for $\Theta \in \mathbf{\Theta_\Delta}$ if and only if either of the following two conditions holds.

- $\|\Theta^{\frac{1}{2}} M \Theta^{-\frac{1}{2}}\| < 1$ is satisfied for some $\Theta \in \mathbf{P\Theta_\Delta}$.

- The operator inequality

$$M^* \Theta M - \Theta < 0,$$

  is feasible over $\Theta \in \mathbf{P\Theta_\Delta}$.

The above equivalence is analogous to that of conditions (i) through (iii) in Proposition 8.6, and can be established with similar methods, with one important exception: here the set $P\Theta_\Delta$ is *not* in general a subset of Θ_Δ. This makes one of the above steps more involved; we will not pursue this here, however, since we will not rely on the above equivalence at this level of generality. Later on we will remark on this issue for the case of time invariant uncertainty, which we are now ready to discuss.

8.4   Time invariant uncertainty

Having laid out the general framework of the structured singular value, we now return to a specific uncertainty model, where the perturbations are assumed to be time invariant in addition to spatially structured. Define the perturbation set Δ_{TI} by

$$\Delta_{TI} = \{\Delta = \mathrm{diag}(\Delta_1 \ldots, \Delta_d) \in \mathcal{L}(L_2) : \Delta \text{ time invariant and } \|\Delta\| \leq 1\}.$$

Notice that this set is simply the intersection of the LTI operators on $L_2[0, \infty)$ with the set Δ_a. Bringing in the Laplace transform from Chapter 3 we see that each $\Delta \in \Delta_{TI}$ can be identified with its transfer function $\hat{\Delta} \in H_\infty$, which inherits the corresponding spatial structure.

We now introduce the relation $\mathcal{R}_{TI}$ that goes along with the operator set Δ_{TI}. It is clear, given an element q in $L_2[0, \infty)$, that

$$q = \Delta p, \text{ for some } \Delta \in \Delta_{TI}, \text{ if and only if } (p,q) \in \mathcal{R}_{TI},$$

where

$$\mathcal{R}_{TI} = \{(p,q) \in L_2 : \hat{\Delta}_k \in H_\infty, \|\hat{\Delta}_k\|_\infty \leq 1, |\hat{\Delta}_k(j\omega)\hat{p}_k(j\omega)| = |\hat{q}_k(j\omega)|\}.$$

From this we see that

$$\mathcal{R}_{TI} \subset \{(p,q) \in L_2 : |\hat{p}_k(j\omega)| \geq |\hat{q}_k(j\omega)|, \text{ for almost every } \omega\}.$$

As with $\mathcal{R}_a$ we have established a constraint between the sizes of the various components of p and q. However, the description is now much tighter since it is imposed at (almost) every frequency. In contrast $\mathcal{R}_a$ can be written as

$$\mathcal{R}_a = \left\{(p,q) \in L_2 : \int_0^\infty |\hat{p}_k(j\omega)|^2 d\omega \geq \int_0^\infty |\hat{q}_k(j\omega)|^2 d\omega, \ k = 1, \ldots, d\right\},$$

which clearly shows that the relation $\mathcal{R}_a$ only imposes quadratic constraints over frequency in an aggregate manner.

As a modeling tool, time invariant uncertainty is targeted at describing dynamics which are fundamentally linear and time invariant, but which we do not desire to model in detail in the nominal description M. Thus M could be a low-order approximation to the linearized system, leaving out high-dimensional effects which arise, for instance, from small-scale spatially distributed dynamics. Instead of simply neglecting these dynamics, time invariant perturbations provide a formalism for including or containing them in $\mathcal{R}_{TI}$ using frequency domain bounds.

8.4.1   Analysis of time invariant uncertainty

We will now discuss questions of robust well-connectedness for uncertain systems with linear time invariant uncertainty. What we will find in the sequel is that robust well-connectedness in this particular setting can be reduced to a matrix test at each point in the frequency domain. Before setting out to show this and the related results we look at an illuminating example. The following example points out the difference from the case of arbitrary structured uncertainty we might expect.

Example:

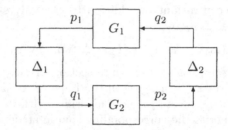

Figure 8.4. Example system

The uncertain system of Figure 8.4 is composed of two fixed linear time invariant systems G_1, G_2, and two uncertainty blocks Δ_1, Δ_2. All blocks are single input and output. We are interested in whether this configuration is well-connected in the sense of $I - \Delta_1 G_1 \Delta_2 G_2$ being nonsingular. We can routinely redraw this system in the standard configuration of Figure 8.1, with

$$M = M_{11} = \begin{bmatrix} 0 & G_1 \\ G_2 & 0 \end{bmatrix} \text{ and } \Delta = \begin{bmatrix} \Delta_1 & 0 \\ 0 & \Delta_2 \end{bmatrix}.$$

Clearly $I - \Delta_1 G_1 \Delta_2 G_2$ is singular if and only if $I - M\Delta$ is singular. Suppose $\|\Delta_1\| \leq 1$ and $\|\Delta_2\| \leq 1$. To investigate the well-connectedness of this system we can invoke the Small gain theorem and impose the sufficient condition for robust well-connectedness given by

$$\|\Delta_1 G_1 \Delta_2 G_2\| < 1.$$

If the uncertainty is time invariant, then Δ_2 and G_1 commute and we can use the submultiplicative inequality to obtain the stronger sufficient condition

$$\|G_1 G_2\| = \|\hat{G}_1 \hat{G}_2\|_\infty = \operatorname*{ess\,sup}_{\omega \in \mathbb{R}} |\hat{G}_1(j\omega)\hat{G}_2(j\omega)| < 1 \qquad (8.19)$$

for robust well-connectedness. If instead Δ_1 and Δ_2 are arbitrary contractive operators, we are not allowed to commute the operators, so we can only write the small gain condition

$$\|G_1\|\|G_2\| = \|\hat{G}_1\|_\infty \|\hat{G}_2\|_\infty = (\text{ess}\sup_{\omega \in \mathbb{R}} |\hat{G}_1(j\omega)|)(\text{ess}\sup_{\omega \in \mathbb{R}} |\hat{G}_2(j\omega)|) < 1.$$

$$(8.20)$$

These conditions are different in general: frequently (8.20) can be more restrictive than (8.19) since values of $|\hat{G}_1(j\omega)|$ and $|\hat{G}_2(j\omega)|$ at *different* frequencies can give a larger product. For instance,

$$\hat{G}_1(s) = \frac{\alpha}{s+1}, \quad \hat{G}_2(s) = \frac{s}{s+1}$$

will satisfy (8.19) for $\alpha < 2$, but (8.20) only for $\alpha < 1$. An extreme case would be if $\hat{G}_1(j\omega)$ and $\hat{G}_2(j\omega)$ had disjoint support.

To interpret these frequency domain conditions, notice that for $\Delta \in \mathbf{\Delta}_{TI}$, the transfer functions $\hat{\Delta}_1(j\omega)$ and $\hat{\Delta}_2(j\omega)$ are contractive at every ω; thus the small gain analysis can be decoupled in frequency, which makes (8.19) sufficient for well-connectedness. When the perturbation class is $\mathbf{\Delta}_a$ the operators Δ_1 and Δ_2 are still contractive, but they are allowed to "transfer energy" between the frequencies where $\hat{G}_1$ and $\hat{G}_2$ achieve their maximum gain, making well-connectedness harder to achieve.

It turns out that these conditions are also *necessary* for robust well-connectedness in each respective case. While this can be shown directly in the configuration of Figure 8.4, it is illustrative to write it in the standard form of Figure 8.1 and apply the robustness analysis techniques of this chapter. This is left as an exercise. □

We now proceed with the analysis of uncertain systems over $\mathbf{\Delta}_{TI}$. A standing assumption throughout the rest of this section is that the nominal operator M is finite dimensional and LTI; plainly, it has a transfer function $\hat{M}(s) \in RH_\infty$.

Our main tool will be the structured singular value μ, defined as in Definition 8.13 for the present uncertainty set $\mathbf{\Delta}_{TI}$. Following the general approach outlined in the previous section, the first task is to identify the operators Θ which commute with the structure $\mathbf{\Delta}_{TI}$. Notice that such an operator must commute with the delay operator S_τ, for every $\tau \geq 0$, since the delay S_τ is itself a member of $\mathbf{\Delta}_{TI}$. Thus we see that such a Θ is necessarily time invariant. If we also take into account the spatial structure, then it is shown in the exercises that the nonsingular commuting set is

$$\mathbf{\Theta}_{TI} =$$

$$\{\Theta \in \mathcal{L}(L_2): \ \Theta \text{ nonsingular, LTI and } \hat{\Theta}(s) = \text{diag}(\hat{\theta}_1(s)I, \ldots, \hat{\theta}_d(s)I)\}.$$

In other words, at every value of s in $\bar{\mathbb{C}}^+$ the matrix $\hat{\Theta}(s)$ must have the block-diagonal structure of (8.4), but it is no longer restricted to be constant. Thus the commuting set has grown as we reduced the perturbation set from $\mathbf{\Delta}_a$ to $\mathbf{\Delta}_{TI}$. Notice also that the inverse Θ^{-1} is automatically LTI, or equivalently $\hat{\Theta}^{-1} \in H_\infty$.

The following result provides a summary of the robustness conditions we have developed so far.

Theorem 8.17.

$$\mu(M, \mathbf{\Delta}_{TI}) \leq \inf_{\Theta \in \Theta_{TI}} \|\hat{\Theta} \hat{M} \hat{\Theta}^{-1}\|_\infty \leq \inf_{\Theta \in \Theta_a} \|\Theta M \Theta^{-1}\| = \mu(M, \mathbf{\Delta}_a)$$

Proof. The first inequality is a direct application of Proposition 8.16 to the set $\mathbf{\Delta}_{TI}$. The second inequality is clear since we are taking infimum over a smaller set. The third equality is a restatement of Theorem 8.12 and is exactly Proposition 8.15. ∎

The immediate question is whether the first inequality above is also an equality: is there a counterpart of Theorem 8.12 for the time invariant uncertainty case? Unfortunately we will find that the answer is negative, except in some specific cases. That is, the inequality is usually strict. The rest of the section will focus on this problem and pursue the analysis of $\mathbf{\Delta}_{TI}$ in more detail.

To provide more generality, at this point we will extend the spatial structure of our uncertainty class; consider the cone of complex matrices

$$C\mathbf{\Delta}_{s,f} =$$
$$\{\operatorname{diag}(\delta_1 I_{m_1}, \ldots, \delta_s I_{m_s}, \Delta_{s+1}, \ldots, \Delta_{s+f}) : \ \delta_k \in \mathbb{C} \ \text{and} \ \Delta_k \in \mathbb{C}^{m_k \times m_k}\},$$

and denote its unit ball by $\mathbf{\Delta}_{s,f} = \{\Delta \in C\mathbf{\Delta}_{s,f} : \bar{\sigma}(\Delta) \leq 1\}$. Now define the time invariant uncertainty set as

$$\mathbf{\Delta}_{TI} = \{\Delta \in \mathcal{L}(L_2) : \ \Delta \text{ is LTI and } \hat{\Delta}(s) \in \mathbf{\Delta}_{s,f}, \text{ for every } \operatorname{Re}(s) \geq 0\}.$$

In addition to the full blocks Δ_k considered before, we are introducing repeated scalar blocks of the form $\delta_k I$ in the uncertainty. This means that the same uncertainty operator acts on a number of scalar channels. One motivation for such repetition is to describe the rational dependence of a transfer function on an uncertain parameter, which can be expressed in this way (see the exercises); another motivation will be seen at the end of this chapter.

The next definition defines the matrix structured singular value, and is entirely analogous to Definition 8.13. This definition will be central in our reduction of LTI well-connectedness to a test on matrices.

Definition 8.18. *Given a matrix $Q \in \mathbb{C}^{m \times m}$, we define the structured singular value of Q with respect to $\mathbf{\Delta}_{s,f}$ by*

$$\mu(Q, \mathbf{\Delta}_{s,f}) = \frac{1}{\min\{\bar{\sigma}(\Delta) : \ \Delta \in C\mathbf{\Delta}_{s,f} \ \text{and} \ I - Q\Delta \ \text{is singular}\}}$$

when this minimum is defined, and $\mu(Q, \mathbf{\Delta}_{s,f}) = 0$ otherwise.

The following properties of μ with respect to the matrix set $\mathbf{\Delta}_{s,f}$ are left as an exercise:

- $\mu(\alpha Q, \boldsymbol{\Delta}_{s,f}) = |\alpha|\mu(Q, \boldsymbol{\Delta}_{s,f})$;
- $\mu(Q, \boldsymbol{\Delta}_{s,f}) = \max\{\rho(Q\Delta) : \Delta \in \boldsymbol{\Delta}_{s,f}\}$.

The second property implies that $\mu(Q, \boldsymbol{\Delta}_{s,f})$ is a continuous function of Q. This follows from the fact that the spectral radius function $\rho(\cdot)$ is continuous on the space of matrices, and the above maximum is over a compact set. While we have not explicitly reviewed these topological facts in this course, it is safe to assume they are known by readers studying this material.

The next major objective we have is to show how robustness analysis over $\boldsymbol{\Delta}_{TI}$ can be converted into a matrix structured singular value test at each frequency. Before we can accomplish this we first need one more property of the matrix structured singular value, in addition to the ones above, namely, that it satisfies a maximum principle over matrix functions that are analytic in a complex domain.

To obtain this result we need a few preliminary facts regarding polynomials with complex variables. The first is a continuity property of the roots of a complex polynomial as a function of its coefficients. In plain language this result says that if all the coefficients of two polynomials are sufficiently close to each other, then their respective roots must also be near each other.

Lemma 8.19. *Let* $p(\xi) = \alpha(\xi - \xi_1)(\xi - \xi_2)\cdots(\xi - \xi_n)$ *be an n-th order polynomial over* $\mathbb{C}$ *($\alpha \neq 0$). If* $p^{(k)}(\xi)$ *is a sequence of n-th order polynomials with coefficients converging to those of* $p(\xi)$, *then there exists a root* $\xi_1^{(k)}$ *of* $p^{(k)}(\xi)$ *such that* $\xi_1^{(k)} \to \xi_1$.

Proof. At each k, we define the factorization

$$p^{(k)}(\xi) = \alpha^{(k)}(\xi - \xi_1^{(k)})(\xi - \xi_2^{(k)})\cdots(\xi - \xi_n^{(k)})$$

such that $\xi_1^{(k)}$ is the root closest to ξ_1 (i.e. $|\xi_1 - \xi_1^{(k)}| \leq |\xi_1 - \xi_i^{(k)}|$ for every $i = 1 \ldots n$).

Since the coefficients of $p^{(k)}(\xi)$ converge to those of $p(\xi)$, we can evaluate at $\xi = \xi_1$ and see that $p^{(k)}(\xi_1) \to p(\xi_1) = 0$ as $k \to \infty$. Also, $\alpha^{(k)} \to \alpha \neq 0$. Therefore

$$|\xi_1 - \xi_1^{(k)}|^n \leq \prod_{i=1}^{n} |\xi_1 - \xi_i^{(k)}| = \left|\frac{p^{(k)}(\xi_1)}{\alpha^{(k)}}\right| \longrightarrow 0 \qquad \text{as } k \to \infty.$$

∎

The next lemma concerns the roots of a polynomial of *two* complex variables. It states that a particular type of root must always exist for such polynomials, one in which the moduli of the arguments are equal.

Lemma 8.20. *Let* $p(\xi, \zeta)$ *be a non-constant polynomial* $p : \mathbb{C}^2 \to \mathbb{C}$. *Define*

$$\beta = \min\{\max(|\xi|, |\zeta|) : p(\xi, \zeta) = 0\} \qquad (8.21)$$

Then there exist $\xi^\star, \zeta^\star$ *such that* $p(\xi^\star, \zeta^\star) = 0$ *and* $|\xi^\star| = |\zeta^\star| = \beta$.

Proof. We start with ξ^*, ζ^* that achieve the minimum in (8.21). The only case we need to consider is when $\beta > 0$ and one of the above numbers has magnitude less than β. Take, for instance, $|\zeta^*| = \beta$, $|\xi^*| < \beta$.

By setting $\zeta = \zeta^*$ and grouping terms in the powers of ξ, we write

$$p(\xi, \zeta^*) = \sum_{n=0}^{N} a_n(\zeta^*)\xi^n. \tag{8.22}$$

Suppose that the above polynomial in ξ is not identically zero. If we replace ζ^* by $(1-\epsilon)\zeta^*$, its coefficients are perturbed continuously, so we know from Lemma 8.19 that as $\epsilon \to 0+$, the perturbed polynomial will have a root ξ_ϵ converging to ξ^*. Thus, for small enough ϵ, we have

$$|\xi_\epsilon| < \beta, \qquad |(1-\epsilon)\zeta^*| < \beta, \qquad p(\xi_\epsilon, (1-\epsilon)\zeta^*) = 0.$$

This contradicts the definition of β; therefore the only alternative is that the polynomial (8.22) must be identically zero in ξ. But then the value of ξ can be set arbitrarily, in particular to have magnitude β, yielding a root with the desired property. ∎

We are now in a position to state the maximum principle for the matrix structured singular value. Just like the maximum modulus theorem for scalar analytic functions, this theorem asserts that structured singular value of a function that is analytic in the right half-plane achieves its supremum on the boundary of the half-plane; we will actually treat the case of a rational function.

Theorem 8.21. *Let $\hat{M}(s)$ be a function in RH_∞, and $\boldsymbol{\Delta}_{s,f}$ be a perturbation structure in the set of complex matrices. Then*

$$\sup_{Re(s)\geq 0} \mu(\hat{M}(s), \boldsymbol{\Delta}_{s,f}) = \sup_{\omega \in \mathbb{R}} \mu(\hat{M}(j\omega), \boldsymbol{\Delta}_{s,f}).$$

Proof. We first convert the problem to a maximization over the unit disk, by means of the change of variables

$$s = \frac{1 + \xi}{1 - \xi}.$$

This linear fractional transformation maps $\{|\xi| \leq 1, \xi \neq 1\}$ to $\{Re(s) \geq 0\}$, and the point $\xi = 1$ to $s = \infty$. Also the boundary $|\xi| = 1$ of the disk maps to the imaginary axis.

Noticing that $\hat{M}(s) \in RH_\infty$ has no poles in $Re(s) \geq 0$ and has a finite limit at $s = \infty$, we conclude that the rational function

$$\hat{Q}(\xi) = \hat{M}\left(\frac{1 + \xi}{1 - \xi}\right)$$

has no poles over the closed unit disk $|\xi| \leq 1$. With this change, our theorem reduces to showing that

$$\mu_0 := \sup_{|\xi| \leq 1} \mu(\hat{Q}(\xi), \boldsymbol{\Delta}_{s,f}) = \sup_{|\xi| = 1} \mu(\hat{Q}(\xi), \boldsymbol{\Delta}_{s,f}).$$

Notice that by continuity of the matrix structured singular value, the supremum μ_0 on the left is a maximum, achieved at some ξ_0, $|\xi_0| \leq 1$; we must show it occurs at the disk boundary.

It suffices to consider the case $\mu_0 > 0$. By definition, we know that $1/\mu_0 = \bar{\sigma}(\Delta_0)$, where

$$\Delta_0 := \arg\min\{\bar{\sigma}(\Delta) : \Delta \in \boldsymbol{C\Delta}_{s,f} \text{ and } I - \hat{Q}(\xi_0)\Delta \text{ is singular }\}.$$

Now consider the equation

$$\det[I - \hat{Q}(\xi)\zeta\Delta_0] = 0$$

in the two complex variables ξ, ζ. Since $\hat{Q}(\xi)$ is rational with no poles in $|\xi| \leq 1$, we can eliminate the common denominator and obtain a polynomial equation

$$p(\xi, \zeta) = 0$$

equivalent to the above for $|\xi| \leq 1$. We claim that

$$1 = \min\{\max(|\xi|, |\zeta|) : p(\xi, \zeta) = 0\}.$$

In fact, the choice $\xi = \xi_0$, $\zeta = 1$ gives $p(\xi_0, 1) = 0$ and $\max(|\xi_0|, 1) = 1$. If we found $p(\xi, \zeta) = 0$ for some $|\xi| < 1$, $|\zeta| < 1$, we would have

$$I - \hat{Q}(\xi)\zeta\Delta_0 \quad \text{singular}, \quad \zeta\Delta_0 \in \boldsymbol{C\Delta}_{s,f}, \quad \bar{\sigma}(\zeta\Delta_0) < 1/\mu_0.$$

This would give

$$\mu(\hat{Q}(\xi), \boldsymbol{\Delta}_{s,f}) > \mu_0$$

contradicting the definition of μ_0.

Thus we have proved our claim, which puts us in a position to apply Lemma 8.20 and conclude there exists a root (ξ^*, ζ^*) of $p(\xi, \zeta)$ with $|\xi^*| = |\zeta^*| = 1$. Consequently,

$$I - \hat{Q}(\xi^*)\zeta^*\Delta_0$$

is singular, with $\bar{\sigma}(\zeta^*\Delta_0) = 1/\mu_0$. So we conclude that, for $|\xi^*| = 1$,

$$\mu(\hat{Q}(\xi^*), \boldsymbol{\Delta}_{s,f}) = \mu_0.$$

■

At this point we have assembled all the results required to prove the following theorem, which is a major step in our analysis of time invariant uncertainty. This theorem converts the well-connectedness of $(M, \boldsymbol{\Delta}_{TI})$ to a pure matrix test at each point in the frequency domain. It therefore has clear computational implications for the analysis of time invariant uncertainty.

Theorem 8.22. *Assume M is a time invariant bounded operator, with its transfer function in RH_∞. The following are equivalent:*

(a) $(M, \mathbf{\Delta_{TI}})$ is robustly well-connected;

(b) $\mu(M, \mathbf{\Delta_{TI}}) < 1$;

(c) $\sup_{\omega \in \mathbb{R}} \mu(\hat{M}(j\omega), \mathbf{\Delta}_{s,f}) < 1$.

Furthermore, the left-hand side quantities in parts (b) and (c) are equal.

Proof. It will be convenient to extend $\hat{M}$ to the compactified right half-plane $\bar{\mathbb{C}}^+ \cup \{\infty\}$, defining $\hat{M}(\infty) = D$. Thus $\hat{M}(s)$ is a continuous function on a compact set, and so is $\mu(\hat{M}(s), \mathbf{\Delta}_{s,f})$.

The fact that (b) implies (a) was already established in the previous section.

To show that (a) implies (c), suppose that (c) does not hold, so

$$\mu(\hat{M}(j\omega_0), \mathbf{\Delta}_{s,f}) = \sup_{\omega \in \mathbb{R}} \mu(\hat{M}(j\omega), \mathbf{\Delta}_{s,f}) \geq 1$$

for some $\omega_0 \in \mathbb{R} \cup \{\infty\}$. Therefore we can find $\Delta_0 \in \mathbf{C\Delta}_{s,f}$ such that $\bar{\sigma}(\Delta_0) \leq 1$ and $I - \hat{M}(j\omega_0)\Delta_0$ is singular. Choosing the constant LTI perturbation $\hat{\Delta}(s) = \Delta_0$, clearly it belongs to $\mathbf{\Delta_{TI}}$ and $I - M\Delta$ is singular, contradicting condition (a).

To show that (c) implies (b), set

$$\eta = \sup_{s \in \mathbb{C}^+} \mu(\hat{M}(s), \mathbf{\Delta}_{s,f}).$$

From condition (c), invoking Theorem 8.21, we know that $\eta < 1$. Fix any scalar β satisfying $1 < \beta < \eta^{-1}$. By the definition of the structured singular value we have that

$$\beta < \min\{\bar{\sigma}(\Delta_0) : \Delta_0 \in \mathbf{C\Delta}_{s,f} \text{ and } I - M(s)\Delta_0 \text{ is singular}\},$$

for all s in $\bar{\mathbb{C}}^+ \cup \{\infty\}$. Therefore $I - \hat{M}(s)\Delta_0$ is an invertible matrix for every s in $\bar{\mathbb{C}}^+ \cup \{\infty\}$, and every matrix $\Delta_0 \in \mathbf{C\Delta}_{s,f}$ with $\bar{\sigma}(\Delta_0) \leq \beta$.

Since both s and Δ_0 are varying over compact sets, we can uniformly bound the norm of the matrix inverse, leading to the conclusion that

$$\bar{\sigma}(\{I - \hat{M}(s)\Delta_0\}^{-1}) \text{ is uniformly bounded.}$$

Now for any time invariant operator $\Delta \in \mathbf{C\Delta_{TI}}$, with $\|\Delta\| \leq \beta$ we conclude that $I - M\Delta$ is a nonsingular operator. Therefore from Definition 8.13 we find that $\mu(M, \mathbf{\Delta_{TI}}) \leq \beta^{-1} < 1$.

Finally note that both $\mu(M, \mathbf{\Delta_{TI}})$ and $\sup_{\omega \in \mathbb{R}} \mu(\hat{M}(j\omega), \mathbf{\Delta}_{s,f})$ are quantities which scale linearly with M. Thus if they were different, we could re-scale M to make one of them less than 1 and the other greater than 1, contradicting their equivalence with condition (a). Therefore they must be equal. ∎

As a first conclusion from the above theorem, we see that the test $\mu(M, \mathbf{\Delta}_{TI}) < 1$ is necessary and sufficient for robust well-connectedness. Recalling our general discussion in §8.3, we have avoided here the difficulty that was mentioned in establishing necessity. The reader should be warned, however, that for this to hold we have allowed in the step "(a) implies (c)" the use of a constant, complex perturbation Δ_0. This is legitimate since we have been working with complex signals and systems, but it would not be allowed if we wished to constrain time domain signals to be real valued, as is often done; in this case the necessity direction will not always hold, as shown in the references at the end of the chapter.

As a second, important conclusion, we see that the test $\mu(M, \mathbf{\Delta}_{TI}) < 1$ reduces exactly to a matrix structured singular value condition over the frequency axis. Henceforth we concentrate exclusively on the latter problem since it provides a direct avenue for computing well-connectedness.

8.4.2   The matrix structured singular value and its upper bound

In this subsection we provide tools to examine and compute the structured singular value of a matrix M. What we will find is that it is not amenable to computation by convex methods except in a number of special cases.

As usual, we begin by considering the set of matrices Θ which commute with the perturbations, that is,

$$\Theta\Delta = \Delta\Theta \quad \text{for all} \quad \Delta \in \mathbf{\Delta}_{s,f}.$$

The set of nonsingular matrices satisfying this property is defined to be $\mathbf{\Theta}_{s,f}$, and we restrict ourselves to using its subset of positive definite matrices

$$\mathbf{P\Theta}_{s,f} =$$
$$\{\text{diag}(\Theta_1, \ldots, \Theta_s, \theta_{s+1}I, \ldots, \theta_{s+f}I) : \Theta_k \in \mathbb{H}^{m_k}, \Theta_k > 0 \text{ and } \theta_k > 0\},$$

where as usual $\mathbb{H}^{m_k}$ denotes the set of Hermitian, $m_k \times m_k$ matrices. These full blocks Θ_k appear in correspondence with the repeated scalar blocks $\delta_k I$. We have the following result, which provides an upper bound.

Proposition 8.23.

(a) $\mu(M, \mathbf{\Delta}_{s,f}) \leq \inf_{\Theta \in \mathbf{P\Theta}_{s,f}} \bar{\sigma}(\Theta^{\frac{1}{2}} M \Theta^{-\frac{1}{2}})$;

(b) *A matrix* $\Theta \in \mathbf{P\Theta}_{s,f}$ *satisfies* $\bar{\sigma}(\Theta^{\frac{1}{2}} M \Theta^{-\frac{1}{2}}) < 1$ *if and only if*

$$M^*\Theta M - \Theta < 0.$$

Clearly the above proposition is just a specialization to the matrix case of our work in §8.3. An interesting point is, however, that in this case both conditions $\Theta \in \mathbf{P\Theta}_{s,f}$ and $M^*\Theta M - \Theta < 0$ of (b) are linear matrix inequalities in the blocks of Θ; therefore they lead directly to an efficient computation.

Also by combining part (a) with Theorem 8.22, we obtain

$$\mu(M, \mathbf{\Delta}_{TI}) = \sup_{\omega \in \mathbb{R}} \mu(\hat{M}(j\omega), \mathbf{\Delta}_{s,f})$$

$$\leq \sup_{\omega \in \mathbb{R}} \inf_{\Theta_\omega \in \mathbf{P\Theta}_{s,f}} \bar{\sigma}\left(\Theta_\omega^{\frac{1}{2}} \hat{M}(j\omega)\Theta_\omega^{-\frac{1}{2}}\right)$$

$$\leq \inf_{\hat{\Theta} \in \mathbf{\Theta}_{TI}} \|\hat{\Theta}\hat{M}\hat{\Theta}^{-1}\|_\infty.$$

In the first inequality above, notice that we allow the scaling matrix Θ_ω to be frequency dependent. If in particular we choose this dependence to be of the form $\Theta_\omega = \hat{\Theta}^*(j\omega)\hat{\Theta}(j\omega)$ where $\Theta \in \mathbf{\Theta}_{TI}$, then we obtain the second inequality, which in fact re-derives the bound of Theorem 8.17 for this generalized spatial structure. Now the intermediate bound appears to be potentially sharper; in fact it is not difficult to show both bounds are equal and we return to this issue in the next chapter.

Another important comment is that verification of the test

$$\sup_{\omega \in \mathbb{R}} \inf_{\Theta_\omega \in \mathbf{P\Theta}_{s,f}} \bar{\sigma}\left(\Theta_\omega^{\frac{1}{2}} \hat{M}(j\omega)\Theta_\omega^{-\frac{1}{2}}\right) < 1$$

amounts to a *decoupled* LMI problem over frequency, which is particularly attractive for computation.

We still have not addressed, however, the conservatism of this bound. The rest of the section is devoted to this issue; our approach will be to revisit the method from convex analysis used in §8.2.2, which was based on the language of *quadratic forms*, and see how far the analysis can be extended. We will find that most of the methods indeed have a counterpart here, but the extension will fail at one crucial point; thus the bound will in general be conservative. However, this upper bound is equal to the structured singular value in a number of cases and the tools we present here can be used to demonstrate this.

We begin by characterizing the quadratic constraints equivalent to the relation $q = \Delta p$, $\Delta \in \mathbf{\Delta}_{s,f}$. The first part of the following lemma is matrix version of Lemma 8.4; the second extends the method to repeated scalar perturbations.

Lemma 8.24. *Let p and q be two complex vectors. Then*

(a) There exists a matrix Δ, $\bar{\sigma}(\Delta) \leq 1$ such that $\Delta p = q$ if and only if

$$p^* p - q^* q \geq 0.$$

(b) There exists a matrix δI, $|\delta| \leq 1$, such that $\delta I p = q$ if and only if

$$pp^* - qq^* \geq 0.$$

Proof. Assume $p \neq 0$; otherwise the result is trivial.

(a) Clearly if $\Delta p = q$, $\bar{\sigma}(\Delta) \leq 1$, then $|p| \geq |q|$ or $p^*p - q^*q \geq 0$. Conversely, if $|p| \geq |q|$ we can choose the contractive rank one matrix

$$\Delta = \frac{qp^*}{|p|^2}.$$

(b) If $\delta p = q$, $|\delta| \leq 1$, then $pp^* - qq^* = pp^*(1 - |\delta|^2) \geq 0$. Conversely, let $pp^* - qq^* \geq 0$; this implies $\operatorname{Ker} p^* \subset \operatorname{Ker} q^*$ and therefore $\operatorname{Im} q \subset \operatorname{Im} p$. So necessarily there exists a complex scalar δ which solves

$$q = \delta p.$$

Now $0 \leq pp^* - qq^* = pp^*(1 - |\delta|^2)$ implies $|\delta| \leq 1$.

$\blacksquare$

Having established a quadratic characterization of the uncertainty blocks, we now apply it to the structured singular value question.

Suppose the matrix $I - M\Delta$ is singular, with $\Delta \in \mathbf{\Delta}_{s,f}$. It will be more convenient to work with $I - \Delta M$, which is also singular (see Lemma 3.12). Let $q \in \mathbb{C}^m$ be nonzero, satisfying $(I - \Delta M)q = 0$, so $q = \Delta(Mq)$. Given the structure of Δ, we can use the previous lemma to write quadratic constraints for the components of q and Mq. We proceed analogously to the operator case in §8.2.2. First, we write these components as $E_k q$ and $E_k M q$, where

$$E_k = \begin{bmatrix} 0 & \cdots & 0 & I & 0 & \cdots 0 \end{bmatrix}.$$

Next we define the quadratic functions

$$\Phi_k(q) = E_k M q q^* M^* E_k^* - E_k q q^* E_k^*,$$
$$\phi_k(q) = q^* M^* E_k^* E_k M q - q^* E_k^* E_k q .$$

Finally we bring in the sets

$$\nabla_{s,f} := \{ (\Phi_1(q), \ldots, \Phi_s(q), \phi_{s+1}(q), \ldots, \phi_{s+f}(q)) : q \in \mathbb{C}^m, |q| = 1 \}$$
$$\Pi_{s,f} := \{ (R_1, \ldots, R_s, r_{s+1}, \ldots, r_{s+f}), R_k = R_k^* \geq 0, \ r_k \geq 0 \}.$$

The latter are subsets of $\mathbb{V} = \mathbb{H}^{m_1} \times \cdots \times \mathbb{H}^{m_s} \times \mathbb{R} \times \cdots \times \mathbb{R}$, which is a real vector space with the inner product

$$\langle Y, R \rangle = \sum_{k=1}^{s} \operatorname{Tr}(Y_k R_k) + \sum_{k=s+1}^{s+f} y_k r_k.$$

The linear functionals on $\mathbb{V}$ are characterized in Proposition 1.6.

The following characterization is the counterpart of Proposition 8.9.

Proposition 8.25. *The following are equivalent:*

(a) $\mu(M, \mathbf{\Delta}_{s,f}) < 1$;

(b) the sets $\nabla_{s,f}$ *and* $\Pi_{s,f}$ *are disjoint.*

Proof. The sets $\nabla_{s,f}$ and $\Pi_{s,f}$ intersect if and only if there exists q, with $|q| = 1$ satisfying

$$\Phi_k(q) \geq 0, \quad k = 1, \ldots, s,$$
$$\phi_k(q) \geq 0, \quad k = s+1, \ldots, s+f.$$

Using Lemma 8.24, this happens if and only if there exist contractive δ_k, $k = 1, \ldots, s$, and Δ_k, $k = s+1, \ldots, s+f$ satisfying

$$\delta_k E_k M q = E_k q, \quad k = 1, \ldots, s$$
$$\Delta_k E_k M q = E_k q, \quad k = s+1, \ldots, s+f.$$

Putting these blocks together, the latter is equivalent to the existence of $\Delta \in \mathbf{\Delta}_{s,f}$ and q such that $\bar{\sigma}(\Delta) \leq 1$, $|q| = 1$ and

$$\Delta M q = q,$$

which is equivalent to $(I - M\Delta)$ being singular for some $\Delta \in \mathbf{\Delta}_{s,f}$, with $\bar{\sigma}(\Delta) \leq 1$. By definition, this is the negation of (a). ∎

Having characterized the structured singular value in terms of properties of the set $\nabla_{s,f}$, we now do the same with the upper bound of Proposition 8.23. Once again, the parallel with §8.2 carries through and we have the following counterpart of Proposition 8.10.

Proposition 8.26. *The following are equivalent:*

(a) There exists $\Theta \in \mathbf{P\Theta}_{s,f}$ satisfying $\bar{\sigma}(\Theta^{\frac{1}{2}} M \Theta^{-\frac{1}{2}}) < 1$;

(b) the convex hull $\mathrm{co}(\nabla_{s,f})$ is disjoint with $\Pi_{s,f}$;

(c) there exists a hyperplane in $\mathbb{V}$ which strictly separates $\nabla_{s,f}$ and $\Pi_{s,f}$; that is, there exists $\bar{\Theta} \in \mathbb{V}$ and $\alpha, \beta \in \mathbb{R}$ such that

$$\langle \bar{\Theta}, Y \rangle \leq \alpha < \beta \leq \langle \bar{\Theta}, R \rangle \quad \text{for all } Y \in \nabla_{s,f}, R \in \Pi_{s,f}. \tag{8.23}$$

Proof. Notice that $\Pi_{s,f}$ is convex, so (b) implies (c) by the hyperplane separation theorem in finite dimensional space. Conversely, if a hyperplane strictly separates two sets it strictly separates their convex hulls, so (c) implies (b). It remains to show that these conditions are equivalent to (a).

Starting from (a), we consider the LMI equivalent $M^*\Theta M - \Theta < 0$, for some $\Theta \in \mathbf{P\Theta}_{s,f}$. For every vector q of $|q| = 1$, we have

$$q^*(M^*\Theta M - \Theta)q < 0.$$

Now we write

$$\Theta = \sum_{k=1}^{s} E_k^* \Theta_k E_k + \sum_{k=s+1}^{s+f} \theta_k E_k^* E_k$$

that leads to the inequality

$$\sum_{k=1}^{s}(q^*M^*E_k^*\Theta_k E_k Mq - q^*E_k^*\Theta_k E_k q)+$$

$$\sum_{k=s+1}^{s+f}(\theta_k q^*M^*E_k^*E_k Mq - \theta_k q^*E_k^*E_k q) < 0$$

for $|q| = 1$. Taking a trace, we rewrite the inequality as

$$\sum_{k=1}^{s}\text{Tr}\,(\Theta_k(E_k Mqq^*M^*E_k^* - E_k qq^*E_k^*))+$$

$$\sum_{k=s+1}^{s+f}\theta_k(q^*M^*E_k^*E_k Mq - q^*E_k^*E_k q) < 0$$

which we recognize as $\langle\Theta, Y\rangle < 0$, with

$$Y = (\Phi_1(q), \ldots, \Phi_s(q), \phi_{s+1}(q), \ldots, \phi_{s+f}(q)) \text{ and}$$
$$\bar{\Theta} = (\Theta_1, \ldots, \Theta_s, \theta_{s+1}, \ldots, \theta_{s+f}).$$

Also since $\Theta_k > 0$, $\theta_k > 0$ we conclude that

$$\langle\bar{\Theta}, R\rangle \geq 0 \text{ for all } R \in \Pi_{s,f}.$$

so we have shown (8.23). The converse implication follows in a similar way and is left as an exercise.   ∎

The last two results have characterized the structured singular value and its upper bound in terms of the sets $\nabla_{s,f}$ and $\Pi_{s,f}$; if in particular the set $\nabla_{s,f}$ were *convex*, we would conclude that the bound is exact. This was exactly the route we followed in §8.2.2. However, in the matrix case the set $\nabla_{s,f}$ is *not* convex, except in very special cases. Recalling the proof of Lemma 8.11, the key feature was the ability to shift in time, which has no counterpart in the current situation. In fact, only for structures with a small number of blocks can the bound be guaranteed to be exact. Such structures are called μ-simple. We have the following classification.

Theorem 8.27.

$$\mu(M, \Delta_{s,f}) = \inf_{\Theta \in P\Theta_{s,f}} \bar{\sigma}(\Theta^{\frac{1}{2}}M\Theta^{-\frac{1}{2}})$$

holds for all matrices M if and only if the block specifiers s and f satisfy

$$2s + f \leq 3.$$

For the alternatives $(s, f) \in \{(0,1), (0,2), (0,3), (1,0), (1,1)\}$ that satisfy $2s + f \leq 3$, the proof of sufficiency can be obtained using the tools introduced in this section. In particular, an intricate study of the set $\nabla_{s,f}$ is required in each case; proofs are provided in Appendix C. Counterexamples exist for all cases with $2s + f > 3$. We remark that the equality may hold if M has special structure.

As an interesting application of the theorem for the case $(s, f) = (1, 1)$, we invite the reader to prove the following discrete time version of the KYP Lemma. We remark, in this case the δI block appears due to the frequency variable z, not to uncertainty; this indicates another application of the structured singular value methods, which also extends to the consideration of multidimensional systems, as will be discussed in Chapter 11.

Proposition 8.28. *Given state space matrices A, B, C, and D, the following are equivalent.*

(a) *The eigenvalues of A are in the open unit disc $\mathbb{D} = \{z \in \mathbb{C} : |z| < 1\}$, and*

$$\sup_{z \in \bar{\mathbb{D}}} \bar{\sigma}(C(I - zA)^{-1}zB + D) < 1 \; ;$$

(b) *There exists a symmetric matrix $X > 0$ such that*

$$\begin{bmatrix} A & B \\ C & D \end{bmatrix}^* \begin{bmatrix} X & 0 \\ 0 & I \end{bmatrix} \begin{bmatrix} A & B \\ C & D \end{bmatrix} - \begin{bmatrix} X & 0 \\ 0 & I \end{bmatrix} < 0 \; .$$

This completes our investigation of time invariant uncertainty and in fact our study of models for uncertain systems in this chapter; in Chapter 10 we will examine uncertainty descriptions in more generality. We now move to the next chapter, where we use our results and framework to investigate stability and performance of uncertain feedback systems.

Exercises

8.1. This example is a follow-up to Exercise 5.6 (with a slight change in notation). As described there, a natural way to perturb a system is to use *coprime factor* uncertainty of the form

$$\hat{P}_\Delta = (\hat{N} + \hat{\Delta}_N)(\hat{D} + \hat{\Delta}_D)^{-1}.$$

Here $\hat{P}_0 = \hat{N}\hat{D}^{-1}$ is a nominal model, expressed as a normalized coprime factorization, and

$$\Delta = \begin{bmatrix} \hat{\Delta}_N \\ \hat{\Delta}_D \end{bmatrix}$$

is a perturbation with a given norm bound.

(a) Find M such that algebraically we have $P_\Delta = \bar{S}(M, \Delta)$. Is your M always a bounded operator?

(b) Repeat the questions of part (a) to describe the closed loop mapping from d_1, d_2 to q in the diagram given in Exercise 5.6.

8.2. Consider the partitioned operators

$$N = \begin{bmatrix} N_{11} & N_{12} \\ N_{21} & N_{22} \end{bmatrix}; \quad H = \begin{bmatrix} H_{11} & H_{12} \\ H_{21} & H_{22} \end{bmatrix}.$$

Find the operator M such that Figure 8.1 represents

(a) The cascade of Figure 8.2, that is,

$$\bar{S}(M, \Delta) = \bar{S}(N, \Delta_1)\bar{S}(H, \Delta_2) \quad \text{with } \Delta = \text{diag}(\Delta_1, \Delta_2).$$

(b) The composition $\bar{S}(M, \Delta) = \bar{S}(H, \bar{S}(N, \Delta))$, where we assume that $I - N_{22}H_{11}$ has a bounded inverse. Draw a diagram to represent this composition.

(c) The inverse $\bar{S}(M, \Delta) = \bar{S}(N, \Delta)^{-1}$, where we assume that N_{22} has a bounded inverse.

8.3. Commutant sets.

(a) Let $\Theta \in \mathbb{C}^{n \times n}$. Prove that if $\Theta\Delta = \Delta\Theta$ for all $\Delta \in \mathbb{C}^{n \times n}$, then $\Theta = \theta I_n$, $\theta \in \mathbb{C}$.

(b) Let $\Theta \in \mathbb{C}^{n \times n}$. Show that if $\Theta\Delta = \Delta\Theta$ for all structured $\Delta = \text{diag}(\Delta_1, \ldots, \Delta_d)$, then $\Theta = \text{diag}(\theta_1 I, \ldots, \theta_d I)$, where $\theta_i \in \mathbb{C}$ and the identity matrices have appropriate dimensions.

(c) Let $\Theta \in \mathbb{C}^{n \times n}$. Show that if $\Theta\Delta = \Delta\Theta$ for all $\Delta \in \mathbf{\Delta}_{s,f}$, then $\Theta \in \mathbf{\Theta}_{s,f}$.

(d) Characterize the commutant of $\mathbf{\Delta}_{TI}$ in $\mathcal{L}(L_2)$ (i.e. the set $\mathbf{\Theta}_{TI}$ of operators that commute with all members of $\mathbf{\Delta}_{TI}$).

e) Characterize the commutant of $\mathbf{\Delta}_a$ in $\mathcal{L}(L_2)$.

8.4. Derive Proposition 8.15 from Theorem 8.12.

8.5. Consider the example of §8.4.1 (note all systems are both single input and single output). It is shown there that

(i) If Δ_1, Δ_2 are arbitrary operators, then $\|G_1\|\|G_2\| < 1$ is sufficient for robust well-connectedness.

(ii) If Δ_1, Δ_2 are LTI operators, then $\|G_1 G_2\|_\infty < 1$ is sufficient for robust well-connectedness.

Show that both of the above conditions are necessary as well as sufficient, by reducing them to the standard (M, Δ) form and applying the theory of this chapter. What happens when only one of the Δ_i is known to be LTI?

8.6. Show that

(a) $\mu(\alpha Q, \mathbf{\Delta}_{s,f}) = |\alpha|\mu_{\mathbf{\Delta}_{s,f}}(Q)$;

(b) $\mu(Q, \mathbf{\Delta}_{s,f}) = \max\{\rho(Q\Delta) : \Delta \in \mathbf{\Delta}_{s,f}\}$.

8.7. Rational functions in linear fractional form.

(a) Given the polynomial $p(\delta) = p_n\delta^n + \cdots + p_1\delta + p_o$, find M such that $\bar{S}(M, \delta I) = p(\delta)$ for an identity matrix of the appropriate size.

(b) Using Problem 2 above, show that a similar representation is available for a *rational* function $r(\delta) = \frac{p(\delta)}{q(\delta)}$ as long as $q(0) \neq 0$.

(c) Discuss how to generalize the above to matrices of rational functions, and to rational functions of many variables $\delta_1, \ldots \delta_d$.

(d) Apply to the following modeling problem. Consider the second-order transfer function

$$\hat{H}(s) = \frac{1}{s^2 + s + \omega_n^2}$$

Assume the natural frequency ω_n is only known up to some error, i.e., $\omega_n = \omega_0 + k\delta$, $\delta \in [-1, 1]$. Find $\hat{M}(s)$ and the corresponding dimension of the identity matrix, such that $\hat{H}(s) = \bar{S}(\hat{M}(s), \delta I)$.

8.8. **Analysis with real parametric uncertainty.** From the preceding exercise we see that rational dependence of a system on a parameter can be represented in our setup of Figure 8.1 by an appropriate block δI, $\delta \in \mathbb{R}$. This leads us to consider the *mixed parametric/LTI uncertainty* structure

$$\boldsymbol{\Delta} =$$
$$\{\operatorname{diag}(\delta_1 I, \ldots, \delta_s I, \Delta_{s+1}, \ldots, \Delta_{s+f}) \in \boldsymbol{\Delta}_{TI} : \delta_k \in [-1, 1], 1 \leq k \leq r\},$$

which is a subset of $\mathcal{L}(L_2)$. In other words, the first r blocks ($r \leq s$) in the general LTI structure correspond to real parameters. For this structure it is possible to generalize Theorem 8.22 and reduce the analysis over frequency to the structured singular value test

$$\sup_{\omega \in \mathbb{R}} \mu(\hat{M}(j\omega), \boldsymbol{\Delta}_{s,f}^r) < 1,$$

where the structure $\boldsymbol{\Delta}_{s,f}^r \subset \mathbb{C}^{m \times m}$ is defined as

$$\boldsymbol{\Delta}_{s,f}^r =$$
$$\{\operatorname{diag}(\delta_1 I, \ldots, \delta_s I, \Delta_{s+1}, \ldots, \Delta_{s+f}) \in \boldsymbol{\Delta}_{s,f} : \delta_k \in [-1, 1], 1 \leq k \leq r\}.$$

We focus on a fixed frequency ω, let $Q = \hat{M}(j\omega)$. Since $\boldsymbol{\Delta}_{s,f}^r \subset \boldsymbol{\Delta}_{s,f}$ we know that the LMI condition

$$Q^*\Theta Q - \Theta < 0, \quad \Theta \in \boldsymbol{P\Theta}_{s,f}$$

is sufficient for $\mu(Q, \boldsymbol{\Delta}_{s,f}^r) < 1$. We wish to refine this bound by exploiting the real blocks of Δ. Consider the additional matrix scaling set

$$\mathcal{G} = \{G = \operatorname{diag}(G_1, \ldots, G_r, 0, \ldots, 0) : G_k = G_k^*\} \subset \mathbb{C}^{m \times m}.$$

(a) Show that if $q \in \mathrm{Ker}\,(I - \Delta Q)$, with $\Delta \in \Delta^r_{s,f}$, then

$$q^*(GQ - Q^*G)q = 0 \quad \text{for all } G \in \mathcal{G}.$$

(b) Deduce that the LMI test in Θ and G,

$$Q^*\Theta Q - \Theta + j(GQ - Q^*G) < 0, \quad \Theta \in \mathbf{P}\Theta_{s,f}, \quad G \in \mathcal{G},$$

is sufficient for $\mu(Q, \Delta^r_{s,f}) < 1$.

8.9. Extension of maximum principle for μ.

  (a) Extend Lemma 8.20 to polynomials of several complex variables.
  (b) Given a matrix M and a complex uncertainty structure $\Delta_{s,f}$, show that if

$$\Delta_0 = \arg \min_{\Delta \in \Delta_{s,f}} \{\bar{\sigma}(\Delta) : I - M\Delta \text{ is singular}\},$$

  then without loss of generality Δ_0 can be taken to be of the form αU, with $\alpha > 0$ and U unitary.
  (c) Deduce that

$$\mu_{\Delta_{s,f}}(M) = \max_{\Delta \in \Delta_{s,f}} \rho(M\Delta) = \max_{U \in \Delta_{s,f}, U \text{ unitary}} \rho(MU).$$

  (d) Let $p(\delta_1, \ldots, \delta_s)$ be a multinomial in several complex variables. Use the above to show the following maximum modulus principle:

$$\max_{|\delta_i| \leq 1} |p(\delta_1, \ldots, \delta_s)| = \max_{|\delta_i| = 1} |p(\delta_1, \ldots, \delta_s)|.$$

8.10. Extension of set Δ_a. In this problem we will generalize the robust well-connectedness results of §8.2 by expanding the set Δ_a to include "diagonal" elements similar to those we saw when studying the matrix structured singular value. Define $\Delta_a^{s,f}$ to be the set

$$\Delta_a^{s,f} := \{\mathrm{diag}(\delta_1 I_{m_1}, \ldots, \delta_s I_{m_s}, \Delta_{s+1}, \ldots, \Delta_{s+f}) :$$
$$\delta_k \in \mathcal{L}(L_2^1), \ \Delta_k \in \mathcal{L}(L_2^{m_k}), \text{ and } \|\Delta\| < 1\}.$$

Here the notation $\delta_k I_{m_k}$ is shorthand for the operator $\mathrm{diag}(\delta_k, \ldots, \delta_k)$ which acts on $L_2^{m_k}$; note that δ_k itself acts on the space of scalar functions L_2^1. Realize that if these new types of blocks are not present we have the usual definition of Δ_a from our work in §8.2.

  (a) Find the commutant set of this expanded perturbation set $\Delta_a^{s,f}$, and therefore the corresponding scaled small gain theorem.
  (b) Here we will derive a generalization of Proposition 8.7, which will relate well-connectedness to an infeasibility condition on a new set of quadratic forms.
    (i) Suppose $x_1, x_2, y_1, y_2 \in \mathbb{R}^n$, and that x_1 and x_2 are linearly independent. Prove: there exists a matrix $A \in \mathbb{R}^{n \times n}$ with

$\bar{\sigma}(A) \leq 1$ such that

$$y_1 = Ax_1 \quad \text{and} \quad y_2 = Ax_2$$

if and only if the matrix

$$\begin{bmatrix} y_1^* \\ y_2^* \end{bmatrix} [y_1 \ \ y_2] - \begin{bmatrix} x_1^* \\ x_2^* \end{bmatrix} [x_1 \ \ x_2] \leq 0.$$

(ii) Suppose p_1, p_2, q_1, q_2 are scalar valued functions in L_2^1, and that p_1 and p_2 are linearly independent. Show that there exists an operator $\delta \in \mathcal{L}(L_2^1)$ with $\|\delta\| \leq 1$ such that

$$q_1 = \delta p_1 \quad \text{and} \quad q_2 = \delta p_2$$

if and only if the matrix

$$\int_0^\infty \begin{bmatrix} q_1(t) \\ q_2(t) \end{bmatrix} [q_1(t) \ \ q_2(t)] - \begin{bmatrix} p_1(t) \\ p_2(t) \end{bmatrix} [p_1(t) \ \ p_2(t)] \, dt \leq 0.$$

(iii) Now prove the generalization: given two $\mathbb{R}^m$-valued functions p, $q \in L_2^m$, there exists an operator $\delta \in \mathcal{L}(L_2^1)$ with $\|\delta\| \leq 1$ such that

$$q = \delta I_m p$$

if and only if the matrix

$$\int_0^\infty q(t)q^*(t) - p(t)p^*(t) \, dt \leq 0.$$

Note: here there are no independence conditions as there were above.

(iv) Define the matrices

$$E_k = \begin{bmatrix} 0 & \cdots & 0 & I & 0 & \cdots & 0 \end{bmatrix}, \text{ for } 1 \leq l \leq s+f,$$

which are defined so that $\sum_{l=1}^{s+f} E_l^* E_l = I$, and we have

$$E_k^* \mathrm{diag}(\delta_1 I_{m_1}, \ldots, \delta_s I_{m_s}, \Delta_{s+1}, \ldots, \Delta_{s+f}) E_k = \begin{cases} \delta_k, & 1 \leq k \leq s \\ \Delta_k, & s+1 \leq k \leq s+f \end{cases}$$

for every $\Delta \in \boldsymbol{\Delta}_a$. Use these to define the quadratic forms $\Phi_k : L_2 \to \mathbb{H}^{m_k}$ and $\phi_k : L_2 \to \mathbb{R}$ by

$$\Phi_k(p) = \int_0^\infty \{E_k q(t)q^*(t)E_k^* - E_k p(t)p^*(t)E_k^*\} \, dt,$$

for $1 \leq k \leq s$ where $q := Mp$, and

$$\phi_k(p) = \int_0^\infty \{q^*(t)E_k^* E_k q(t) - p^*(t)E_k^* E_k p(t)\} \, dt,$$

for $s+1 \leq k \leq s+f$. Prove that if there exists a nonzero p in L_2 so that $\Phi_k(p) \geq 0$ for $1 \leq k \leq s$, and $\phi_k(p) \geq 0$ for $s+1 \leq k \leq s+f$, then $(M, \Delta_a^{s,f})$ is *not* robustly well-connected. Observe this is a generalization of Proposition 8.7.

(c) Use the quadratic forms in (b, iv) to generalize the work in §8.2 to the set $(M, \Delta_a^{s,f})$, and thus provide a more general version of Theorem 8.12. The work in §8.4.2 will be helpful.

8.11. **Loop shifting.** Suppose Q and R are operators in $\mathcal{L}(L_2)$, that $Q^*Q = QQ^* = I$, and that Q_{12} is surjective. Also assume that the system in the figure below is well-connected.

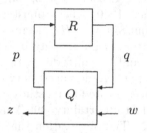

(a) Prove the norm $\|R\| \leq 1$ if and only if $\|\bar{S}(Q,R)\| \leq 1$. Consulting the figure may be helpful.

(b) Suppose A, B, C, D is a realization for the state space system P, where $\bar{\sigma}(D) < 1$ and A is Hurwitz. Show that $\|\hat{P}\|_\infty \leq 1$, if and only if, $\|\hat{P}_Q\|_\infty \leq 1$. Here P_Q is the state space system with realization

$$A_Q = A + B(I - D^*D)^{-1}D^*C$$
$$B_Q = B(I - D^*D)^{-\frac{1}{2}}$$
$$C_Q = (I - DD^*)^{-\frac{1}{2}}C$$
$$D_Q = 0$$

Hint: exploit part (a) by setting $\hat{R} = \hat{P}$ and

$$Q = \begin{bmatrix} D^* & (I - D^*D)^{\frac{1}{2}} \\ (I - DD^*)^{\frac{1}{2}} & -D \end{bmatrix}.$$

Notice: this question converts a question about a system with a D-term to one without a D-term. This procedure is called *loop-shifting*.

(c) Prove that A_Q in (b) must be Hurwitz if $\|\hat{P}\|_\infty \leq 1$ and A is Hurwitz.

(d) Generalize these results to strict inequalities: that is, systems that are strictly contractive.

Notes and references

The contemporary methods of robustness analysis are rooted in the work of the 1960s on feedback stability for nonlinear systems from an input–output perspective. In this category fall the Small gain theorem [144, 183] and methods based on passivity, particularly emphasized by the Russian school, where they were termed "absolute stability theory" [124, 176]. In this literature we already find "multiplier" methods for stability analysis of various component nonlinearities which are direct precursors of the Θ-scalings considered in this chapter.

The observation that these methods could be used to study *uncertain* systems can be traced back to [183], but the subject was largely not pursued until the late 1970s when robustness began to be emphasized (see [140]) and the connection to H_∞ norms was pointed out [184]. In particular the formalization of uncertainty in terms of Δ blocks appeared in the papers [37, 137], and [37] introduced the matrix structured singular value for studying time invariant uncertainty. Subsequently a great deal of research activity was devoted to its study, and to its extension to mixed real parametric and LTI uncertainty [49]. Much of this research is summarized in [111, 113, 182]. The observation that the test $\mu(M, \mathbf{\Delta_{TI}}) < 1$ can be conservative when dealing with real valued signals can be found in [159]. Computing the structured singular value for an arbitrary number of blocks appeared to be a vexing task, and in [22, 160] it was demonstrated that the problem is NP-hard. In reference [161] it is shown that the gap between the structured singular value and its upper bound is not uniformly bounded with respect to the number of blocks.

Given these difficulties, it was welcome news when it was discovered in the early 1990s that convex upper bounds had an interpretation in their own right. The first result came from the parallel theory of robust control using L_∞ signal norms, which we have not covered in this course (see [28], and also [7] for structured singular value methods in this context). It was shown in [89] that scaled small gain conditions in the L_∞-induced norm, analogous to Theorem 8.12, were exact for the analysis of arbitrary time-varying operators. These results led to the study of the analogous question for the case of L_2 norms, leading to the proof of Theorem 8.12 in [101, 148] (see also the related work [14]). The proof we presented is based on [103], which brought in and extended the S-procedure [177] that had been developed in parallel in the absolute stability literature. An extension of this viewpoint, which makes IQCs [102] the central object, will be discussed in a later chapter. In a sense, robust control has come full-circle to reinterpret the tools of input–output stability analysis: instead of a known nonlinearity we are dealing with an unknown ball of operators, and thus we obtain necessary as well as sufficient conditions.

In [123] these results are extended to show that the test with frequency varying scalings, presented in Theorem 8.17, has a similar interpretation as an exact test for the analysis of structured *slowly varying* perturbations. The monograph [45] studies these structured robustness problems for hybrid sampled-data systems, and shows they can have unique properties in this setting.

A treatment of star products, and loop shifting, for general operators can be found in [133].

At the beginning of this chapter sets of systems were defined in terms of explicit perturbations to a nominal plant. Another way to obtain a set of systems is to put a topology on the space of all plants, and consider open sets around the nominal system. This approach was initiated in [165] and [185] where the graph topology and gap metric were defined, respectively. For details on this research direction see, for instance, [60, 168] and the references therein; at present these approaches do not explicitly consider subsystem and component uncertainty.

9
Feedback Control of Uncertain Systems

In this chapter we bring together the separate threads of synthesis of feedback controllers in the absence of uncertainty, and analysis of uncertain systems, into a common problem involving both uncertainty and control. This problem is represented by the diagram shown in Figure 9.1, where G is the generalized plant as in earlier chapters, but now also describes dependence on system uncertainty. The perturbation Δ belongs to a structured ball, and K represents the controller.

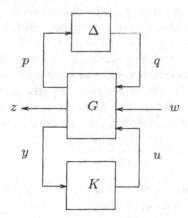

Figure 9.1. Feedback system with uncertainty

We will assume *throughout* this chapter that G and K are standard state space systems, and they are therefore LTI and causal, and have associated initial conditions. The mapping Δ will vary in a set of structured, bounded operators, and is only described in input output terms with no initial conditions[1].

While the above figure exhibits an attractive symmetry, we should immediately note that the upper and lower feedback loops have been used in the preceding chapters in two different settings. In Chapters 5 through 7 we studied the feedback interconnection of causal, LTI, finite dimensional systems, motivated by the modeling of an actual physical interconnection. In contrast, in Chapter 8 we used the feedback configuration as a mathematical tool to parametrize a set of models. Our goal is to combine these viewpoints for the study of analysis and synthesis in uncertain feedback systems. Before embarking on this detailed investigation, however, it is worthwhile to informally overview our objectives in regard to the diagram of Figure 9.1.

A first main objective of our feedback is that it must achieve *stability* in the interconnection. This notion has not yet been defined for such general systems, a task we undertake below. As a first approximation, however, we state the following requirements:

- For any $w \in L_2[0, \infty)$, and any initial conditions there must exist unique solutions in L_2 for the functions z, p, q, u, and y.

- If $w = 0$, the states of G and K should tend asymptotically to zero from any initial conditions.

- For zero initial conditions, the five maps on L_2 from the input w to the functions z, p, q, u, and y should all be bounded.

These conditions ensure that (a) the equations which describe the feedback system always have meaning; (b) deviations in the initial states of G and K away from equilibrium are innocuous; and (c) the effect of w on the system is bounded.

The requirement of stability should be *robust*, that is it should hold for *every* Δ in the uncertainty set. Since we always assume that this set contains the element $\Delta = 0$, we see that in particular we are requiring the internal stability of the interconnection of G and K, as defined in Chapter 5. This is called *nominal stability*. An immediate consequence of nominal stability is that it makes the star product

$$M = \underline{S}(G, K)$$

a bounded LTI operator with transfer function in RH_∞. Thus we are led to consider the configuration of Figure 9.2 below.

---

[1]This modeling assumption implies in particular that all unstable dynamics should be included in G; while restrictive, this kind of assumption is essential for the analysis that follows.

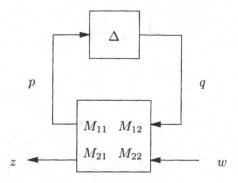

Figure 9.2. Robustness analysis setup

Now this setup loops exactly like the arrangement from Chapter 8. Indeed our strategy is to study the robust stability of the overall system Figure 9.1 using the approach and results of Chapter 8, which revolve around the study of nonsingularity of the operator $I - M_{11}\Delta$. Also, given this requirement, we will study the system *performance*, as usual measured in terms of keeping the mapping

$$w \mapsto z = \bar{S}(M, \Delta) = M_{22} + M_{21}\Delta(I - M_{11}\Delta)^{-1}M_{12} \qquad (9.1)$$

small in some appropriate sense. In this chapter we will focus on the induced norm as measure of size and impose the condition

$$\|\bar{S}(M, \Delta)\| < 1$$

for every perturbation Δ as a notion of *robust performance*.

At this point it is apparent our work in Chapter 8 will pay large dividends provided that we can rigorously connect the interconnection in Figure 9.1 to the simplified setup in Figure 9.2. Although the diagrams appear to make this obvious, we must be careful to ensure that the our diagrams are backed by well-defined mathematical operations.

In particular, notice that the above procedure of closing first the loop on K and subsequently studying the dependence on Δ, is in effect the opposite of the order in which the problem is naturally formulated. It is in fact the interconnection of G and Δ, depicted in Figure 9.3 below, which must first be well-defined, since this is the uncertain open-loop system we desire to control.

Focusing on the latter picture, we notice a difficulty: all of our studies in Chapter 8 assumed that Δ was an operator on L_2; for the above model to be meaningful under such circumstances would require that the signal p be an element of L_2. However before interconnection with the controller, it is not natural to impose such restriction. In fact the plant G may well be unstable in the open loop, and one of the purposes of control may be stabilization. In other words, the fact that signals are in L_2 should be a *conclusion*, not

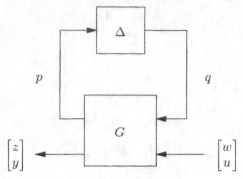

Figure 9.3. Uncertain plant

an assumption, of our stability analysis, and our uncertain models should be meaningful even if stabilization has not yet been performed.

Thus to make our study rigorous will require the extension of our uncertain models to signals which are not necessarily elements of L_2. This is tackled in the following section.

9.1   Extended spaces and stability

Following up on the preceding discussion, we begin the section by emphasizing again that the state space plant G does not necessarily define a bounded LTI operator on L_2, nor does the state space controller K. In particular, their transfer functions $\hat{G}(s)$ and $\hat{K}(s)$ are not necessarily in RH_∞.

More explicitly, consider a general state space system described by a minimal realization (A, B, C, D). From Chapter 2, the input–output mapping R defined by this system is given by

$$(Ru)(t) = C \int_0^t e^{A(t-\tau)} Bu(\tau)\, d\tau + Du(t).$$

If A is not Hurwitz, the fact that $u(t) \in L_2$ will not imply that Ru is in L_2; therefore R does not define a bounded mapping on L_2. However, it is easily verified that, for any $T > 0$, the inequality

$$\int_0^T |(Ru)(t)|^2\, dt < \infty \tag{9.2}$$

is satisfied. So although Ru may not be an element of L_2, the energy of $(Ru)(t)$ is bounded on any finite interval, provided the same is true of $u(t)$.

This observation motivates the definition of the space of functions L_2-extended[2] denoted by L_{2e}. Define

$$L_{2e} = \{\ w : \mathbb{R} \to \mathbb{C}^n \text{ satisfying } P_T w \in L_2[0, \infty), \text{ for all } T > 0\}.$$

Recall that P_T is the mapping which truncates the support of any function to the interval $(-\infty, T]$; further note that, by definition, $P_T w$ only will have support on $[0, T]$ when w is in L_{2e}. Thus this is just the space of functions whose energy is finite on any finite interval. The space L_{2e} is a vector space by pointwise addition, where we equate two elements v and g if they satisfy $\|P_T(v - g)\| = 0$, for all $T \geq 0$. Thus it contains L_2 as a subspace; however, L_{2e} itself is not a normed space.

Clearly since we have a vector space structure we can talk about linear mappings on L_{2e}. Now it follows from this definition that any state space system such as R defined above satisfies

$$R : L_{2e} \to L_{2e}.$$

Namely, any state space system defines a linear mapping on L_{2e}.

We make some more observations about these mappings. First, if $R : L_{2e} \to L_{2e}$ is bijective the inverse R^{-1} is defined in the usual sense. Also a mapping R on L_{2e} will be called *causal* if it satisfies the familiar relationship

$$P_T R = P_T R P_T, \text{ for all } T \geq 0.$$

In particular the mapping defined by a state space system is causal. Finally, we say that a mapping on L_{2e} is *bounded* if it defines a bounded linear mapping when restricted to L_2.

We can now use these notions to study the stability of Figure 9.1. We make the following assumptions:

- all signals w, p, q, u, y, and z are assumed a *priori* to be in L_{2e};

- G and K are state space systems;

- the operator Δ is well-defined on L_{2e}.

The first assumption says that we set up our stability problem in the space L_{2e}. This is not a very restrictive scenario, however it does a priori preclude issues of finite escape time which arise in some nonlinear models; a detailed discussion is beyond our scope here. The second is our customary assumption on generalized plant and controller, and guarantees, in particular, that G and K are well-defined over the signal space L_{2e} under consideration. The final assumption postulates the same property for our uncertainty class Δ; we will later discuss conditions on Δ that make this extension possible. We are now ready to give our main definition.

---

[2]This space is also called L_2-local, but the term L_2-extended is more apt here.

Definition 9.1. *Consider the system of Figure 9.1, under the above assumptions. We say the system is stable if:*

(i) *Given any w in L_{2e}, and initial states of G and K, there exist unique solutions in L_{2e} for p, q, u, y, and z. Furthermore if w is in L_2, so are the other five functions.*

(ii) *The states of G and K tend asymptotically to zero, from any initial condition, when $w = 0$.*

(iii) *The maps w to p, q, u, y and z are all bounded on L_2, when the initial states of G and K are zero.*

If stability holds for all Δ in an uncertainty class $\mathbf{\Delta}$, we say the system is robustly stable.

9.1.1   A general result on stability

Armed with our new definition, we now return to our problem of studying stability for the Figure 9.1 by means of the simplified setup of Figure 9.2. We have the following basic result.

Theorem 9.2. *Pertaining to Figure 9.1, suppose that*

(a) *Δ is a linear mapping on L_{2e} and is bounded on L_2;*

(b) *the state space system G is internally stabilized by the state space controller K;*

(c) *denoting $M = \underline{S}(G, K)$, the inverse $(I - M_{11}\Delta)^{-1}$ exists as a mapping on L_{2e} and is bounded on the normed space L_2.*

Then the interconnection of Figure 9.1 is stable.

Proof. We must prove conditions (i), (ii) and (iii) in Definition 9.1. We start with (i). Choose w in L_{2e} and an initial state $(x(0), x_K(0))$ for the state space systems G and K. Now the equations governing this system are the state space equations for G and K, and the operator equation $q = \Delta p$. We initially suppose that solutions for p, q, u, y, and z exist in L_{2e}, and we will show that they are unique. Now by routine *algebraic* manipulations we can write the equations that govern the feedback loops of this system as

$$q = \Delta p$$
$$\dot{x}_{cl} = A_{cl}x_{cl} + B_{cl1}q + B_{cl2}w, \text{ where } x_{cl}(0) = (x(0), x_K(0))$$
$$p = C_{cl1}x_{cl} + D_{cl11}q + D_{cl12}w.$$

Here A_{cl}, B_{cl}, C_{cl}, and D_{cl} is the state space representation for the interconnection of G and K, and thus we know from hypothesis (b) that A_{cl} is Hurwitz. Clearly this is a state space realization for the operator M. From this we see that

$$p = M_{11}q + M_{12}w + C_{cl0}\tilde{A}_{cl}x_{cl}(0),$$

where $\tilde{A}_{cl} : \mathbb{C}^n \to L_2$ via the relationship $\tilde{A}_{cl}x_{cl}(0) = e^{A_{cl}t}x_{cl}(0)$. Now applying assumption (c) above we have that

$$p = (I - M_{11}\Delta)^{-1}M_{12}w + (I - M_{11}\Delta)^{-1}C_{cl0}\tilde{A}_{cl}x_{cl}(0). \qquad (9.3)$$

Thus we see that p must be uniquely determined since w and $x_{cl}(0)$ are fixed, and then the function q is uniquely specified by p. Since the inputs q, w, and the initial conditions of the *state space* systems G and K are all specified, we know from (b) and the work in Chapter 5 that u, y, and z are unique and in L_{2e}.

To complete our demonstration of (i) we must show the existence of solutions. To see this simply start by defining p from (9.3), given any $w \in L_{2e}$ and initial condition $x_{cl}(0)$, and essentially reverse the above argument. Also if w is in L_2, then by the boundedness of M, $(I - M_{11}\Delta)^{-1}$, and $\tilde{A}_{cl}$ we see that p must be in L_2. Immediately it follows that q must be in L_2 since Δ is bounded, and again by the nominal stability in (b) we have now that the other functions are also in L_2.

We now prove (ii). From above we know that if $w = 0$ we have

$$p = (I - M_{11}\Delta)^{-1}C_{cl0}\tilde{A}_{cl}x_{cl}(0),$$

and so $q = \Delta p$ is in L_2 since A_{cl} is Hurwitz and Δ is bounded. Continuing under these conditions we have

$$\dot{x}_{cl} = A_{cl}x_{cl} + B_{cl0}q, \text{ with some initial condition } x_{cl}(0).$$

Since q is in L_2 and A_{cl} is Hurwitz, it follows from the exercises of Chapter 3 that $x_{cl}(t) \to 0$ as $t \to \infty$.

Finally we need to show that the maps from w to the other functions are bounded on L_2. From above we already know that the maps from w to p and q are bounded on L_2. This means that the map from w to the state x_{cl} must be bounded on L_2 since A_{cl} is Hurwitz. Finally the functions z, u, and y are all simply given by matrices times w, x_{cl}, and q and so the maps from w to them must also be bounded on L_2.   ∎

We remark that the mapping w to z is given by (9.1) follows routinely from the above formula for p, and the definition of the operator M.

The preceding theorem describes what we should aim for if we intend to apply the tools of Chapter 8 to our stability problem. Namely, we must ensure that our uncertainty operator Δ, and the inverse $(I - M_{11}\Delta)^{-1}$ are not only defined and bounded over L_2 but that they admit suitable extensions to L_{2e}. The topic of when such extensions exist will be tackled next.

Before doing this we make an additional remark. We have only shown the *sufficiency* of the conditions in Theorem 9.2 for stability. In fact, to show these conditions are also necessary would require a modification to our definition of stability, requiring that the system be tolerant to disturbances injected directly into the loop between G and Δ. Since this loop is not "physical" such an added condition may not be natural, and in this sense our method may be conservative. However, such injected disturbances may be appropriate for the modeling of nonzero initial conditions in the dynamics described by the perturbations.

9.1.2 Causality and maps on L_2-extended

From our discussion so far it is clear that any bounded mapping on L_{2e} defines an element of $\mathcal{L}(L_2)$ simply by restricting its domain. We would now like to examine the converse problem: given $Q \in \mathcal{L}(L_2)$, when can the domain of Q can be extended to L_{2e} while maintaining the linearity of Q? Consider the following example.

Example:

Define the convolution operator Q on L_2 by

$$(Qu)(t) = \int_t^\infty e^{(t-\tau)} u(\tau) d\tau,$$

which can be shown to have an L_2-induced norm of one. However we see that it is not possible to make this integral meaningful for some inputs in L_{2e}; for instance set $u(\tau) = e^\tau$, which yields a diverging integral, for *every* $t \geq 0$.   □

The above difficulty is strongly related to the non-causality of the operator Q. In contrast, we now see that causal operators on L_2 can always be extended to L_{2e}. To this end observe that given any causal operator $Q \in \mathcal{L}(L_2)$, we can define $Qu = v$, for $u \in L_{2e}$, by

$$P_T v = P_T Q u := P_T Q P_T u,$$

for each T. We leave it to the reader to verify that this provides a valid definition, namely, that if $P_{T_1} v$ and $P_{T_2} v$ are defined as above for $T_1 < T_2$, then $P_{T_1} P_{T_2} v = P_{T_1} v$; this follows from the causality of Q in L_2. Summarizing, we have —

Proposition 9.3. *Suppose Q is a causal operator in $\mathcal{L}(L_2)$. Then Q defines a linear causal mapping on L_{2e}, which is bounded on L_2.*

To complete this discussion it is natural to inquire whether causal operators are the only ones that extend to L_{2e}. In other words, is existence of an extension to L_{2e} any more general than causality? The following example shows the answer is affirmative.

Example:

Let us define the mapping Q on L_2 by

$$(Qu)(t) = u(2t), \text{ for } t \geq 0.$$

Clearly this mapping is linear. It is straightforward to verify that $2\|Qu\|^2 = \|u\|^2$, for any u in L_2, and so Q is a bounded operator with an induced norm of $\frac{1}{\sqrt{2}}$. This operator is *anti-causal* since the value of $(Qu)(t)$ is equal to the value of u at time $2t$ in the future. From the above formula for Q we see that it can be immediately extended to a linear mapping on L_{2e}. $\square$

Therefore we see that postulating an extension to L_{2e} is in principle less restrictive than imposing causality. Nevertheless, causal operators constitute the simplest setting in which such an extension is guaranteed, and for this reason we concentrate on causal operators from now on. In particular the following corollary to Theorem 9.2, while somewhat more restrictive, is general enough for most purposes. Its proof is immediate from Theorem 9.2 by applying Proposition 9.3.

Theorem 9.4. *Suppose that* $\Delta \in \mathcal{L}(L_2)$ *is causal, and the assumption of nominal stability holds. If* $(I - M_{11}\Delta)^{-1}$ *exists as a bounded, causal mapping on* L_2, *then the feedback system in Figure 9.1 is stable.*

Thus we find that a theory of uncertain systems suitable for questions of stability can be naturally obtained if we use *causal* perturbations in our uncertainty modeling, and insist, in addition to well-connectedness, that the inverse operator $(I - M_{11}\Delta)^{-1}$ be causal. This will be our setting for the remainder of the chapter.

Before proceeding, we briefly remark on the implications of the causality restriction from the point of view of uncertainty modeling. The usual argument in favor of causal models is physical realizability; in this regard, it is natural that a component such as K, which directly models a physical controller, be taken as causal. However the situation is less obvious in the case of a perturbation Δ, which often does not directly model a physical system, but instead is a part of a *parametrization* of a relation (e.g., $\mathcal{R}_a$) which is the actual model of a physical component. In these cases the main reason to choose a causal model is mathematical convenience: this allows us to isolate Δ as though it were a component, and in this way to standardize a large class of uncertainty models into a single prototypical form. While this might restrict somewhat our modeling possibilities, experience seems to indicate that the restriction is not severe. Furthermore in the next section we will see that the restriction to causal operators has no effect on the robustness conditions we are able to obtain.

A final question in regard to causality is whether we must explicitly check the causality of $(I - M_{11}\Delta)^{-1}$, or is it automatically guaranteed once M_{11} and Δ are assumed to be causal. In the exercises it is shown by counterexample that no such guarantees exist.

Nevertheless, we will now see that the causality of $(I - M_{11}\Delta)^{-1}$ can be assured if Δ is causal and a spectral radius condition is satisfied. This result is significant for our robustness analysis because it shows that if $(I - M_{11}\Delta)^{-1}$ is not causal, then there exists a complex scalar $|\alpha| \leq 1$, such that $I - \alpha M_{11}\Delta$ is singular. In words, if the above inverse is not causal, then there exists a *smaller* perturbation of the same structure which causes our well-connectedness condition to fail. Now for the result.

Proposition 9.5. *Let M_{11} and Δ be causal operators in $\mathcal{L}(L_2)$. If $\rho(M_{11}\Delta) < 1$, then $(I - M_{11}\Delta)^{-1}$ exists in $\mathcal{L}(L_2)$ and is causal.*

Proof. Clearly $I - M_{11}\Delta$ is invertible since its spectrum is by assumption inside the unit disc. Also by this assumption we know from Chapter 3 that its inverse has a series expansion

$$(I - M_{11}\Delta)^{-1} = \sum_{k=0}^{\infty} (M_{11}\Delta)^k.$$

Now since M_{11}, Δ, are causal, so is $(M_{11}\Delta)^k$ for every k, and thus also the series which gives $(I - M_{11}\Delta)^{-1}$.   ∎

Having clearly established the relationship between feedback systems involving uncertainty and the configuration of Chapter 8, we are now ready to apply its tools to analyze robust stability and performance.

9.2   Robust stability and performance

In this section we extend the machinery of Chapter 8 to confront robustness for the general feedback control setup of Figure 9.1. We concentrate on the *analysis* of robust stability and performance for a given controller K, and the main objective is to adapt our work on well-connectedness to include the causality requirements of Theorem 9.4.

In this regard, we assume that K is nominally stabilizing so that $M = \underline{S}(G, K)$ is an LTI, causal system with transfer function in RH_∞. Also we define our uncertainty set to be of the form

$$\Delta^c = \{\Delta \in \mathbf{\Delta}, \ \Delta \text{ is causal } \} \subset \mathcal{L}(L_2)$$

where $\mathbf{\Delta}$ is one of the uncertainty classes considered in Chapter 8. Namely the subset of causal operators with the same spatial and dynamic structure. We now give the following definition, which is motivated by Theorem 9.4, and can be interpreted in terms of Figure 9.2.

- The uncertain system $(M_{11}, \mathbf{\Delta}^c)$ is *robustly stable* if $(I - M_{11}\Delta)^{-1}$ exists in $\mathcal{L}(L_2)$ and is causal, for each Δ in $\mathbf{\Delta}^c$;

- The uncertain system $(M, \mathbf{\Delta}^c)$ has *robust performance* if $(M_{11}, \mathbf{\Delta}^c)$ is robustly stable and $\|\bar{S}(M, \Delta)\| < 1$, for every $\Delta \in \mathbf{\Delta}^c$.

Stability here is identical to our definition of well-connectedness in Chapter 8, except that now we demand that the relevant inverse be causal. We define robust performance to be stability plus contractiveness of the closed-loop, which again is identical to the performance criterion we imposed in Chapter 8.

By virtue of Theorem 9.4, it is clear that robust stability of (M_{11}, Δ^c) directly implies the stability of the configuration of Figure 9.1 for every $\Delta \in \Delta^c$. Furthermore, robust performance implies the stability plus contractiveness of the mapping from w to z in Figure 9.1, once again for every $\Delta \in \Delta^c$.

The remainder of this section is devoted to methods of analyzing robust stability and performance. As we will see, the results will be completely parallel to those on robust well-connectedness in the previous chapter.

9.2.1 Robust stability under arbitrary structured uncertainty

We first consider the uncertainty structure

$$\Delta_a^c = \{\mathrm{diag}(\Delta_1, \ldots, \Delta_d) : \Delta_k \in \mathcal{L}(L_2), \Delta_k \text{ causal }, \|\Delta_k\| \leq 1\},$$

which is obtained by imposing causality on our class Δ_a from Chapter 8. We will assume nominal stability, so M in Figure 9.2 will be a causal, bounded LTI system.

The main result, stated below, tells us that robust stability can be studied analogously to the corresponding robust well-connectedness problem. The set Θ_a is the commutant set corresponding to Δ_a, and is used here with the new set Δ_a^c. Also $P\Theta_a$ is the set of positive elements associated with Θ_a.

Theorem 9.6. *Suppose that M is a causal, bounded LTI operator. Then the following are equivalent:*

(i) *The uncertain system (M_{11}, Δ_a^c) is robustly stable.*

(ii) *The inequality $\inf_{\Theta \in \Theta_a} \|\Theta M_{11} \Theta^{-1}\| < 1$ holds.*

Proof. The sufficiency direction (ii) implies (i) is essentially identical to our work in Chapter 8. From the small gain condition and the fact that Θ commutes with Δ we have

$$1 > \|\Theta M_{11} \Theta^{-1}\| \geq \rho(\Theta M_{11} \Theta^{-1} \Delta) = \rho(M_{11}\Delta).$$

Thus $I - M_{11}\Delta$ must be nonsingular, and $(I - M_{11}\Delta)^{-1}$ is causal by virtue of Proposition 9.5.

For the necessity direction (i) implies (ii), we can take advantage of most of our work in §8.2.2. Referring back to this section (in particular to the definitions of the sets ∇ and Π), we see that the following step, analogous to Proposition 8.9, is all that is required.

Proposition 9.7. *Suppose that* $(M_{11}, \mathbf{\Delta}_a^c)$ *is robustly stable. Then the sets* Π *and* ∇ *are strictly separated, i.e.,*

$$D(\Pi, \nabla) := \inf_{r \in \Pi, y \in \nabla} |r - y| > 0.$$

In fact, given this proposition we can then follow the rest of §8.2.2, leading to condition (ii). Proposition 9.7 stated above is proved in Appendix B. ∎

We now turn to time invariant uncertainty.

9.2.2 Robust stability under LTI uncertainty

For the case of LTI uncertainty we recall the definition of the structure $\mathbf{\Delta}_{TI}$ from Chapter 8.

$$\mathbf{\Delta}_{TI} = \{\Delta = \operatorname{diag}(\Delta_1 \ldots, \Delta_d) \in \mathcal{L}(L_2) : \Delta \text{ time invariant and } \|\Delta\| \leq 1\}.$$

Now each Δ in $\mathbf{\Delta}_{TI}$ is LTI on the space of functions $L_2[0, \infty)$. By Theorem 3.30 we see that this immediately means that Δ must be *causal*. That is, our LTI uncertainty set is already composed of operators that are causal.

Having made this observation, we can state the following theorem.

Theorem 9.8. *Let* M *be a causal, bounded LTI system with transfer function* $\hat{M} \in RH_\infty$. *Then the following are equivalent:*

(i) The uncertain system $(M_{11}, \mathbf{\Delta}_{TI})$ *is robustly stable.*

(ii) The structured singular value satisfies $\sup_{\omega \in \mathbb{R}} \mu(\hat{M}_{11}(j\omega), \mathbf{\Delta}_{s,f}) < 1.$

Proof. By Theorem 8.22 it is sufficient to show that robust well-connectedness of $(M_{11}, \mathbf{\Delta}_{TI})$ is equivalent to robust stability. Since by definition the latter condition implies the former, we need only prove the converse.

Thus we need to show that if $I - M_{11}\Delta$ is nonsingular, where $\Delta \in \mathbf{\Delta}_{TI}$, then $(I - M_{11}\Delta)^{-1}$ is causal. This follows directly from frequency domain analysis. The nonsingularity condition implies that $(I - \hat{M}(s)\hat{\Delta}(s))^{-1}$ is in H_∞, and thus by Theorem 3.30 we know it represents a causal operator. ∎

Thus we see that robust stability of $(M_{11}, \mathbf{\Delta}_{TI})$ is equivalent to robust well-connectedness; they both amount to a complex structured singular value condition parametrized over frequency. Let us move forward to robust performance.

9.2.3 Robust performance analysis

Having characterized robust stability, we now turn to the robust performance problem for the configuration of Figure 9.2, with the performance measured by the induced norm of the map from w to z.

In Chapter 8 we explored a similar issue in the context of abstract L_2 operators, and saw how a restriction on the norm of $\bar{S}(M, \Delta)$ could be turned into a robust well-connectedness problem with an augmented uncertainty structure. We now see that the procedure carries through in the context of causal uncertainties for the classes treated in the preceding sections.

Proposition 9.9. *Suppose M is a causal LTI operator on $L_2[0, \infty)$. Define the perturbation set*

$$\boldsymbol{\Delta}^c_{a,p} = \left\{ \begin{bmatrix} \Delta_u & 0 \\ 0 & \Delta_p \end{bmatrix} : \Delta_u \in \boldsymbol{\Delta}^c_a, \ \Delta_p \in \mathcal{L}(L_2) \text{ causal and } \|\Delta_p\| \leq 1 \right\}.$$

The following are equivalent:

(a) *The uncertain system $(M, \boldsymbol{\Delta}^c_a)$ satisfies robust performance: it is robustly stable and $\|\bar{S}(M, \Delta_u)\| < 1$, for every $\Delta_u \in \boldsymbol{\Delta}^c_a$.*

(b) *The uncertain system $(M, \boldsymbol{\Delta}^c_{a,p})$ is robustly stable.*

(c) *There exists $\Theta \in \boldsymbol{\Theta}_a$ such that*

$$\left\| \begin{bmatrix} \Theta & 0 \\ 0 & I \end{bmatrix} M \begin{bmatrix} \Theta & 0 \\ 0 & I \end{bmatrix}^{-1} \right\| < 1. \tag{9.4}$$

Proof. The fact that (a) implies (b) follows analogously to Proposition 8.3 in Chapter 8. Here we have by hypothesis that $(I - M_{11}\Delta_u)$ has a causal, bounded inverse, and the identity (8.2) can be written:

$$\begin{bmatrix} I - M_{11}\Delta_u & -M_{12}\Delta_p \\ -M_{21}\Delta_u & I - M_{22}\Delta_p \end{bmatrix} = $$

$$\begin{bmatrix} I & 0 \\ -M_{21}\Delta_u(I - M_{11}\Delta_u)^{-1} & I \end{bmatrix} \begin{bmatrix} I - M_{11}\Delta_u & -M_{12}\Delta_p \\ 0 & I - \bar{S}(M, \Delta_u)\Delta_p \end{bmatrix}.$$

By Proposition 9.5, the hypothesis implies that $I - \bar{S}(M, \Delta_u)\Delta_p$ has a causal, bounded inverse, so the same happens with $I - M\Delta$.

The step from (b) to (c) is a direct application of Theorem 9.6, once we recognize that the scaling set corresponding to $\boldsymbol{\Delta}^c_{a,p}$ is

$$\boldsymbol{\Theta}_{ap} = \left\{ \begin{bmatrix} \Theta & 0 \\ 0 & \theta_p I \end{bmatrix} : \Theta \in \boldsymbol{\Theta}_a, \theta_p \neq 0 \right\}$$

and we realize that θ_p can be normalized to one without loss of generality.

Finally to prove (c) to (a), first note that the top-left block of (9.4) implies $\|\Theta M_{11}\Theta^{-1}\| < 1$ and hence robust stability. Next, we rewrite (9.4) in operator inequality form,

$$\Psi := M^* \begin{bmatrix} \Theta_0 & 0 \\ 0 & I \end{bmatrix} M - \begin{bmatrix} \Theta_0 & 0 \\ 0 & (1 - \epsilon)I \end{bmatrix} < 0,$$

where $\Theta_0 = \Theta^*\Theta$ and by continuity we may introduce a suitably small $\epsilon > 0$ in the bottom block.

Applying the signals q and w from Figure 9.2, we have

$$
\begin{aligned}
0 &\geq \left\langle \begin{bmatrix} q \\ w \end{bmatrix}, \Psi \begin{bmatrix} q \\ w \end{bmatrix} \right\rangle \\
&= \left\langle \begin{bmatrix} p \\ z \end{bmatrix}, \begin{bmatrix} \Theta_0 & 0 \\ 0 & I \end{bmatrix} \begin{bmatrix} p \\ z \end{bmatrix} \right\rangle - \left\langle \begin{bmatrix} q \\ w \end{bmatrix}, \begin{bmatrix} \Theta_0 & 0 \\ 0 & (1-\epsilon)I \end{bmatrix} \begin{bmatrix} q \\ w \end{bmatrix} \right\rangle \\
&= \sum_{k=1}^{d} \theta_k (\|p_k\|^2 - \|q_k\|^2) + \|z\|^2 - (1-\epsilon)\|w\|^2 \\
&\geq \|z\|^2 - (1-\epsilon)\|w\|^2.
\end{aligned}
$$

In the last step, the fact that $q_k = \Delta_k p_k$ and $\|\Delta_k\| \leq 1$ was used. The final inequality says that the map from w to z has induced norm less than one. ∎

We now give an analogous result for the time invariant case.

Proposition 9.10. *Suppose the configuration of Figure 9.2 is nominally stable, and the uncertainty class is $\boldsymbol{\Delta_{TI}}$. Define the perturbation set*

$$
\boldsymbol{\Delta_{TI,p}} = \left\{ \begin{bmatrix} \Delta_u & 0 \\ 0 & \Delta_p \end{bmatrix} : \Delta_u \in \boldsymbol{\Delta_{TI}}, \ \Delta_p \in \mathcal{L}(L_2) \ LTI, \ causal, \ \|\Delta_p\| \leq 1 \right\}.
$$

Notice that $\hat{\Delta}(s)$ takes values in $\boldsymbol{\Delta_{s,f+1}}$ for any $\Delta \in \boldsymbol{\Delta_{TI,p}}$ (i.e., there is one extra full block with respect to the uncertainty structure $\boldsymbol{\Delta_{s,f}}$). The following are equivalent:

(a) *The uncertain system $(M, \boldsymbol{\Delta_{TI}})$ satisfies robust performance: it is robustly stable and $\|\bar{S}(M, \Delta_u)\| < 1$, for every $\Delta_u \in \boldsymbol{\Delta_{TI}}$.*

(b) *The uncertain system $(M, \boldsymbol{\Delta_{TI,p}})$ is robustly stable.*

(c)

$$
\sup_{\omega \in \mathbb{R}} \mu(\hat{M}(j\omega), \boldsymbol{\Delta_{s,f+1}}) < 1. \tag{9.5}
$$

The proof follows along similar lines as that of Proposition 9.9 and is left as an exercise. Notice that here the so-called performance block Δ_p can be taken to be LTI.

This concludes our work on *analysis* of robust stability and performance in causal feedback systems. We now turn our attention to the *synthesis* of controllers.

9.3   Robust controller synthesis

In this section we tackle the problem of synthesizing feedback controllers for robust performance in the presence of uncertainty. In other words, going back to our original setup of Figure 9.1 our objective is to design the controller K as in earlier chapters; however, we are no longer satisfied with nominal stability and performance, but wish these properties to be maintained throughout the causal uncertainty class $\mathbf{\Delta}^c$.

As in previous sections, we will concentrate here on the L_2-induced norm from w to z as a measure of performance. The advantage of this choice is that, as just seen, robust performance can be studied with completely analogous methods as robust stability. We thus focus our discussion on the *robust stabilization* problem: given an LTI state space system G, and an uncertainty class $\mathbf{\Delta}^c$, find a controller K such that

- K internally stabilizes G;

- the uncertain system $(M, \mathbf{\Delta}^c)$ is robustly stable, where $M :=$ $\underline{S}(G, K)$.

We will only consider perturbation structures for which we have developed the necessary analysis tools. In this way our problem turns into the search for a controller K that satisfies a precise mathematical condition. The simplest case is when we have unstructured uncertainty, i.e., when

$$\mathbf{\Delta}^c = \{\Delta \in \mathcal{L}(L_2), \ \Delta \text{ causal} , \|\Delta\| \leq 1\}.$$

In this case a controller K will be robustly stabilizing if and only if it satisfies

$$\|\underline{S}(G, K)\|_\infty < 1;$$

this small gain property follows, for instance, from our more general result of Theorem 9.6. So our problem reduces to H_∞ synthesis; in fact, this observation is the main motivation behind H_∞ control as a design method, completing the discussion which was postponed from Chapter 7.

What this method does not consider is uncertainty structure, which as we have seen can arise in two ways:

- Uncertainty models derived from interconnection of more simple component structures.

- Performance block, added to account for a performance specification in addition to robust stability.

The remainder of this chapter is dedicated to the robust synthesis problem under the structured uncertainty classes $\mathbf{\Delta}_a^c$ and $\mathbf{\Delta}_{TI}$ which we have studied in detail.

9.3.1 Robust synthesis against $\mathbf{\Delta}_a^c$

We begin our discussion with the class $\mathbf{\Delta}_a^c$. Let $\mathcal{K}$ be the set of state space controllers K which internally stabilize G. Then robust synthesis is reduced, via Theorem 9.6, to the optimization problem

$$\inf_{\Theta \in P\Theta_a, K \in \mathcal{K}} \|\Theta^{\frac{1}{2}} \underline{S}(G, K)\Theta^{-\frac{1}{2}}\|,$$

which is H_∞ synthesis under constant scaling matrices; we have constrained the latter to the positive set $P\Theta_a$, which is a slight change from Theorem 9.6 but is clearly inconsequential.

Robust stabilization is achieved if and only if the above infimum is less than one, since the set Δ_a^c contains only contractive operators. We therefore have a recipe to tackle the problem. The main question is whether the above optimization can be reduced to a tractable computational method. Note that we have already considered and answered two restrictions of this problem:

- for fixed Θ we have H_∞ synthesis;

- for fixed K we have robustness analysis over Δ_a^c.

As seen in previous chapters, both these subproblems can be reduced to LMI computation, and the new challenge is the joint search over the matrix variable Θ and the system variable K.

The first result is encouraging. It says that we can restrict our search for a controller to those which have the same dynamic order as the plant.

Proposition 9.11. *Let G be an LTI system of order n. If*

$$\inf_{\Theta \in P\Theta_a, K \in \mathcal{K}} \|\Theta^{\frac{1}{2}} \underline{S}(G, K)\Theta^{-\frac{1}{2}}\| < 1, \tag{9.6}$$

then there exits a controller $K \in \mathcal{K}$ of order at most n, such that

$$\|\Theta^{\frac{1}{2}} \underline{S}(G, K)\Theta^{-\frac{1}{2}}\| < 1.$$

The proposition states that if the the robust stabilization problem can be solved, then it can be solved using a controller K of state dimension no greater than the plant G. Figure 9.4 illustrates the constructions used in the proof.

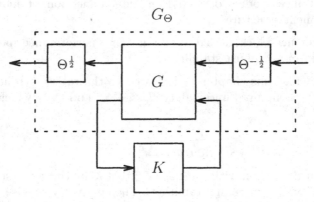

Figure 9.4. Scaled synthesis problem

Proof. Consider Θ, K satisfying $\|\Theta^{\frac{1}{2}}\underline{S}(G,K)\Theta^{-\frac{1}{2}}\| < 1$. Defining

$$G_\Theta := \begin{bmatrix} \Theta^{\frac{1}{2}} & 0 \\ 0 & I \end{bmatrix} G \begin{bmatrix} \Theta^{-\frac{1}{2}} & 0 \\ 0 & I \end{bmatrix},$$

(see Figure 9.4) we have

$$\Theta^{\frac{1}{2}}\underline{S}(G,K)\Theta^{-\frac{1}{2}} = \underline{S}(G_\Theta,K).$$

Also G_Θ has order n, so from Chapter 7 we know that the condition

$$\|\underline{S}(G_\Theta,K)\| < 1$$

can be achieved with K of order no greater than n. ∎

Therefore we see that for purposes of robust synthesis over $\mathbf{\Delta}_a^c$, we can confine our attention to controllers of the order of the generalized plant G. In particular, our optimization problem is *finite*, in the sense that it involves a finite set of variables; they are, Θ, and the controller state space matrices A_K, B_K, C_K, D_K of a fixed order. This is substantially simpler than the problem we initially started with, namely, arbitrary state space controllers and uncertainties in an abstract operator space. One could search over this finite number of variables with a variety of computational tools.

Unfortunately this still does not mean that the problem is computationally tractable by the standards we have employed in this course, namely, problems which can be globally optimized in a moderate amount of computational time, such as LMI optimization problems. Can a further simplification be achieved? To explore this issue it is natural to combine our robust stability condition from §9.2.1, which resulted from robustness analysis, with our LMI method for H_∞ synthesis described in Chapter 7. This is done next.

We begin with a state space realization for the generalized plant

$$\hat{G}(s) = \left[\begin{array}{c|cc} A & B_1 & B_2 \\ \hline C_1 & D_{11} & D_{12} \\ C_2 & D_{21} & 0 \end{array} \right],$$

and incorporate the scalings $\Theta^{\frac{1}{2}}$ and $\Theta^{-\frac{1}{2}}$ to obtain

$$\hat{G}_\Theta(s) = \left[\begin{array}{c|cc} A & B_1\Theta^{-\frac{1}{2}} & B_2 \\ \hline \Theta^{\frac{1}{2}}C_1 & \Theta^{\frac{1}{2}}D_{11}\Theta^{-\frac{1}{2}} & \Theta^{\frac{1}{2}}D_{12} \\ C_2 & D_{21}\Theta^{-\frac{1}{2}} & 0 \end{array} \right].$$

Next, we apply Theorem 7.10 to obtain necessary and sufficient conditions for the existence of a controller satisfying $\|\underline{S}(G_\Theta,K)\|_\infty < 1$. These take the form

$$\begin{bmatrix} N_o & 0 \\ 0 & I \end{bmatrix}^* \begin{bmatrix} A^*X + XA & XB_1\Theta^{-\frac{1}{2}} & C_1^*\Theta^{\frac{1}{2}} \\ \Theta^{-\frac{1}{2}}B_1^*X & -I & \Theta^{-\frac{1}{2}}D_{11}^*\Theta^{\frac{1}{2}} \\ \Theta^{\frac{1}{2}}C_1 & \Theta^{\frac{1}{2}}D_{11}\Theta^{-\frac{1}{2}} & -I \end{bmatrix} \begin{bmatrix} N_o & 0 \\ 0 & I \end{bmatrix} < 0, \quad (9.7)$$

$$\begin{bmatrix} N_c & 0 \\ 0 & I \end{bmatrix}^* \begin{bmatrix} AY + YA^* & YC_1^*\Theta^{\frac{1}{2}} & B_1\Theta^{-\frac{1}{2}} \\ \Theta^{\frac{1}{2}}C_1Y & -I & \Theta^{\frac{1}{2}}D_{11}\Theta^{-\frac{1}{2}} \\ \Theta^{-\frac{1}{2}}B_1^* & \Theta^{-\frac{1}{2}}D_{11}^*\Theta^{\frac{1}{2}} & -I \end{bmatrix} \begin{bmatrix} N_c & 0 \\ 0 & I \end{bmatrix} < 0, \quad (9.8)$$

$$\begin{bmatrix} X & I \\ I & Y \end{bmatrix} \geq 0, \quad (9.9)$$

where N_o and N_c are full rank matrices whose images satisfy

$$\operatorname{Im}N_o = \operatorname{Ker} \begin{bmatrix} C_2 & D_{21}\Theta^{-\frac{1}{2}} \end{bmatrix},$$
$$\operatorname{Im}N_c = \operatorname{Ker} \begin{bmatrix} B_2^* & D_{12}^*\Theta^{\frac{1}{2}} \end{bmatrix}.$$

So we see that the robust stabilization problem is solvable if and only if conditions (9.7–9.9) in the variables X, Y, and Θ are satisfied. This once again emphasizes the finite dimensionality of the problem.

To get these conditions into a more transparent form, it is useful to redefine the outer multiplication factors so that they are independent of Θ. Define $\bar{N}_o$, $\bar{N}_c$ to be full rank matrices whose images satisfy

$$\operatorname{Im}\bar{N}_o = \operatorname{Ker} \begin{bmatrix} C_2 & D_{21} \end{bmatrix},$$
$$\operatorname{Im}\bar{N}_c = \operatorname{Ker} \begin{bmatrix} B_2^* & D_{12}^* \end{bmatrix}.$$

Then $\bar{N}_o$, $\bar{N}_c$ are constant, and clearly we can take

$$N_o = \begin{bmatrix} I & 0 \\ 0 & \Theta^{\frac{1}{2}} \end{bmatrix} \bar{N}_o, \quad (9.10)$$

$$N_c = \begin{bmatrix} I & 0 \\ 0 & \Theta^{-\frac{1}{2}} \end{bmatrix} \bar{N}_c. \quad (9.11)$$

Substituting (9.10) into (9.7) gives

$$\begin{bmatrix} \bar{N}_o & 0 \\ 0 & I \end{bmatrix}^* \begin{bmatrix} A^*X + XA & XB_1 & C_1^*\Theta^{\frac{1}{2}} \\ B_1^*X & -\Theta & D_{11}^*\Theta^{\frac{1}{2}} \\ \Theta^{\frac{1}{2}}C_1 & \Theta^{\frac{1}{2}}D_{11} & -I \end{bmatrix} \begin{bmatrix} \bar{N}_o & 0 \\ 0 & I \end{bmatrix} < 0.$$

As a final simplification, we can multiply the last row and column of the preceding LMI by $\Theta^{\frac{1}{2}}$. This gives the condition

$$\begin{bmatrix} \bar{N}_o & 0 \\ 0 & I \end{bmatrix}^* \begin{bmatrix} A^*X + XA & XB_1 & C_1^*\Theta \\ B_1^*X & -\Theta & D_{11}^*\Theta \\ \Theta C_1 & \Theta D_{11} & -\Theta \end{bmatrix} \begin{bmatrix} \bar{N}_o & 0 \\ 0 & I \end{bmatrix} < 0. \quad (9.12)$$

An analogous procedure combining (9.8) and (9.11) leads to

$$\begin{bmatrix} \bar{N}_c & 0 \\ 0 & I \end{bmatrix}^* \begin{bmatrix} AY + YA^* & YC_1^* & B_1\Theta^{-1} \\ C_1Y & -\Theta^{-1} & D_{11}\Theta^{-1} \\ \Theta^{-1}B_1^* & \Theta^{-1}D_{11}^* & -\Theta^{-1} \end{bmatrix} \begin{bmatrix} \bar{N}_c & 0 \\ 0 & I \end{bmatrix} < 0. \qquad (9.13)$$

To summarize, we have now reduced robust stabilization to conditions (9.9), (9.12), and (9.13). Condition (9.12) is an LMI on the variables X, Θ, but unfortunately we find that (9.13) is an LMI on Y and Θ^{-1}, not on Θ. This means that the conditions, as written, are not jointly convex on the variables X, Y, and Θ. Thus we have *not* been able to reduce robust synthesis to LMI computation. In particular, examples can be given where the allowable set of Θ-scalings is even disconnected.

Of course at this point we may wonder whether some further manipulations and possibly additional changes of variables may not yield an equivalent problem which is convex. No such conversion is known, but at the time of writing no mathematical proof to the contrary is known either. However, insights gained from computational complexity theory (see remarks below) seem to indicate that such a conversion is impossible.

Since our approach does not provide a convex answer for the general stabilization problem, the next question is whether there are special cases where the conclusion is different. Clearly if somehow we could get rid of one of the two conditions (9.12) and (9.13), an LMI would result. One such case is the so-called full information control problem where the measurements are

$$y = \begin{bmatrix} x \\ q \end{bmatrix}.$$

That is the controller has direct access to the states and all the outputs of the uncertainty blocks. Here

$$C_2 = \begin{bmatrix} I \\ 0 \end{bmatrix}, \quad D_{21} = \begin{bmatrix} 0 \\ I \end{bmatrix};$$

therefore the kernel of $[C_2\ D_{21}]$ is trivial, so the constraint (9.12) disappears completely. Consequently the variable X can be eliminated and the robust synthesis problem reduces to (9.13) and $Y > 0$, hence an LMI problem in Y and Θ^{-1}.

A dual situation which is also convex is the so-called full control case, where (9.13) disappears. A few other instances of simplification in synthesis are mentioned in the references at the end of the chapter. However, these cases are very special and in general robust design under structured uncertainty remains a difficult problem. An alternative viewpoint which reinforces this conclusion, is work on *bilinear* matrix inequalities (BMIs): these are feasibility problems of the form

$$f(X, Y) < 0,$$

where f is matrix valued and bilinear in the variables X, Y. Our robust synthesis problem can indeed be rewritten in this form, as is shown in the references. This is unfortunately not very helpful, since BMIs do not share the attractive computational features of LMIs; rather, the general BMI problem falls in an intractable computational complexity class.[3]

Given this complexity we are led to consider heuristic methods for optimization; these will be mentioned after discussing synthesis issues associated with time invariant uncertainty.

9.3.2   Robust synthesis against $\mathbf{\Delta}_{TI}$

We now turn our attention to the uncertainty class $\mathbf{\Delta}_{TI}$. From the preceding theory the robust stabilization problem reduces in this case to the optimization

$$\inf_{K \in \mathcal{K}} \sup_{\omega \in \mathbb{R}} \mu\big(\underline{S}(\hat{G}, \hat{K})(j\omega),\, \mathbf{\Delta}_{s,f}\big).$$

This synthesis problem is, however, even harder than the one studied just now in §9.3.1, since the function μ to be optimized is difficult to evaluate. In other words we are starting with a weaker result from the analysis side, so the additional search over K can only make matters worse. For this reason the above optimization is rarely attempted, and what usually goes by the name of μ-synthesis is the minimization based on the upper bound for the structured singular value. This is the optimization

$$\inf_{K \in \mathcal{K}} \sup_{\omega \in \mathbb{R}} \inf_{\Theta_\omega \in \boldsymbol{P}\boldsymbol{\Theta}_{s,f}} \bar{\sigma}\Big(\Theta_\omega^{\frac{1}{2}} \underline{S}(\hat{G}, \hat{K})(j\omega)\Theta_\omega^{-\frac{1}{2}}\Big). \qquad (9.14)$$

If the above infimum is made less than one, we have a robustly stabilizing controller from the analysis theory. One should note, however, that except for μ-simple structures, the converse is not true, that is the previous method might fail even when robust stabilization is achievable.

At this point we take the opportunity to discuss the relationship between the above problem and the one obtained by using scaling operators from $\boldsymbol{\Theta}_{TI}$, namely,

$$\inf_{K \in \mathcal{K}, \Theta \in \boldsymbol{\Theta}_{TI}} \|\hat{\Theta}\underline{S}(\hat{G}, \hat{K})\hat{\Theta}^{-1}\|_\infty. \qquad (9.15)$$

This topic pertains in fact to the analysis problem for a fixed K, but has been postponed from Chapter 8, since it is for the synthesis step that this issue acquires the most relevance.

We will now argue that (9.14) and (9.15) give the same optimal value. Clearly, if we find a controller $K \in \mathcal{K}$ and $\Theta \in \boldsymbol{\Theta}_{TI}$ satisfying

$$\|\hat{\Theta}\underline{S}(\hat{G}, \hat{K})\hat{\Theta}^{-1}\|_\infty < 1, \qquad (9.16)$$

---

[3]To be precise it is provably NP-hard, in the terminology of complexity theory.

then the answer to (9.14) will also be less than one, because we can always pick $\Theta_\omega = \hat{\Theta}^*(j\omega)\hat{\Theta}(j\omega) \in P\Theta_{s,f}$ at each ω.

What is less obvious is the converse implication. So suppose that we have a fixed stabilizing controller K and a family of matrix scalings $\Theta_\omega \in P\Theta_{s,f}$ satisfying

$$\sup_{\omega \in \mathbb{R}} \bar{\sigma}\left(\Theta_\omega^{\frac{1}{2}}\underline{S}(\hat{G},\hat{K})(j\omega)\Theta_\omega^{-\frac{1}{2}}\right) < 1,$$

or equivalently the LMI form

$$\hat{M}^*(j\omega)\Theta_\omega\hat{M}(j\omega) - \Theta_\omega \leq -\epsilon I \quad \text{for all } \omega \in \mathbb{R},$$

where $\epsilon > 0$ and $M := \underline{S}(G,K)$.

Now a topological argument, outlined in the exercises at the end of the chapter, shows that since $\hat{M}(s)$ is in RH_∞, then in the above Θ_ω can be replaced by a *rational* function $\hat{\Upsilon}(j\omega)$, satisfying —

- $\hat{\Upsilon}(j\omega) \in RL_\infty$.
- $\hat{\Upsilon}(j\omega) = \hat{\Upsilon}^*(j\omega) \in P\Theta_{s,f}$ for all ω.
- For a fixed $\epsilon > 0$,
$$\hat{M}^*(j\omega)\hat{\Upsilon}(j\omega)\hat{M}(j\omega) - \hat{\Upsilon}(j\omega) \leq -\epsilon I \quad \text{for all } \omega \in \mathbb{R}.$$

What remains to be done, in order to show the equivalence of the two bounds, is a *spectral factorization* step where one writes

$$\hat{\Upsilon}(j\omega) = \hat{\Theta}^*(j\omega)\hat{\Theta}(j\omega) \tag{9.17}$$

with $\hat{\Theta} \in RH_\infty$, and $\hat{\Theta}(s) \in \Theta_{s,f}$ for each s in the right half-plane. This final step follows by observing that each of the diagonal blocks of $\hat{\Upsilon}(j\omega)$ is lower bounded by ϵI at every frequency, and invoking the following result.

Proposition 9.12. *Given a rational function $\hat{Q} \in RL_\infty$, such that $\hat{Q}(j\omega) \geq \epsilon I > 0$ for all ω, then there exists $\hat{F} \in RH_\infty$, with $\hat{F}^{-1} \in RH_\infty$, such that*

$$\hat{F}^*(j\omega)\hat{F}(j\omega) = \hat{Q}(j\omega) \text{ holds for all } \omega \in \mathbb{R}.$$

This spectral factorization theorem is covered in the exercises of Chapter 7. Invoking this result for each block, we construct the desired $\hat{\Theta}(s) \in RH_\infty$, with $\hat{\Theta}(s) \in \Theta_{s,f}$. In other words, $\hat{\Theta}$ represents an operator in Θ_{TI}. Now the inequality

$$\hat{M}^*(j\omega)\hat{\Theta}^*(j\omega)\hat{\Theta}(j\omega)\hat{M}(j\omega) - \hat{\Theta}^*(j\omega)\hat{\Theta}(j\omega) \leq -\epsilon I \quad \text{for all } \omega \in \mathbb{R}$$

implies that $\|\hat{\Theta}\hat{M}\hat{\Theta}^{-1}\|_\infty < 1$ as required.

Remark 9.13. *In the above process we have in fact learned about the structure of the positive scaling set $P\Theta_{TI}$ which goes along with the commutant Θ_{TI}. We recall from §8.3 that such set is defined by*

$$P\Theta_{TI} = \{\Theta \in \mathcal{L}(L_2) : \Theta = \tilde{\Theta}^*\tilde{\Theta} \text{ and } \tilde{\Theta} \in \Theta_{TI}\},$$

and we remarked there that such sets are not always contained in the cor-responding commutant. Here, in fact, operators in $P\Theta_{TI}$ are self-adjoint over $L_2[0,\infty)$, so in general they are not LTI; in other words, they are not represented by a transfer function in H_∞, except when they are memoryless. Now the above factorization provides in effect a characterization of $P\Theta_{TI}$ in terms multiplication operators defined from $\hat{L}_\infty$. To be more precise, the operator $\Upsilon_0 = \Theta^*\Theta \in \mathcal{L}(L_2)$ that corresponds to $\Theta \in \Theta_{TI}$ would be given by

$$\Upsilon_0 = P_+ \Upsilon \Big|_{L_2[0,\infty)},$$

where Υ is the LTI operator on $L_2(-\infty,\infty)$ associated with the $\hat{L}_\infty$ func-tion $\hat{\Upsilon}(j\omega) = \hat{\Theta}^*(j\omega)\hat{\Theta}(j\omega)$, and P_+ is the projection from $L_2(-\infty,\infty)$ to $L_2[0,\infty)$.

We are now in a position to return to the controller synthesis problem, which we have reformulated as (9.15); further note from the above argument that we can restrict attention to rational scaling functions; explicitly our optimization problem is to infimize

$$\|\hat{\Theta}\underline{S}(\hat{G}, \hat{K})\hat{\Theta}^{-1}\|_\infty \qquad (9.18)$$

over the stabilizing controllers $K \in \mathcal{K}$, and scaling functions $\hat{\Theta} \in RH_\infty$ with $\hat{\Theta}^{-1} \in RH_\infty$ and $\hat{\Theta}(s) \in \Theta_{s,f}$ for every s.

As expected from the discussion in the previous section, this robust syn-thesis problem is difficult as well, and we will not be able to provide a computationally tractable solution to the optimization.[4] Thus we are led to consider heuristic algorithms which search for a minimum, without global guarantees; the most commonly used one, known as D-K iteration, will be introduced in the next section.

As a further complication, in the case of (9.18) we will not be able to bound a priori the order of the optimizing controller. To discuss this, sup-pose we knew the correct scaling function $\hat{\Theta}(s)$ and it had order n_d. Then $\hat{\Theta}(s)^{-1}$ can be realized with the same order and we see from an analogous argument to that in Proposition 9.11 that the controller order need not be greater than $n + 2n_d$. However, there is no way to know the required n_d: this issue will become more transparent once we look at the typical methods for searching for scaling functions.

Suppose we have a fixed controller K and thus a fixed $\hat{M} = \underline{S}(\hat{G}, \hat{K}) \in RH_\infty$. There are two main ways of finding a scaling $\hat{\Theta}$ which satisfies (9.16):

(a) *Frequency-by-frequency analysis plus curve fitting:* In this alterna-tive, for a given M we minimize, at each ω

$$\bar{\sigma}\big(\hat{\Theta}(j\omega)M(j\omega)\hat{\Theta}(j\omega)^{-1}\big)$$

---

[4] Some results exist for special cases; see the chapter references.

over $\hat{\Theta}(j\omega)$. For example, we can check for the feasibility of the LMI

$$\hat{M}^*(j\omega)\hat{\Upsilon}(j\omega)\hat{M}(j\omega) - \gamma^2\hat{\Upsilon}(j\omega) < 0 \qquad (9.19)$$

in $\hat{\Upsilon}(j\omega) = \hat{\Theta}^*(j\omega)\hat{\Theta}(j\omega)$, and search over γ. In practice this is done over a discrete grid of frequencies. Subsequently, a curve fitting step is done to find a rational $\Theta(s)$ that approximately satisfies (9.17). The tighter the approximation, the higher order of $\Theta(s)$ which will be required, and the step gives no guarantees for frequencies between grid points.

(b) *Basis function approach:* Here $\hat{\Theta}(j\omega)$ is constrained a priori to lie on a certain finite dimensional space of transfer functions; for instance, one can take $\hat{\Upsilon}(j\omega) = \hat{\Theta}^*(j\omega)\hat{\Theta}(j\omega)$ parametrized by

$$\hat{\Upsilon}(j\omega) = \begin{bmatrix} (j\omega I - A_\Theta)^{-1}B_\Theta \\ I \end{bmatrix}^* Q \begin{bmatrix} (j\omega I - A_\Theta)^{-1}B_\Theta \\ I \end{bmatrix},$$

where A_Θ and B_Θ are fixed and Q is allowed to vary. Endowing A_Θ, B_Θ, and Q with appropriate spatial structure, a family of transfer functions $\hat{\Upsilon}(j\omega) \in \boldsymbol{P\Theta}_{s,f}$ is parametrized, as discussed in the exercises. It is also shown there how the search over $\hat{\Upsilon}$ in this family satisfying (9.19) over frequency can be reduced to state space LMI computation. This method involves some conservatism over the unconstrained scalings $\hat{\Theta}(j\omega)$; as a counterpart, the frequency search and curve fitting is avoided.

More details on both approaches are given in the references at the end of the chapter. Having reviewed these approaches, we see that in both the order of the scaling (determined either by the curve fitting step or by the order of the basis expansion) can be chosen to be arbitrarily high. Unless some result could be given showing we need not go higher than a certain bound, there will no bound on the resulting n_d and thus on the order of the optimizing controller. No such result exists, and in fact examples show that approximating the infimum in our optimization may indeed require controllers of arbitrary high order.

We are now ready to move on to a design heuristic.

9.3.3   D-K iteration: A synthesis heuristic

In the previous two sections we have seen that robust synthesis for our standard setup in Figure 9.1 can be recast as the finding the infimum of

$$\|\Theta \underline{S}(G,K)\Theta^{-1}\|,$$

where K ranges over the stabilizing class $\mathcal{K}$, and Θ ranges over:

- the set of constant matrices $\boldsymbol{\Theta}_a$ for the case of uncertainty in $\boldsymbol{\Delta}_a^c$;
- the set of rational functions in $\boldsymbol{\Theta}_{TI}$ for uncertainty in $\boldsymbol{\Delta}_{TI}^c$.

We have also mentioned that this is not in general solvable by convex, finite dimensional methods. This reality leads us to consider heuristic algorithms, and this section is devoted to introducing the most commonly used approach, which is termed *D-K iteration*. The idea is to split the problem of robust synthesis, which we cannot solve, into two simpler problems which we can solve. The two simpler problems are (a) synthesizing an H_∞ controller; and (b) finding a scaling that infimizes a scaled gain. We have the following algorithm, which differs for each of these problems only in the set from which Θ is chosen.

1. Set $\Theta_1 = I$, $\eta_0 = \infty$ and the counter $k = 1$.

2. Solve for K_k in the H_∞ synthesis $\inf_{K_k} \|\Theta_k \underline{S}(G, K_k)\Theta_k^{-1}\|$; let η_k denote the achieved norm.

3. Compare η_k with η_{k-1}; if these are approximately equal stop, and set $K = K_k$ as the final controller. Otherwise continue to next step.

4. Solve for Θ_{k+1} in the scaled gain problem

$$\inf_{\Theta_{k+1}} \|\Theta_{k+1}\underline{S}(G, K_k)\Theta_{k+1}^{-1}\|;$$

increment k and go to step 2.

This is the *D-K* iteration algorithm, which is so named because the scaling variable Θ is often denoted by D, and thus the algorithm iterates between finding solutions for D and K. Starting with the trivial scaling $\Theta = I$, the algorithm begins by performing an H_∞ synthesis in step (2); later in step (4) a new scaling Θ is found, which can be chosen from either $\boldsymbol{\Theta}_a$ or $\boldsymbol{\Theta}_{TI}$ depending on which type of uncertainty is involved. Then these scalings are included for a new H_∞ synthesis; the algorithm stops when there is no significant improvement in the scaled norm. Alternatively, we can choose to stop when $\eta_k < 1$, i.e., when robust stabilization is achieved.

What are the properties of this algorithm? First the achieved performance η_k at any step forms a nonincreasing sequence up to the tolerance employed in the infimization steps (an exercise). Also if we are dealing with the uncertainty set $\boldsymbol{\Delta}_a^c$ the scaling Θ_k is a constant matrix, and therefore the controller K_k is never required to have order larger than that of G. In contrast, when dealing with the uncertainty set $\boldsymbol{\Delta}_{TI}$ we must find rational scalings by any of the two methods discussed in §9.3.2; in general the scaling Θ_k may need to be of arbitrarily high order, and thus so must K_k. Now notice that the scaled problem always involves the original generalized plant G; this means that the Θ_k are modified, not accumulated, in the process. Therefore if we impose a restriction on the order of the scalings to be fit, this automatically forces our algorithm to search only over controllers up to a certain order.

However, there is no guarantee that this process converges to the global minimum, in fact not even to a local minimum: the iteration can get "stuck" in values which are minima with respect to each variable separately, but not for both of them at a time. This kind of difficulty is not surprising given the lack of convexity of the problem. Consequently these methods should be viewed as ways of searching the controller space to improve an initial design, rather than global solutions to the robust synthesis problem.

Exercises

9.1. Complete the discussion in §9.1.2 by providing the details for the proof of Proposition 9.3. Also show that if W and W^{-1} are causal and in $\mathcal{L}(L_2)$, then their extensions to L_{2e} preserve the inverse.

9.2. Construct an example where M_{11} and Δ are causal and bounded, and $(I - M_{11}\Delta)^{-1}$ exists and is bounded, but is not causal. *Hint:* Exploit the anti-causal operator

$$(Qu)(t) = u(2t)$$

considered in §9.1.2 and its inverse.
Note: Such examples are impossible in discrete time when starting at time zero.

9.3. Small gain and nonlinearity. Consider the static function $f : \mathbb{R} \to \mathbb{R}$ given by

$$f(x) = \begin{cases} \frac{1}{2}|x| \text{ for } |x| \geq 1; \\ 0 \text{ for } |x| < 1. \end{cases}$$

Clearly, $|f(x)| \leq \frac{1}{2}x$ so f has small gain. If I is the identity, is the function $I - f$ invertible?

9.4. Prove Proposition 9.10.

9.5. Consider the robust synthesis problem under a constant scaling Θ and static state feedback, i.e.

$$\hat{G}(s) = \left[\begin{array}{c|cc} A & B_1 & B_2 \\ \hline C_1 & D_{11} & D_{12} \\ I & 0 & 0 \end{array} \right]$$

and $K(s) = F$, a static matrix. Show that the search for F and Θ reduces to a bilinear matrix inequality (BMI) condition in the variables F and (X, Θ), where X is a square state space matrix as in the KYP Lemma.

9.6. Show that the sequence η_k in the D-K iteration of §9.3.3 is non-increasing. Thus the performance of the controllers K_k generated by the algorithm will improve, or remain unchanged, as the iteration proceeds. Discuss the implications of a curve fitting step in this regard.

9.7. Basis function approach to uncertainty scalings.

(a) Let $v_0, v_2 \in \mathbb{R}$, $v_1 \in \mathbb{C}$. Consider the scalar transfer function

$$\hat{v}(j\omega) = v_0 + v_1 \frac{1}{1+j\omega} + v_1^* \frac{1}{1-j\omega} + v_2 \frac{1}{1+\omega^2},$$

which is real valued for every ω. Find fixed A_0, B_0, and Q affine in (v_0, v_1, v_2), so that

$$\hat{v}(j\omega) = \begin{bmatrix} (j\omega I - A_0)^{-1} B_0 \\ I \end{bmatrix}^* Q \begin{bmatrix} (j\omega I - A_0)^{-1} B_0 \\ I \end{bmatrix}. \qquad (9.20)$$

(b) Discuss how to modify the previous construction to describe:

(i) A Hermitian matrix function

$$\hat{\Upsilon}(j\omega) = \Upsilon_0 + \Upsilon_1 \frac{1}{1+j\omega} + \Upsilon_1^* \frac{1}{1-j\omega} + \Upsilon_2 \frac{1}{1+\omega^2}$$

with a desired spatial structure.

(ii) Terms of higher order in $\frac{1}{1+j\omega}$ and its conjugate.

(c) Given a scaling $\hat{\Upsilon}(j\omega)$ of this form (9.20), and

$$\hat{M}(s) = \left[\begin{array}{c|c} A & B \\ \hline C & D \end{array} \right],$$

find fixed matrices $\mathcal{A}$, $\mathcal{B}$ and a matrix Ψ, affinely dependent on Q, such that

$$\hat{M}^*(j\omega)\hat{\Upsilon}(j\omega)\hat{M}(j\omega) - \hat{\Upsilon}(j\omega) =$$
$$\begin{bmatrix} (j\omega I - \mathcal{A})^{-1}\mathcal{B} \\ I \end{bmatrix}^* \Psi \begin{bmatrix} (j\omega I - \mathcal{A})^{-1}\mathcal{B} \\ I \end{bmatrix}.$$

(d) Explain how to reduce condition (9.19) to an LMI.

9.8. Comparison of frequency scaling approaches.

(a) Suppose that a rational scaling from Θ_{TI} is required to minimize the gain $\|\Theta \underline{S}(G, K)\Theta^{-1}\|$ subject to the constraint of having a *fixed* maximum order. Can each of the frequency scaling methods of §9.3.2 guarantee that such a solution will be found?

(b) Discuss the computational tradeoffs between the two methods.

9.9. Continuous frequency scalings.
Given $\hat{M} \in RH_\infty$ we wish to show that if there exists a family of matrices $\Upsilon_\omega \in \boldsymbol{P\Theta}_{s,f}$ such that

$$\hat{M}^*(j\omega)\Upsilon_\omega\hat{M}(j\omega) - \Upsilon_\omega \leq -\epsilon I \qquad \text{for all } \omega \in \mathbb{R},$$

then such a Υ_ω exists which is continuous on $\omega \in \mathbb{R} \cup \{\infty\}$. A sketch of a proof is provided below, fill in the details.

(a) At each frequency $\bar{\omega}$ in $\mathbb{R} \cup \{\infty\}$, there exist a fixed matrix $\Upsilon_{\bar{\omega}} \in \boldsymbol{P\Theta}_{s,f}$ and an open interval $I_{\bar{\omega}}$ containing $\bar{\omega}$, such that

$$\hat{M}^*(j\omega)\Upsilon_{\bar{\omega}}\hat{M}(j\omega) - \Upsilon_{\bar{\omega}} < 0 \text{ for all } \omega \in I_{\bar{\omega}}.$$

(b) Using the matrices and intervals from a) show that there exist a finite number of intervals I_k and matrices $\Upsilon_k \in \boldsymbol{P\Theta}_{as,f}$ such that

$$\hat{M}^*(j\omega)\Upsilon_k\hat{M}(j\omega) - \Upsilon_k < 0 \text{ for all } \omega \in I_k, \text{ and}$$

$\cup_{k=1}^{N} I_k = \mathbb{R} \cup \{\infty\}$ holds.

(c) From this show that the desired function Υ_ω, continuous in $\mathbb{R} \cup \{\infty\}$ can be constructed.

Note: it is a fact that any bounded function Υ_ω which is continuous on $\mathbb{R} \cup \{\infty\}$ can be approximated as closely as desired by a function in RL_∞. This proves the claim made in §9.3.2 when arguing that rational scalings $\Theta \in \boldsymbol{\Theta}_{TI}$ can always be constructed.

Notes and references

The input–output perspective for the study of stability for possibly nonlinear or time varying systems goes back to the 1960, in particular to [183] on the small gain theorem, and the absolute stability theory of [124, 176]. Standard references on this early theory, in particular covering the nonlinear case, are [34] and [172]. The case of structured perturbations was already considered in [176], but mostly gained attention in the late 1970s [138]. The emphasis up to this point was on sufficient conditions; for subsequent work on necessity of small gain conditions including nonlinear nominal plants, see [147] and references therein. As mentioned before, necessity theorems for the structured case were developed in [89, 101, 148].

The fact that robust performance problems could be reduced to robust stability by addition of a fictitious block is from [41]. The $D - K$ iteration for robust synthesis was proposed in [43], including curve fitting in the D-step; the alternative basis function approach is from [139].

The observation that full-information and full-control problem gives convex robust synthesis is due to [114]. Other special problems with this property include the "rank one" case studied in [130], and problems where uncertainty is represented by a certain type of L_2 constraints [30]. Also the synthesis for robust sensitivity under additive LTI uncertainty has been reduced in [186, 110] to a convex, yet infinite dimensional problem. A recent survey of the BMI approach for the general (non-convex) case is [104].

10
Further Topics: Analysis

At this point we have achieved the major goals of our course — the detailed study of the topics in Chapters 1 through 9. This chapter and the next are devoted to broadening and deepening our background by considering a number of additional topics. Our approach will be that of a technical overview, stressing the main ideas and technical machinery, with a somewhat reduced emphasis on formal demonstrations.

This chapter will consider two important topics in robustness analysis. The first is a more sophisticated tool for the modeling of uncertainty. The second involves a combination of uncertainty with the H_2 performance criterion studied in Chapter 6.

10.1 Analysis via integral quadratic constraints

In previous chapters we pursued the modeling of uncertain systems in terms of spatially structured perturbation balls, and developed the corresponding methods for robustness analysis. It should be clear by now that the above modeling strategy is general enough to handle a wide variety of situations by manipulating the structure of these operator parametrizations. The corresponding robustness analysis conditions take the form of generalizations of the Small gain theorem.

Other strategies are, however, available for this class of problems. For instance, a widely used alternative is to make system *passivity* (see a definition below) the central concept instead of contractiveness, which leads to parallel robustness analysis tools. In this section we will provide a unified viewpoint which directly incorporates these classical robustness analysis methods, and further leads to useful generalizations. This will be accomplished by an analysis framework based on integral quadratic constraints.

We have already incorporated some elements of this framework in our presentation of the previous chapters, namely when we exploited *quadratic forms* over L_2 as a tool to recast some of the robustness analysis conditions. To help motivate the generalizations to follow, we begin by revisiting these results to provide a slightly different perspective. The next step will be to generalize the method to arbitrary quadratic characterizations of uncertainty, and develop robustness analysis theorems that go along with these characterizations. Finally, at the end of the section we address the question of computation of the resulting analysis conditions.

We begin by defining the relation, in $L_2[0, \infty) \times L_2[0, \infty)$, associated with an uncertainty set $\mathbf{\Delta} \subset \mathcal{L}(L_2)$:

$$\mathcal{R}_{\mathbf{\Delta}} := \left\{ \begin{bmatrix} p \\ q \end{bmatrix} \in L_2 \times L_2 : q = \Delta p, \text{ for some } \Delta \in \mathbf{\Delta} \right\}.$$

Recall from Chapter 8 that such relations can be taken as a natural starting point for uncertainty modeling, and from this perspective the operator set $\mathbf{\Delta}$ is merely a parametrization. As compared to the prior chapters, the tools of this section place much greater emphasis on the underlying relation. In particular if we focus on the uncertainty set $\mathbf{\Delta}_a$, the associated relation $\mathcal{R}_a$ can be more explicitly written as

$$\mathcal{R}_a = \left\{ \begin{bmatrix} p \\ q \end{bmatrix} \in L_2 \times L_2 : \|E_k p\| \geq \|E_k q\|, \text{ for } k = 1, \dots, d \right\},$$

where the matrices $E_k = \begin{bmatrix} 0 & \cdots & 0 & I & 0 & \cdots & 0 \end{bmatrix}$.

By introducing the quadratic forms

$$\psi_k \left(\begin{array}{c} p \\ q \end{array} \right) = \|E_k p\|^2 - \|E_k q\|^2,$$

we rewrite this characterization as

$$\mathcal{R}_a = \left\{ \begin{bmatrix} p \\ q \end{bmatrix} \in L_2 \times L_2 : \psi_k \left(\begin{array}{c} p \\ q \end{array} \right) \geq 0, \text{ for } k = 1, \dots, d \right\}.$$

Notice that the forms ψ_k act on the *product* space $L_2[0, \infty) \times L_2[0, \infty)$, which is slightly different from the forms ϕ_k of Chapter 8. We can further express the quadratic forms ψ_k in terms of the inner product

$$\psi_k \left(\begin{array}{c} p \\ q \end{array} \right) = \left\langle \begin{bmatrix} p \\ q \end{bmatrix}, \Psi_k \begin{bmatrix} p \\ q \end{bmatrix} \right\rangle,$$

involving the memoryless operators Ψ_k defined by

$$\Psi_k = \begin{bmatrix} E_k^* E_k & 0 \\ 0 & -E_k^* E_k \end{bmatrix}.$$

Now let us combine the constraints given in terms of the ψ_k using the multipliers $\theta_k > 0$: define

$$\psi\begin{pmatrix} p \\ q \end{pmatrix} := \sum_{k=1}^{d} \theta_k \psi_k \begin{pmatrix} p \\ q \end{pmatrix} = \left\langle \begin{bmatrix} p \\ q \end{bmatrix}, \Psi \begin{bmatrix} p \\ q \end{bmatrix} \right\rangle, \tag{10.1}$$

where the operator

$$\Psi = \sum_{k=1}^{d} \theta_k \Psi_k = \begin{bmatrix} \Theta & 0 \\ 0 & -\Theta \end{bmatrix}. \tag{10.2}$$

Recall that

$$\Theta = \sum_{k=1}^{d} \theta_k E_k^* E_k = \mathrm{diag}(\theta_1 I, \ldots, \theta_d I)$$

is an element of $\boldsymbol{P\Theta_a}$, the set of positive scaling matrices from Chapters 8 and 9.

From the above discussion, for any p and q related by $q = \Delta p$, for some $\Delta \in \boldsymbol{\Delta_a}$, we have the inequality

$$\psi\begin{pmatrix} p \\ q \end{pmatrix} \geq 0 \tag{10.3}$$

satisfied. To express this in geometric language, we use the quadratic form ψ to define the set (a cone in the product space)

$$\mathcal{C}_\psi^+ = \left\{ \begin{bmatrix} p \\ q \end{bmatrix} \in L_2 \times L_2 : \psi\begin{pmatrix} p \\ q \end{pmatrix} \geq 0 \right\}. \tag{10.4}$$

Then we see that the condition in (10.3) is compactly written as

$$\mathcal{R}_a \subset \mathcal{C}_\psi^+.$$

That is the relation defined by the uncertainty set $\boldsymbol{\Delta_a}$ is a subset of $\mathcal{C}_\psi^+$.

Next we turn our attention to our earlier robustness analysis results for the uncertainty set $\boldsymbol{\Delta_a}$. It was shown in the previous chapters that the feasibility of the operator inequality

$$M^* \Theta M - \Theta < 0,$$

for $\Theta \in \boldsymbol{P\Theta_a}$, is both necessary and sufficient for robust well-connectedness of $(M, \boldsymbol{\Delta_a})$, and also for robust stability of $(M, \boldsymbol{\Delta_a^c})$ when M is causal.

The above condition is easily rewritten as

$$\begin{bmatrix} M \\ I \end{bmatrix}^* \begin{bmatrix} \Theta & 0 \\ 0 & -\Theta \end{bmatrix} \begin{bmatrix} M \\ I \end{bmatrix} < 0,$$

which means that, for some $\epsilon_0 > 0$, the inequality

$$\left\langle q, \begin{bmatrix} M \\ I \end{bmatrix}^* \begin{bmatrix} \Theta & 0 \\ 0 & -\Theta \end{bmatrix} \begin{bmatrix} M \\ I \end{bmatrix} q \right\rangle \leq -\epsilon_0^2 \|q\|^2 \quad \text{holds for all } q \in L_2.$$

From the definition of ψ in (10.2) we see that the above simply says

$$\psi \left(\begin{array}{c} Mq \\ q \end{array} \right) \leq -\epsilon_0 \|q\|^2, \text{ for every } q \in L_2. \tag{10.5}$$

Notice that the set

$$\mathcal{R}_M := \left\{ \begin{bmatrix} Mq \\ q \end{bmatrix} : q \in L_2 \right\} \subset L_2 \times L_2$$

is the relation or *graph* defined by the operator M. Consequently, we can interpret the above results in geometric terms inside the space $L_2 \times L_2$. From ψ define the set

$$\mathcal{C}_\psi^{-\epsilon} = \left\{ \begin{bmatrix} p \\ q \end{bmatrix} \in L_2 \times L_2 : \psi \left(\begin{array}{c} p \\ q \end{array} \right) \leq -\epsilon \left\| \begin{array}{c} p \\ q \end{array} \right\|^2 \right\}, \tag{10.6}$$

for $\epsilon > 0$ fixed. Then the condition in (10.5) will be satisfied whenever

$$\mathcal{R}_M \subset \mathcal{C}_\psi^{-\epsilon} \tag{10.7}$$

is satisfied for some appropriately chosen $\epsilon > 0$.

Thus (10.7) states that $\mathcal{R}_M$ is contained in a cone where our quadratic form is strictly negative definite, whereas (10.3) states that the uncertainty relation $\mathcal{R}_a$ is in the non-negative cone. Thus we say the two relations are *quadratically separated* by the form ψ. In particular notice that the two cones $\mathcal{C}_\psi^+$ and $\mathcal{C}_\psi^{-\epsilon}$ only intersect at zero, and furthermore the separation is strict in a sense made precise in Lemma 10.4 below.

The results of Chapters 8 and 9 imply that the robust well-connectedness of $(M, \boldsymbol{\Delta}_a)$, or respectively the robust stability of $(M, \boldsymbol{\Delta}_a^c)$, are equivalent to the existence of an operator Ψ in the parametrization given by (10.2), such that the resulting form ψ provides this quadratic separation.

Can we convert our earlier the analysis results over the time invariant uncertainty set $\boldsymbol{\Delta}_{TI}$ to this language? The exact structured singular value test is difficult to express, however the sufficient convex upper bound test can be formulated in terms of quadratic separation.

To see the latter, let Ψ be an LTI, self-adjoint operator on $L_2(-\infty, \infty)$, characterized by the frequency domain $\hat{L}_\infty$ function

$$\hat{\Psi}(j\omega) = \begin{bmatrix} \hat{\Theta}(j\omega) & 0 \\ 0 & -\hat{\Theta}(j\omega) \end{bmatrix},$$

where $\hat{\Theta}(j\omega) \in \boldsymbol{P\Theta}_{s,f}$, corresponding to the spatial structure $\Delta_{s,f}$ of $\boldsymbol{\Delta}_{TI}$. The quadratic form

$$\psi(v) = \langle v, \Psi v \rangle = \frac{1}{2\pi} \int_{-\infty}^{\infty} \hat{v}(j\omega)^* \hat{\Psi}(j\omega) \hat{v}(j\omega) d\omega$$

is thus defined on $L_2(-\infty, \infty)$, and can be restricted to $L_2[0, \infty)$. It follows similarly to the above that

$$\mathcal{R}_{\Delta_{TI}} \subset \mathcal{C}_\psi^+.$$

Also, the condition

$$\sup_{\omega \in \mathbb{R}} \bar{\sigma}(\hat{\Theta}(j\omega)^{\frac{1}{2}} \hat{M}(j\omega) \hat{\Theta}(j\omega)^{-\frac{1}{2}}) < 1$$

is once again equivalent to (10.7). So here we can also interpret the above convex analysis test in terms of quadratic separation. Notice that in this case the test is only sufficient for robust well-connectedness of (M, Δ_{TI}).

The above discussion suggests a generalization of our procedure to more general uncertainty sets Δ: instead of constraining ourselves to operators Ψ of the special structures that we have encountered above, what if we are allowed complete freedom in our choice? In other words, we pose the question of finding a suitable self-adjoint Ψ so that $\mathcal{R}_\Delta$ is contained in $\mathcal{C}_\psi^+$, and the graph of the nominal system satisfies $\mathcal{R}_M \subset \mathcal{C}_\psi^{-\epsilon}$. If such a Ψ is found, we might have a general means of establishing robustness properties of the (M, Δ) system. To this end we make the following definition.

Definition 10.1. *Let Ψ be a self-adjoint operator on $L_2 \times L_2$. The uncertainty set Δ is said to satisfy the integral quadratic constraint (IQC) defined by Ψ if*

$$\mathcal{R}_\Delta \subset \mathcal{C}_\psi^+.$$

Remark 10.2. *In the preceding definition Ψ is allowed to be any self-adjoint operator. However, the typical situation is that the quadratic form ψ can be written as*

$$\psi(v) = \frac{1}{2\pi} \int_{-\infty}^{\infty} \hat{v}(j\omega)^* \hat{\Psi}(j\omega) \hat{v}(j\omega) d\omega$$

for a certain $\hat{L}_\infty$ function $\hat{\Psi}(j\omega) = \hat{\Psi}(j\omega)^*$. The fact that the uncertainty is described by a sign constraint on the above integral explains the terminology "IQC."

We now look at different examples of uncertainty properties that can be encompassed by IQCs, in addition to the structured contractiveness which was studied above.

Examples:

Consider the matrix

$$\Psi = \begin{bmatrix} 0 & I \\ I & 0 \end{bmatrix}.$$

A component $q = \Delta p$ satisfying

$$\psi\left(\begin{array}{c} p \\ q \end{array}\right) = 2\mathrm{Re}\langle p, q \rangle \geq 0$$

for every $q \in L_2$ is called *passive*; this property arises naturally in physical (electrical, mechanical) modeling.

Now suppose Δ is a scalar, time-varying real parameter, $q(t) = \delta(t)p(t)$. Then for any anti-Hermitian matrix $\Omega = -\Omega^*$ we have

$$\int_0^\infty \begin{bmatrix} p(t) \\ q(t) \end{bmatrix}^* \begin{bmatrix} 0 & \Omega \\ \Omega^* & 0 \end{bmatrix} \begin{bmatrix} p(t) \\ q(t) \end{bmatrix} = \int_0^\infty \delta(t)[p(t)^*\Omega^* p(t) + p(t)^*\Omega p(t)]dt = 0,$$

so Δ satisfies the IQC defined by

$$\Psi = \begin{bmatrix} 0 & \Omega \\ \Omega^* & 0 \end{bmatrix}.$$

If in addition we have the contractiveness condition $|\delta(t)| \leq 1$, it follows easily that Δ satisfies the IQC defined by

$$\Psi = \begin{bmatrix} \Theta & 0 \\ 0 & -\Theta \end{bmatrix}$$

for any matrix $\Theta > 0$. Now one can always superimpose two IQCs, so we find that a contractive, time-varying parameter gain always satisfies the IQC defined by

$$\Psi = \begin{bmatrix} \Theta & \Omega \\ \Omega^* & -\Theta \end{bmatrix}.$$

Finally, assume the parameter is real, contractive, and also constant over time. Then it follows analogously that the component $q = \delta p$ satisfies the IQC defined by

$$\hat{\Psi}(j\omega) = \begin{bmatrix} \hat{\Theta}(j\omega) & \hat{\Omega}(j\omega) \\ \hat{\Omega}(j\omega)^* & -\hat{\Theta}(j\omega) \end{bmatrix},$$

for any bounded $\hat{\Omega}(j\omega) = -\hat{\Omega}(j\omega)^*$ and $\hat{\Theta}(j\omega) > 0$.   □

As a final remark on modeling, we notice that we have not imposed any a priori restrictions on the allowable choices of IQCs; to be interesting, however, an IQC must allow for a rich enough set of signals to be in $\mathcal{C}_\psi^+$. For instance a negative definite Ψ will mean that $\mathcal{C}_\psi^+$ only contains the zero element, and therefore would not contain any uncertainty relation $\mathcal{R}_\Delta$ as a subset, since the latter always allow the signal p to span the entire space L_2. In what follows we will therefore assume that our IQCs are models for some set Δ of operators; namely, there exists a nonempty uncertainty set such that $\mathcal{R}_\Delta \subset \mathcal{C}_\psi^+$.

10.1.1 Analysis results

Having introduced IQC descriptions of uncertainty, we are now ready to pursue the generalization of our robustness analysis work based on IQCs. The following discussion refers to the setup of Figure 10.1, relevant to robust well-connectedness or robust stability questions. We assume M and Δ are bounded operators on $L_2[0,\infty)$. While robust performance problems can also be included in this setup, they will not be discussed here; interested readers can consult the references at the end of the chapter.

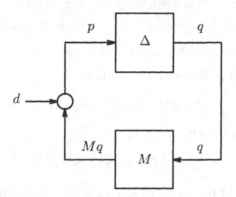

Figure 10.1. Setup for robustness analysis

We will assume we have found an IQC which quadratically separates $\mathcal{R}_\Delta$ and $\mathcal{R}_M$. Namely, for some $\epsilon > 0$, we have

$$\mathcal{R}_\Delta \subset \mathcal{C}_\psi^+ \quad \text{and} \quad \mathcal{R}_M \subset \mathcal{C}_\psi^{-\epsilon}.$$

The first obvious consequence is that $\mathcal{R}_\Delta \cap \mathcal{R}_M = \{0\}$. So if $d = 0$, the only L_2 solutions to the diagram of Figure 10.1 are $p = 0$, $q = 0$. Equivalently, $\mathrm{Ker}\,(I - M\Delta) = \{0\}$; however, something slightly stronger can be said.

Proposition 10.3. *Suppose $\mathcal{R}_\Delta \subset \mathcal{C}_\psi^+$ and $\mathcal{R}_M \subset \mathcal{C}_\psi^{-\epsilon}$. Then there exists a constant $\kappa > 0$ such that*

$$\|p\| \leq \kappa \|(I - M\Delta)p\|$$

is satisfied, for every $\Delta \in \boldsymbol{\Delta}$ and every $p \in L_2$.

To prove this result we will rely on the following lemma about quadratic forms on L_2.

Lemma 10.4. *Let $\psi(v) = \langle v, \Psi v \rangle$, where Ψ is a self-adjoint operator in $\mathcal{L}(L_2)$, and the number $\epsilon > 0$. Then there exists a constant κ, depending only on Ψ and ϵ, such that any z, v in L_2 satisfying*

$$z \in \mathcal{C}_\psi^+, \qquad v \in \mathcal{C}_\psi^{-\epsilon},$$

will satisfy the inequality

$$\kappa \| z - v \| \geq \| z \|.$$

Proof. Without loss of generality we may assume that $\epsilon = 1$, since it can be absorbed into Ψ by scaling. We first write the identity

$$\psi(z) - \psi(v) = \langle v, \Psi(z - v) \rangle + \langle (z - v), \Psi v \rangle + \langle (z - v), \Psi(z - v) \rangle,$$

that leads to the inequality

$$\psi(z) - \psi(v) \leq 2 \| \Psi \| \, \| v \| \, \| z - v \| + \| \Psi \| \, \| z - v \|^2.$$

Since $z \in \mathcal{C}_\psi^+$, $v \in \mathcal{C}_\psi^{-1}$, we have $\| v \|^2 \leq \psi(z) - \psi(v)$ and thus

$$\| v \|^2 \leq 2 \| \Psi \| \, \| v \| \, \| z - v \| + \| \Psi \| \, \| z - v \|^2. \tag{10.8}$$

Separately, we find α large enough[1] so that

$$0 \leq \frac{1}{2} \| v \|^2 - 2 \| \Psi \| \, \| v \| \, \| z - v \| + \alpha^2 \| z - v \|^2. \tag{10.9}$$

Now add (10.8) and (10.9) to obtain

$$\frac{1}{2} \| v \|^2 \leq (\alpha^2 + \| \Psi \|) \| z - v \|^2,$$

which is a bound of the form $\| v \| \leq C \| z - v \|$. Finally, from the triangle inequality we have

$$\| z \| \leq \| v \| + \| z - v \| \leq (C + 1) \| z - v \|,$$

which is the desired bound with $\kappa := C + 1$. ∎

We now return to the main proof.

Proof of Proposition 10.3. First apply the above lemma to obtain the appropriate κ, depending only on ψ, ϵ. Now for any $p \in L_2$, $\Delta \in \boldsymbol{\Delta}$ set $q = \Delta p$,

$$z = \begin{bmatrix} p \\ q \end{bmatrix} \in \mathcal{C}_\psi^+, \text{ and } v = \begin{bmatrix} Mq \\ q \end{bmatrix} \in \mathcal{C}_\psi^{-\epsilon}.$$

Then Lemma 10.4 gives

$$\| p \| \leq \left\| \begin{bmatrix} p \\ q \end{bmatrix} \right\| \leq \kappa \left\| \begin{bmatrix} p \\ q \end{bmatrix} - \begin{bmatrix} Mq \\ q \end{bmatrix} \right\| = \kappa \| p - Mq \| = \kappa \| (I - M\Delta) p \|.$$

∎

The above result provides an important first step for robustness analysis with IQCs. It tells us that the mapping $(I - M\Delta)$ is injective over L_2, and that the inverse mapping defined on $\text{Im}(I - M\Delta)$ has a norm bound of κ. However, to establish that $(I - M\Delta)$ is invertible, as required for

---

[1] One can take $\alpha = \sqrt{2} \| \Psi \|$ (depending only on Ψ) to produce a perfect square.

well-connectedness, we also need to be sure it is surjective. The following example illustrates the difficulty.

Example:

Suppose M is the constant gain $M = 2$, and Δ is the LTI system with transfer function

$$\hat{\Delta}(s) = \frac{s-1}{s+1}.$$

It is easy to see that Δ is isometric, so $\|p\| = \|q\|$ whenever $q = \Delta p$. Therefore Δ satisfies, for example, the IQC

$$\psi\left(\begin{array}{c} p \\ q \end{array}\right) = -\|p\|^2 + 2\|q\|^2 \geq 0.$$

corresponding to

$$\Psi = \begin{bmatrix} -I & 0 \\ 0 & 2I \end{bmatrix}.$$

Also it is clear that

$$\psi\left(\begin{array}{c} Mq \\ q \end{array}\right) = -4\|q\|^2 + 2\|q\|^2 \leq -2\|q\|^2.$$

Applying Proposition 10.3 we see that

$$\|p\| \leq \kappa \|(I - M\Delta)p\|$$

for some constant κ (it is easily verified here that $\kappa = 1$ suffices). However, $Q := I - M\Delta$ has transfer function

$$\hat{Q}(s) = \frac{3-s}{s+1}$$

and does not have a bounded inverse on $L_2[0, \infty)$. In fact it is easy to see that the operator Q, while injective, is not surjective, since the Laplace transform of any element in its image must belong to the set

$$\{\hat{v}(s) \in H_2 : \hat{v}(3) = 0\}.$$

In fact the above subspace of H_2 exactly characterizes the image of Q. □

The above discussion implies that in addition to separation of the graphs of M and Δ by an IQC, some additional property is required to prove the invertibility of $I - M\Delta$. Namely, quadratic separation does not necessarily imply that $I-M\Delta$ is surjective. For the specific IQCs considered in previous chapters, this stronger sufficiency result was provided by the Small gain theorem. We seek an extended argument to establish this for a richer class of IQCs. As a first step we state a property about the image of operators such as those in the previous proposition.

Lemma 10.5. *Suppose $Q \in \mathcal{L}(L_2)$ satisfies*

$$\kappa \|Qp\| \geq \|p\|$$

for all $p \in L_2$. Then $\operatorname{Im} Q$ is a closed subspace of L_2.

Proof. Take a sequence Qp_n which converges to $z \in L_2$. Then Qp_n is a Cauchy sequence so $\|Qp_n - Qp_m\| < \epsilon$ holds for sufficiently large n and m. Now applying the hypothesis we find

$$\|p_n - p_m\| \leq \kappa\epsilon$$

so p_n is also a Cauchy sequence; hence p_n converges to some $p \in L_2$. By continuity of Q, the sequence Qp_n converges to Qp, and therefore $z = Qp$ is in the image of Q.   ∎

In order to state the main result we make the following definition. An uncertainty set $\boldsymbol{\Delta}$ is *radial* if for every $\Delta \in \boldsymbol{\Delta}$ we have

$$\tau\Delta \in \boldsymbol{\Delta} \quad \text{for all } \tau \in [0, 1].$$

That is the uncertainty set is closed under linear homotopy to $\Delta = 0$. Now for the result.

Theorem 10.6. *Suppose the uncertainty set $\boldsymbol{\Delta}$ is radial. If there exist a self-adjoint operator Ψ and $\epsilon > 0$, such that $\mathcal{R}_M \subset \mathcal{C}_\psi^{-\epsilon}$ and $\mathcal{R}_\Delta \subset \mathcal{C}_\psi^+$, then $I - M\Delta$ is invertible over $\boldsymbol{\Delta}$.*

The theorem states that if $\mathcal{R}_\Delta$ and $\mathcal{R}_M$ can be quadratically separated by a single IQC, then $(M, \boldsymbol{\Delta})$ is robustly well-connected.

Proof. Fix $\Delta \in \boldsymbol{\Delta}$. Given Proposition 10.3, it suffices to show that $I - M\Delta$ is surjective, because in that case it follows that $(I - M\Delta)^{-1}$ exists and has norm bounded by κ.

Let us suppose that $\tau_0 \in [0, 1]$ and $\operatorname{Im}(I - M\tau_0\Delta) = L_2$. Then to prove the theorem it suffices to show that $I - M\tau\Delta$ is surjective when

$$|\tau - \tau_0| \leq \frac{1}{\kappa \|M\| \, \|\Delta\|} \quad \text{and } \tau \in [0, 1]. \tag{10.10}$$

To see this, note the above states that $I - M\tau\Delta$ must be surjective for τ in an interval of nonzero length around τ_0; furthermore this length does not depend on the initial τ_0. Thus we can begin with $\tau_0 = 0$ were surjectivity holds, and expand the interval of validity to be $[0, 1]$ by successively enlarging the initial interval via the above result, each time setting τ_0 equal to an interval endpoint; this process converges in a finite number of steps.

We thus focus on the perturbation argument for a given τ_0, and τ satisfying (10.10). By contradiction, suppose

$$\operatorname{Im}(I - M\tau\Delta)$$

is a strict subspace of L_2. Since it is closed by Lemma 10.5, then by the projection theorem it has a non-trivial orthogonal complement. Namely, we can find a function $v \in L_2$, $\|v\| = 1$, such that

$$\langle (I - M\tau\Delta)p, v \rangle = 0, \text{ for all } p \in L_2.$$

Now observe that

$$(I - M\tau_0\Delta) = (I - M\tau\Delta) + M(\tau - \tau_0)\Delta;$$

therefore,

$$\langle (I - M\tau_0\Delta)p, v \rangle = \langle M(\tau - \tau_0)\Delta p, v \rangle, \text{ for all } p \in L_2. \qquad (10.11)$$

Since $I - M\tau_0\Delta$ is surjective we can find p_0 satisfying

$$(I - M\tau_0\Delta)p_0 = v,$$

and furthermore such p_0 has norm bounded by κ. Substitution into (10.11), and applications of the Cauchy-Schwartz and submultiplicative inequalities yield

$$1 = \langle v, v \rangle = \langle M(\tau - \tau_0)\Delta p_0, v \rangle \leq \|M\| \, |\tau - \tau_0| \, \|\Delta\| \, \kappa < 1,$$

which is a contradiction. Therefore $I - M\tau\Delta$ must be surjective as required.
∎

We have thus obtained a fairly general robust well-connectedness test based on IQCs; the extra condition imposed on the uncertainty set Δ is quite mild, since one usually wishes to consider the nominal system ($\Delta = 0$) as part of the family, and thus it is not too restrictive to impose that Δ is radial. The reader can verify that this is true with the IQC models presented in the above examples.

An important comment here is that we have only shown the *sufficiency* of this test for a given IQC. When studying the uncertainty set Δ_a in Chapter 8 we showed as well that the method is non-conservative, in the sense that if the system is robustly well-connected this fact can always be established by an IQC of the form $\Psi = \begin{bmatrix} \Theta & 0 \\ 0 & -\Theta \end{bmatrix}$ with $\Theta \in P\Theta_a$. Notice that such family of IQCs is obtained as the conic hull of a finite number of time invariant IQCs (Ψ_k considered at the beginning of the section), and the uncertainty set is rich enough to be characterized exactly by such family. Only in such quite rare situations can one expect necessity theorems to hold; this subject will not be pursued further here.

Having discussed well-connectedness, we now turn our attention to the question of robust stability of the configuration of Figure 10.1, understood as establishing a relationship in the extended space L_{2e}. The following is a simple immediate corollary of Theorem 10.6.

Corollary 10.7. *Suppose the hypothesis of Theorem 10.6 holds, and that M and the uncertainty class Δ are well defined over L_{2e}. Suppose further* *that $I - M\Delta$ is invertible as a mapping in L_{2e}, for each $\Delta \in \mathbf{\Delta}$. Then the system $(M, \mathbf{\Delta})$ is robustly stable.*

Proof. Since $(I - M\Delta)^{-1}$ is assumed to be well defined over L_{2e}, it only remains to establish boundedness over L_2. But this follows directly from Theorem 10.6.   ∎

The most common situation in which the conditions of the above corollary can be satisfied is when M and Δ are taken to be *causal*, and $I - M\Delta$ has a causal inverse over L_{2e}; the IQC analysis comes in as a way of ensuring boundedness over L_2.

As a final remark in this section, we note that we have developed the theory exclusively for *linear* operators M and Δ, in tune with the style of this course. In fact the analysis applies, with minor modifications to nonlinear components; in a sense, IQCs provide a more sophisticated tool to capture nonlinearities than the sector bounds that we used to motivate our uncertainty $\mathbf{\Delta}_a$. For nonlinear formulations of the analysis, as well as application examples of IQCs for some common nonlinearities, we refer the reader to the references at the end of the chapter.

10.1.2   The search for an appropriate IQC

A question we have not yet addressed explicitly is how one finds an IQC to satisfy the assumptions of the above theory, namely, to separate quadratically the graph $\mathcal{R}_M$ from the uncertainty relation $\mathcal{R}_\Delta$. The success of the method as an analysis technique clearly hinges on this question. There are two aspects to the above issue:

  (i) how to find IQCs ψ that describe a given uncertainty set $\mathbf{\Delta}$;

  (ii) how to search, among the above class, to find one IQC that satisfies the quadratic separation condition $\mathcal{R}_M \subset \mathcal{C}_\psi^{-\epsilon}$.

The first question cannot have a general purpose answer, since it depends strongly on how the uncertainty set $\mathbf{\Delta}$ is presented to us; in particular it may be more or less easy to extract, from the problem formulation, a valid family of IQCs, as was done for instance in the examples of §10.1.1. Whatever method is employed, however, the available family is always *convex*.

To be more precise define the set of operators

$$\mathcal{Q}_\Delta = \{\Psi \in \mathcal{L}(L_2 \times L_2): \ \Psi \text{ self-adjoint and } \mathcal{R}_\Delta \subset \mathcal{C}_\psi^+\}.$$

This set is comprised of all the operators which generate IQCs that are satisfied by the uncertainty set $\mathbf{\Delta}$. An important feature of $\mathcal{Q}_\Delta$ is that it is a *convex* cone: if Ψ_1 and Ψ_2 are members of $\mathcal{Q}_\Delta$, then so is

$$\alpha_1 \Psi_1 + \alpha_2 \Psi_2,$$

for any $\alpha_1, \alpha_2 \geq 0$. Similarly, the set of operators Ψ which satisfy

$$\mathcal{R}_M \subset \mathcal{C}_\psi^{-\epsilon}, \tag{10.12}$$

for some $\epsilon > 0$, is convex. Therefore the search for a quadratically separating IQC is a convex feasibility problem, over a possibly infinite dimensional space.

In more practical terms, one usually considers a *subset* of $\mathcal{Q}_\Delta$, described by some free parameters. Then one searches over the set of parameters to find an operator which satisfies (10.12). If successful, the search is over; if not, it is conceivable that a richer subset of $\mathcal{Q}_\Delta$ may yield a quadratically separating IQC.

How is such a parametrization of IQCs obtained? Clearly, one can be generated from any finite list of elements $\Psi_1, \ldots, \Psi_d$ in $\mathcal{Q}_\Delta$, by means of the relevant convex cone. Sometimes, however, as in the example of scalar parameters discussed above, a matrix parameter such as $\Omega = -\Omega^*$ or $\Theta > 0$ is more convenient. In some cases, most notably for LTI uncertainty, one starts with an infinite family of IQCs characterized in the frequency domain by

$$\hat{\Psi}(j\omega) \text{ in some matrix set } \mathcal{S}_\Psi,$$

at every ω. In this case we have the same options that were discussed in Chapter 9: either keep the parametrization infinite dimensional, or restrict it to a finite dimensional space of transfer functions by writing

$$\hat{\Psi}(j\omega) = \begin{bmatrix} (j\omega I - A_\Psi)^{-1} B_\Psi \\ I \end{bmatrix}^* Q \begin{bmatrix} (j\omega I - A_\Psi)^{-1} B_\Psi \\ I \end{bmatrix}, \tag{10.13}$$

where A_Ψ and B_Ψ are fixed and impose the structure of the relevant set $\mathcal{S}_\Psi$, and Q is a free matrix parameter. The latter form can fairly generally accommodate a finite parametrization of a family of self-adjoint, LTI operators.

The two options are distinguished when we address the second basic question, namely the search for an IQC satisfying (10.12). Assuming M is LTI, the frequency domain characterization of the condition is that

$$\begin{bmatrix} \hat{M}(j\omega) \\ I \end{bmatrix}^* \hat{\Psi}(j\omega) \begin{bmatrix} \hat{M}(j\omega) \\ I \end{bmatrix} \leq -\epsilon I \text{ holds for all } \omega,$$

which in practice implies a test based on one-dimensional gridding. In contrast the finite parametrization of (10.13) reduces to a state space LMI over Q by application of the KYP Lemma, as discussed in the exercises of Chapter 9.

In summary, we see that given a family of IQCs which are satisfied by the uncertainty, the search for an appropriate one to establish robust stability can be handled with our standard tools, and provides an attractive general methodology for a variety of robustness analysis problems. We now move on to discussing a completely new robustness analysis problem.

10.2   Robust H_2 performance analysis

In this section we study a new type of robust performance problem based on uncertainty sets in $\mathcal{L}(L_2)$ combined with the H_2 performance norm — the "robust H_2" problem. Before doing this let us briefly review some of our work so far, in order to provide strong motivation for this new problem. What we will see is that the H_2 norm is a particularly well-motivated metric for performance, whereas the H_∞ norm has the most relevance when viewed as a measure of robust stability. This leads directly to the problem we consider here.

To begin, let us reflect on the two performance measures we studied in Chapters 6 and 7. The H_∞ norm has an interpretation as the maximum system gain when inputs are taken over the input space L_2; as such it directly carries information on the *worst-case* input that can affect our system. This worst-case philosophy in the treatment of disturbances easily blends with our approach to uncertainty modeling, and we took advantage of this in Chapters 8 and 9 when combining performance with robust stability.

However this view of performance is often not natural for the treatment of disturbances or noise, a point that is seen more clearly from a frequency domain perspective. Assume our systems are LTI; worst-case signals for H_∞ concentrate their energy around the frequency of peak system response, the limit case would be a sinusoidal signal at that frequency. By adopting the H_∞ performance measure we are effectively saying that we potentially expect these very systematic disturbances, but we do not know at what frequency they will occur. This may be unnatural for two reasons:

1. If indeed we knew nothing about the spectral content of disturbances, there is also no reason to expect the *worst* one to be exciting our system. Rather, since disturbances are usually "neutral" towards our system it is more natural to consider the average response to various spectra. For instance the average response to sinusoids leads to an H_2 performance specification.

2. Very often we know very approximately what the input spectrum is, either in statistical terms or because it is a known reference signal, and we would like to exploit this information in our problem. As explained in Chapter 6, a weighted H_2 performance measure, with the weight given by the known spectrum, provides the desired measure. A weighted H_∞ measure does not directly achieve this, except in the extreme case of a sinusoid of known frequency. In general, while it is possible to use H_∞ weights to indirectly influence the system frequency response, the choice of these weights is by no means systematic.

For these reasons, the H_2 norm has a more compelling engineering motivation than H_∞ norm as a measure of performance, in most circumstances.

However, as seen from the preceding chapters, an induced norm such as H_∞ is indeed the most natural tool when dealing with robustness to model uncertainty. First of all, for *stability* it is indeed natural to adopt a worst-case philosophy. Secondly, it is the Banach algebra structure available in induced norms that allows us to perform these worst-case evaluations by means of small-gain-type results; such structure is not available in H_2 space.

The above discussion motivates a compromise for robust performance problems, namely to continue to model uncertainty and treat robust stability in terms of L_2-induced norms as before, yet to modify the performance specification to be an H_2 criterion. The search for such a compromise is the topic of this section, and we will see that similar tools to the ones we have developed can be brought to bear on this problem.

10.2.1 *Problem formulation*

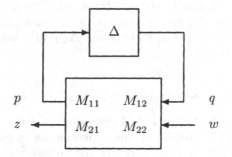

Figure 10.2. Setup for Robust H_2 analysis.

Throughout the section we will consider the uncertain system of Figure 10.2. The nominal map M is taken to be a finite dimensional LTI system with state space realization

$$
\begin{aligned}
\dot{x} &= Ax + B_1 q + B_2 w \\
p &= C_1 x + D_{11} q \\
z &= C_2 x + D_{21} q
\end{aligned}
$$

with no feed-through terms for the input w, which will ensure that the H_2 norm is well-defined. The uncertainty Δ will vary as usual in a block-structured class of operators.

There is however a slight difference in our problem formulation as compared to the preceding chapters: instead of working with signals supported in positive time, we will find it advantageous to set up our problem in the space $L_2(-\infty, \infty)$. We now make some remarks regarding this choice.

The space $L_2[0, \infty)$ in which we have mostly been working is best suited for studying questions that have a natural "starting" time, such as transient response and the issue of stability. In particular our study of stability for uncertain systems relied strongly on the extended signal space L_{2e}; this construction appears to be only possible with a semi-infinite time axis. In contrast, for situations where transient stability has already been established, it is often interesting to study system response in *stationary steady state*, namely to assume the system has been in operation for an arbitrary amount of time. This is in particular natural for problems of disturbance attenuation, such as the ones we consider in this section.

Thus we will assume our signals are elements of $L_2(-\infty, \infty)$, and that the uncertainty operators belongs to the structured class

$$\mathbf{\Delta}_a = \{\text{diag}(\Delta_1, \ldots, \Delta_d) : \Delta_k \in \mathcal{L}(L_2(-\infty, \infty)), \|\Delta_k\| \leq 1\},$$

or the time invariant subset $\mathbf{\Delta}_{TI}$. Notice that the signal *relation* that goes along with $\mathbf{\Delta}_a$ has the form

$$\mathcal{R}_a = \{(p, q) \in L_2(-\infty, \infty) : \|p_k\| \geq \|q_k\|, \text{ where } k = 1, \ldots, d\},$$

analogous to the case of $L_2[0, \infty)$ discussed before, because the basic Lemma 8.4 is also valid in our new space. Moreover in this new setting one can also characterize the LTI subset $\mathbf{\Delta}_{TI}$ in a very concise way by means of the relation

$$\mathcal{R}_{TI} =$$
$$\{(p, q) \in L_2(-\infty, \infty) : |\hat{p}_k(j\omega)| \geq |\hat{q}_k(j\omega)| \text{ a.e., for } k = 1, \ldots, d\}.$$

This follows from the fact that LTI operators in $L_2(-\infty, \infty)$ are exactly characterized (see Chapter 3) by multiplication by an element of $\hat{L}_\infty(j\mathbb{R})$. In each case, we will assume a *priori* the robust well-connectedness of $(M_{11}, \mathbf{\Delta})$, i.e. that $I - M_{11}\Delta$ is invertible over our uncertainty set.

Since we are not discussing stability it is not strictly necessary to enforce causality of perturbations; nevertheless it is natural to discuss the effect of imposing causality since our robust H_2 performance analysis will typically be performed after an analysis of robust stability. We denote the causal subsets of $\mathbf{\Delta}_a$ and $\mathbf{\Delta}_{TI}$ by $\mathbf{\Delta}_a^c$ and $\mathbf{\Delta}_{TI}^c$. Notice that causality of LTI operators is *not* automatically guaranteed in the space $L_2(-\infty, \infty)$.

For LTI uncertainty, the robust H_2 performance analysis problem is to evaluate

$$\sup_{\Delta \in \mathbf{\Delta}_{TI}} \|\bar{S}(M, \Delta)\|_2,$$

where $\bar{S}(M, \Delta)$ as usual is the closed-loop map from w to z in Figure 10.2. Recall from Chapter 6 that the above norm is defined by

$$\|\hat{G}\|_2^2 = \frac{1}{2\pi} \int_{-\infty}^{\infty} \text{Tr}\{\hat{G}(j\omega)^* \hat{G}(j\omega)\} d\omega. \tag{10.14}$$

Unfortunately the above supremum over LTI operators Δ is difficult to compute; this is not unexpected since the analysis of robust stability alone is a structured singular value problem, which we have seen is hard. Thus we are led to seek convex upper bounds for this supremum, like the procedure we followed in Chapter 8, using the set of scaling matrices $P\Theta_{s,f}$. In the current context we have $s = 0$ and $f = d$, and so

$$P\Theta_{s,f} = \{\Theta \in \mathbb{R}^{m \times m} : \Theta = \mathrm{diag}(\theta_1 I_{m_1}, \ldots, \theta_d I_{m_d}), \; \theta_k > 0\},$$

which in fact coincides with the set $P\Theta_a$. We will also study bounds for the perturbation class Δ_a using the set $P\Theta_a$. In this case of perturbations Δ_a the closed loop is no longer necessarily time invariant, and so the definition of (10.14) cannot be used. We will therefore discuss possible generalizations of the H_2 norm and resulting robust performance tests.

10.2.2 Frequency domain methods and their interpretation

In this section we will study a frequency domain characterization of robust H_2 performance that is closely related to our work in Chapter 8. Consider the following optimization:

Problem 10.8. *Find*

$$\mathbf{J_{f,TI}} := \inf \frac{1}{2\pi} \int_{-\infty}^{\infty} \mathrm{Tr}(Y(\omega)) d\omega,$$

subject to $\Theta(\omega) \in P\Theta_{s,f}$, $Y(\omega)$ *Hermitian, and*

$$\hat{M}(j\omega)^* \begin{bmatrix} \Theta(\omega) & 0 \\ 0 & I \end{bmatrix} \hat{M}(j\omega) - \begin{bmatrix} \Theta(\omega) & 0 \\ 0 & Y(\omega) \end{bmatrix} \leq 0, \; \text{for all } \omega. \qquad (10.15)$$

This is an infinite dimensional convex optimization problem, consisting of the minimization of a linear objective subject to a family of LMI constraints over frequency. Observe that its solution amounts to minimizing the trace of $Y(\omega)$, at each frequency, subject to (10.8); in this sense the above is a pointwise optimization in frequency.

The LMI (10.15) is very closely related to the upper bounds for the structured singular value: if we replace $Y(\omega)$ by $\gamma^2 I$ in (10.15), we would impose the robust H_∞ performance specification that the worst-case gain of $\bar{S}(M, \Delta)$, for LTI perturbations, across every frequency and spatial direction is bounded by γ.

Here we have added a *slack variable* $Y(\omega)$ that allows such gain to vary, provided that the accumulated effect over frequency and spatial direction is minimized, reflecting an H_2-type specification. For LTI uncertainty, this argument is easily formalized.

Theorem 10.9. *Suppose the system in Figure 10.2 is robustly well-connected over the class* Δ_{TI}. *Then*

$$\sup_{\Delta \in \Delta_{TI}} \|\bar{S}(M, \Delta)\|_2^2 \leq \mathbf{J_{f,TI}}.$$

Proof. Consider a fixed frequency ω; by introducing signals $\hat{q}(j\omega)$, $\hat{w}(j\omega)$ we obtain from (10.15) the inequality

$$\begin{bmatrix} \hat{q}(j\omega) \\ \hat{w}(j\omega) \end{bmatrix}^* \left(\hat{M}(j\omega)^* \begin{bmatrix} \Theta(\omega) & 0 \\ 0 & I \end{bmatrix} \hat{M}(j\omega) - \begin{bmatrix} \Theta(\omega) & 0 \\ 0 & Y(\omega) \end{bmatrix} \right) \begin{bmatrix} \hat{q}(j\omega) \\ \hat{w}(j\omega) \end{bmatrix} \leq 0,$$

which can be expanded into

$$|\hat{z}(j\omega)|^2 + \sum_{k=1}^{d} \theta_k(\omega)|\hat{p}_k(j\omega)|^2 \leq \hat{w}(j\omega)^* Y(\omega)\hat{w}(j\omega) + \sum_{k=1}^{d} \theta_k(\omega)|\hat{q}_k(j\omega)|^2,$$

using the signal conventions of Figure 10.2. Since Δ is time invariant and contractive we have $|\hat{p}_k(j\omega)| \geq |\hat{q}_k(j\omega)|$, so $\theta_k(\omega) > 0$ implies that

$$|\hat{z}(j\omega)|^2 \leq \hat{w}(j\omega)^* Y(\omega)\hat{w}(j\omega).$$

Now, $\hat{z}(j\omega) = [\bar{S}(M,\Delta)(j\omega)]\hat{w}(j\omega)$, so

$$\hat{w}(j\omega)^* \left[\bar{S}(M,\Delta)(j\omega)^* \bar{S}(M,\Delta)(j\omega) - Y(\omega) \right] \hat{w}(j\omega) \leq 0$$

holds for any $\hat{w}(j\omega)$. Therefore we conclude that the preceding matrix is negative semidefinite and in particular

$$\mathrm{Tr}\{\bar{S}(M,\Delta)(j\omega)^* \bar{S}(M,\Delta)(j\omega)\} \leq \mathrm{Tr}(Y(\omega)).$$

Using definition (10.14), we have

$$\|\bar{S}(M,\Delta)\|_2^2 \leq \frac{1}{2\pi} \int_{-\infty}^{\infty} \mathrm{Tr}(Y(\omega))d\omega.$$

Taking supremum over Δ and infimum over Y, we complete the proof. ∎

The preceding proof is remarkably simple and close in spirit to the methods of the structured singular value theory. In an analogous manner, we can study the conservatism of this bound. Not surprisingly, this issue is related to whether the associated uncertainty structure is μ-simple or not, but there are other issues as well. We state the following result.

Proposition 10.10. *Assume that the number of blocks $d \leq 2$, and that the signal w is scalar. Then*

$$\sup_{\Delta \in \boldsymbol{\Delta}_{TI}} \|\bar{S}(M,\Delta)\|_2^2 = \mathbf{J_{f,TI}}.$$

The proof of this result is a standard application of the structured singular value theory. Since $\hat{w}$ is scalar, so is $Y(\omega)$, and it can be identified as the worst-case gain of $\bar{S}(M,\Delta)(j\omega)$ at that particular frequency, using the classification of μ-simple structures in Theorem 8.27. Choosing the worst case Δ at each frequency results in $\hat{\Delta}(j\omega) \in \hat{L}(j\mathbb{R})$, which achieves the desired bound by integrating over frequency. We leave details as an exercise.

Having identified the non-conservative case, we will now comment on the sources of conservatism for this condition:

- Non-μ-simple structures. Clearly as the number of blocks increases, we expect a similar conservatism as the one we have in the structured singular value.

- Multivariable noise. As we see in the preceding proof, we are imposing the matrix inequality $\bar{S}(M, \Delta)(j\omega)^* \bar{S}(M, \Delta)(j\omega) \leq Y(\omega)$ as a way of constraining the matrix traces. Examples can be given showing this is conservative.

- Causal uncertainty. As we have seen, for the robust stability question it is natural to consider causal perturbations. This information, however, is not used in the preceding bound. We will discuss this issue in more detail below.

In view of these observations, it is natural to inquire what exactly is the bound $\mathbf{J}_{f,TI}$ computing in regard to this problem. It turns out that frequency domain conditions can be interpreted as tests for white noise rejection in a worst-case perspective, as explained below. This approach will apply as well to the case of arbitrary structured uncertainty $\mathbf{\Delta}_a$; in that case the conditions will involve the use of *constant* uncertainty scalings $\Theta \in \mathbf{P\Theta}_a$, analogously to the situation of previous chapters. In particular, we will consider the modified problem

$$\mathbf{J}_{f,a} := \inf \frac{1}{2\pi} \int_{-\infty}^{\infty} \mathrm{Tr}(Y(\omega))d\omega,$$

subject to $\Theta \in \mathbf{P\Theta}_a$, $Y(\omega)$ Hermitian, and

$$\hat{M}(j\omega)^* \begin{bmatrix} \Theta & 0 \\ 0 & I \end{bmatrix} \hat{M}(j\omega) - \begin{bmatrix} \Theta & 0 \\ 0 & Y(\omega) \end{bmatrix} \leq 0, \text{ for all } \omega. \qquad (10.16)$$

We now outline the set-based approach to white noise rejection, directly tailored to the robustness analysis problem. By treating both noise and uncertainty from a worst-case perspective, exact characterizations can be obtained. At first sight, this proposition may seem strange to readers accustomed to a stochastic treatment of white noise; notice, however, that a stochastic process is merely a model for the *generation* of signals with the required statistical spectrum, and other models (e.g., deterministic chaos) are possible. In this formulation we take the standpoint that rather than model the generating mechanism, we can directly characterize a *set* of signals of white spectrum, defined by suitable constraints, and subsequently pursue worst-case analysis over such set.

One way to do this for scalar L_2 signals is to constrain the *cumulative spectrum* by defining the set

$$W_{\eta,B} := \qquad\qquad\qquad\qquad\qquad\qquad\qquad\qquad\qquad\qquad (10.17)$$

$$\left\{ w \in L_2 : \min\left(\frac{\beta}{\pi} - \eta, \frac{B}{\pi} - \eta\right) \leq \int_{-\beta}^{\beta} |\hat{w}(j\omega)|^2 \frac{d\omega}{2\pi} \leq \frac{\beta}{\pi} + \eta, \ \forall \beta > 0 \right\}.$$

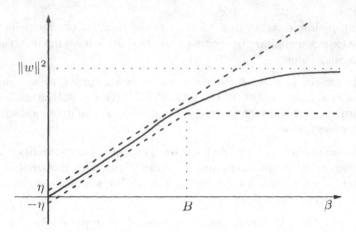

Figure 10.3. Constraints on the cumulative spectrum

This approach is inspired in statistical tests for white noise that are commonly used in the time series analysis literature. The constraints, depicted in Figure 10.3, impose that signals in $W_{\eta,B}$ have approximately unit spectrum (controlled by the accuracy $\eta > 0$) up to bandwidth B, since the integrated spectrum must exhibit approximately linear growth in this region. Notice that this integrated spectrum will have a finite limit as $\beta \to \infty$, for L_2 signals, so we only impose an sublinear upper bound for frequencies above B.

Having defined such approximate sets, the white noise rejection measure will be based on the worst-case rejection of signals in $W_{\eta,B}$ in the limit as $\eta \to 0$ and $B \to \infty$:

$$\|\bar{S}(M,\Delta)\|_{2,wn} := \lim_{\substack{\eta \to 0 \\ B \to \infty}} \sup_{w \in W_{\eta,B}} \|\bar{S}(M,\Delta)w\|.$$

For an LTI system under some regularity assumptions, $\|\cdot\|_{2,wn}$ can be shown to coincide with the standard H_2 norm; see the references for details. The method can be extended to multivariable noise signals, where the components are restricted to have low cross-correlation. Notice that the preceding definition can apply to any bounded operator, and even nonlinear maps.

We are now ready to state a characterization of the frequency domain test.

Theorem 10.11. *Suppose the system in Figure 10.2 is robustly well-connected over the class Δ_a. With the above definitions,*

$$\mathbf{J_{f,a}} = \lim_{\substack{\eta \to 0 \\ B \to \infty}} \sup_{\substack{w \in W_{\eta,B} \\ \Delta \in \Delta_a}} \|\bar{S}(M,\Delta)w\|^2.$$

The proof of this result involves extending the S-procedure method employed in Chapter 8, to accommodate the additional constraints on the signal spectrum. Notice that the constraints (10.17) are indeed IQCs in the signal w, although in contrast to the situation of Chapter 8, we have an infinite family of them, indexed by β. Thus the proof is based on a convex duality argument in infinite dimensional space; for details see the references. An additional comment is that the characterization remains true if $\mathbf{\Delta}_a$ is replaced by the causal subset $\mathbf{\Delta}_a^c$.

We remark that a similar interpretation can be given to the cost function $\mathbf{J_{f,TI}}$, as a test for robust performance over white noise sets in the sense defined above, for uncertainty of *arbitrarily slow* time variation, paralleling results available for H_∞-type performance. See the references for a precise treatment.

In summary, by using the worst-case interpretation of white noise rejection, which in itself entails some conservatism, we are able to obtain a parallel theory to the one obtained in the previous chapters for the H_∞ performance measure.

We end the discussion on these frequency domain characterizations with a few remarks on the computational aspect. Problem 10.8 is infinite dimensional as posed; however, we note that the problem decouples over frequency, into the minimization of $\text{Tr}(Y(\omega))$ subject to (10.15) at each fixed frequency. This decoupling can be exploited to build practical approximation methods, that complement the tools of μ-analysis.

If Θ is constant as in (10.16) the problem is coupled across frequency. However, in this case it can be reduced exactly to a finite dimensional state-space LMI optimization. To simplify the formulas we will assume in the next proposition that D_{11} and D_{21} are zero, but the general case can be treated similarly.

Proposition 10.12. $\mathbf{J_{f,a}} = \inf \text{Tr}(Z)$, *where the variables* Z, X_-, X_+, *and* Θ *are subject to*

$$
\left[\begin{array}{cc} AX_- + X_-A^* + B_1\Theta B_1^* & [X_-C_1^* \; X_-C_2^*] \\ \left[\begin{array}{c} C_1X_- \\ C_2X_- \end{array} \right] & -\left[\begin{array}{cc} \Theta & 0 \\ 0 & I \end{array} \right] \end{array} \right] < 0,
$$

$$
\left[\begin{array}{cc} AX_+ + X_+A^* + B_1\Theta B_1^* & [X_+C_1^* \; X_+C_2^*] \\ \left[\begin{array}{c} C_1X_+ \\ C_2X_+ \end{array} \right] & -\left[\begin{array}{cc} \Theta & 0 \\ 0 & I \end{array} \right] \end{array} \right] < 0,
$$

$$
\left[\begin{array}{cc} Z & B_2^* \\ B_2 & X_+ - X_- \end{array} \right] > 0.
$$

A proof is provided in the references at the end of the chapter. Here we will turn our attention to an alternative state space method.

10.2.3 State space bounds involving causality

The above frequency domain methods have been interpreted as picking the worst among possible "white" disturbances. Now we will present a different bound, that is obtained by focusing on *impulse response* interpretation of the H_2 norm, but in addition applies to average case white noise rejection. As we will see, in these interpretations the causality of Δ plays a larger role.

We state a second convex optimization problem based on the state space description. Once again we work for simplicity in the case where D_{11} and D_{21} are zero.

Problem 10.13. *Find* $\mathbf{J_{s,a}} := \inf \, \mathrm{Tr}(B_2^* X B_2)$, *subject to* $X > 0$, $\Theta \in P\Theta_a$, *and*

$$\begin{bmatrix} A^* X + X A + C_1^* \Theta C_1 + C_2^* C_2 & X B_1 \\ B_1^* X & -\Theta \end{bmatrix} \leq 0.$$

To interpret this problem, we define another generalization of the H_2 norm for time-varying operators. Recall from Chapter 6 that, for LTI systems, the H_2 norm is equal to the sum of the energies of the output responses corresponding to impulses applied at each input channel.

In the case of time-varying systems, such response may vary in time so it is natural in addition to average over the time when the impulse is applied. Denote by δ^τ the Dirac impulse at time τ, and $z_i^\tau = \bar{S}(M, \Delta) \delta^\tau e_i$, where e_i is the i-th coordinate vector in $\mathbb{R}^m$. That is z_i^τ is the system response to an impulse applied in the i-th input channel at time τ. Now define

$$\|\bar{S}(M, \Delta)\|_{2,imp}^2 := \limsup_{\tau \to \infty} \frac{1}{2\tau} \int_{-\tau}^{\tau} \left(\sum_{i=1}^{m} \|z_i^\tau\|^2 \right) d\tau,$$

the average output energy for such impulses. We have the following result.

Theorem 10.14. *Suppose the system of Figure 10.2 is robustly stable against the ball Δ_a^c of structured, contractive, causal operators. Then*

$$\sup_{\Delta \in \Delta_a^c} \|\bar{S}(M, \Delta)\|_{2,imp}^2 \leq \mathbf{J_{s,a}}.$$

Proof. It suffices to show that

$$\sup_{\Delta \in \Delta_a^c} \sum_{i=1}^{m} \|z_i^\tau\|^2$$

is bounded by $\mathbf{J_{s,a}}$ for every τ, since we can then average over τ. For simplicity we prove the above for $\tau = 0$, but the proof is identical for any other value.

Applying the impulse $\delta^0 e_i$ as an input has the effect of "loading" an initial condition $x_0 = B_2 e_i$ in the system, which subsequently responds autonomously. For this reason we first focus on the problem for fixed initial condition and no input,

$$J(x_0) := \sup_{\Delta \in \Delta_a^c, x(0)=x_0} \|z\|^2.$$

We now write the bound

$$J(x_0) \leq \sup_{\substack{q \in L_2[0,\infty), \|p_k\|^2 \geq \|q_k\|^2 \\ x(0)=x_0}} \|z\|^2$$

$$\leq \inf_{\theta_k > 0} \sup_{q \in L_2[0,\infty)} \left(\|z\|^2 + \sum_{k=1}^{d} \theta_k (\|p_k\|^2 - \|q_k\|^2) \right). \qquad (10.18)$$

In the first step the k-th uncertainty block is replaced by the integral quadratic constraint $\|p_k\|^2 \geq \|q_k\|^2$; this constraint would characterize the class of contractive (possibly non-causal) operators. However, by requiring $q \in L_2[0,\infty)$, we are imposing some causality in the problem by not allowing q to anticipate the impulse. This does not, however, impose full causality in the map from p to q, hence the inequality.

Secondly, we are bounding the cost by using the Lagrange multipliers $\theta_k > 0$ to take care of the constraints. It is straightforward to show the stated inequality. This step is closely related to the S-procedure method explained in Chapter 8 when studying the structured well-connectedness problem. In that case we showed the procedure was not conservative; in fact, a slight extension of those results can be used to show there is equality in the second step of (10.18).

To compute the right-hand side of (10.18), observe that for fixed θ_k we have

$$\sup_{q \in L_2[0,\infty)} \int_0^\infty [x(t)^*(C_1^* \Theta C_1 + C_2^* C_2)x(t) - q(t)^* \Theta q(t)] \, dt, \qquad (10.19)$$

where we recall that $\Theta = \text{diag}(\theta_1 I, \ldots, \theta_d I)$.

This is a linear-quadratic optimization problem that can be solved by Riccati equation techniques. The following proposition summarizes some facts which are proved in the exercises of Chapters 6 and 7.

Proposition 10.15. *If the H_∞ norm condition*

$$\left\| \begin{bmatrix} \Theta^{\frac{1}{2}} & 0 \\ 0 & I \end{bmatrix} M \begin{bmatrix} \Theta^{-\frac{1}{2}} \\ 0 \end{bmatrix} \right\|_\infty < 1$$

holds, then the optimal value of (10.19) is given by

$$x_0^* X x_0,$$

where X is the stabilizing solution of the algebraic Riccati equation

$$A^* X + XA + C_1^* \Theta C_1 + C_2^* C_2 + X B_1 \Theta^{-1} B_1^* X = 0.$$

Furthermore, this solution X is the minimizing solution of the LMI

$$\begin{bmatrix} A^*X + XA + C_1^*\Theta C_1 + C_2^*C_2 & XB_1 \\ B_1^*X & -\Theta \end{bmatrix} \leq 0. \qquad (10.20)$$

The above LMI is the same as the one considered in Problem 10.13. To apply this result, first notice that the norm condition is

$$\left\| \begin{bmatrix} \Theta^{\frac{1}{2}} M_{11} \Theta^{-\frac{1}{2}} \\ M_{21} \Theta^{-\frac{1}{2}} \end{bmatrix} \right\| < 1.$$

Given the robust stability assumption, we know by Theorem 9.6 that the norm of the top block can be made less than one by appropriate choice of Θ. Now since Θ can be scaled up, the norm of the bottom block can be made as small as desired, yielding the required condition, and thus the feasibility of (10.20).

Now we wish to combine the solution X with the minimization over Θ. Here is where the LMI (10.20) is most advantageous, since it is jointly affine in Θ and X. We have

$$J(x_0) \leq \inf_{X, \Theta > 0 \text{ satisfying (10.20)}} x_0^* X x_0,$$

that is, a semidefinite programming problem.

The final step is to return to the sum over the impulses applied at the input channels:

$$\sup_{\Delta \in \mathbf{\Delta}_a^c} \sum_{i=1}^{m} \|z_i^0\|^2 \leq \sum_{i=1}^{m} J(B_2 e_i) \qquad (10.21)$$

$$\leq \sum_{i=1}^{m} \inf_{X, \Theta > 0 \text{ satisfying (10.20)}} e_i^* B_2^* X B_2 e_i \qquad (10.22)$$

$$\leq \inf_{X, \Theta > 0 \text{ satisfying (10.20)}} \sum_{i=1}^{m} e_i^* B_2^* X B_2 e_i \qquad (10.23)$$

$$= \inf_{X, \Theta > 0 \text{ satisfying (10.20)}} \text{Tr}(B_2^* X B_2)$$

$$= \mathbf{J_{s,a}}.$$

In the above chain of inequalities, (10.21) follows by interchanging the supremum with the sum, (10.22) is the previous derivation, and (10.23) results from exchanging the sum with the infimum.   ∎

We remark that (10.21) and (10.23) are in principle conservative steps when $m > 1$; more remarks on this conservatism are given later on.

The previous result focuses on the H_2 norm as a measure of transient performance; we immediately wonder if the same bound applies to the other notions that were used to motivate the H_2 norm, in particular in regard to the rejection of stationary white noise. Indeed it can be shown that

$$\sup_{\Delta \in \Delta_a^c} \|\bar{S}(M, \Delta)\|_{2,aov}^2 \leq \mathbf{J_{s,a}}, \tag{10.24}$$

where

$$\|\bar{S}(M, \Delta)\|_{2,aov}^2 := \limsup_{\tau \to \infty} \frac{1}{2\tau} \int_{-\tau}^{\tau} \mathcal{E}|z(t)|^2 dt$$

is the *average output variance* of the time-varying system when the input is stochastic white noise. In fact for linear operators the above quantity coincides with the norm $\|\bar{S}(M, \Delta)\|_{2,imp}^2$; see the references for this equivalence, and also for a direct stochastic proof of (10.24) which also applies to nonlinear uncertainty. Notice, however, that in these interpretations we can only prove a bound, not an exact characterization as with $\mathbf{J_{f,a}}$.

We end the discussion of this method by explaining how the bound can be refined in the case of LTI uncertainty. Returning to the proof we would consider in this case frequency depending scalings $\theta_k(\omega)$, and write

$$J(x_0) \leq \sup_{\substack{q \in L_2[0,\infty), |\hat{p}_k(j\omega)|^2 \geq |\hat{q}_k(j\omega)|^2 \\ x(0)=x_0}} \|z\|^2$$

$$\leq \inf_{\theta_k(\omega)>0} \sup_{q \in L_2[0,\infty)} \left(\|z\|^2 + \sum_{k=1}^{d} \int_{-\infty}^{\infty} \theta_k(\omega)(|\hat{p}_k(j\omega)|^2 - |\hat{q}_k(j\omega)|^2)\frac{d\omega}{2\pi} \right). \tag{10.25}$$

However at this level of generality the restriction $q \in L_2[0,\infty)$ (related to causality) is not easily handled, and thus (10.25) is not yet in a form amenable to computation. In order to extend the previous method we must rely on state-space methods; thus we are led to a *basis function* approach, where we further bound the right-hand side of (10.25) by constraining $\theta_k(\omega)$ to the span of a finite set of rational functions. With this additional restriction, the problem can indeed be reduced to LMI computation, as shown in the references, and thus one generates a family of optimization costs $\mathbf{J_{s,TI}^N}$ of state dimension increasing with number N of elements in the basis, in the limit approaching the optimization (10.25); we will have more to say about this below.

10.2.4 Discussion

We have shown two alternative methods to approach the robust H_2 performance problem; we end the section with a few remarks on the comparison between them. For simplicity, we initially focus on the case of scalar disturbances w.

We first discuss perturbations in $\mathbf{\Delta}_a$ and state the following relationships:

$$
\sup_{\Delta \in \mathbf{\Delta}_a^c} \|\bar{S}(M,\Delta)\|_{2,imp}^2 \leq \sup_{\Delta \in \mathbf{\Delta}_a} \|\bar{S}(M,\Delta)\|_{2,imp}^2
$$

$$
\leq \lim_{\substack{\eta \to 0 \\ B \to \infty}} \sup_{\substack{w \in W_{\eta,B} \\ \Delta \in \mathbf{\Delta}_a}} \|\bar{S}(M,\Delta)w\|^2 \tag{10.26}
$$

$$
= \lim_{\substack{\eta \to 0 \\ B \to \infty}} \sup_{\substack{w \in W_{\eta,B} \\ \Delta \in \mathbf{\Delta}_a^c}} \|\bar{S}(M,\Delta)w\|^2 = \mathbf{J_{f,a}}.
$$

The first inequality is clear; (10.26) follows from the fact that the impulse (or more exactly, an L_2 approximation) is always an element of the "white" set $W_{\eta,B}$. The equalities with $W_{\eta,B}$ were stated before.

Notice that the previous inequality does not transparently relate $\mathbf{J_{f,a}}$ and $\mathbf{J_{s,a}}$, since we only know that $\mathbf{J_{s,a}}$ is an upper bound for the first quantity. Nevertheless, we have the following:

Proposition 10.16. $\mathbf{J_{s,a}} \leq \mathbf{J_{f,a}}$.

Proof. It is possible to provide a direct proof based on the state space version of $\mathbf{J_{f,a}}$. However, here we will restrict ourselves to a more insightful argument for the special case of scalar w. Notice that in the case of a scalar impulse, $x_0 = B_2$ and the right-hand side of (10.18) is directly $\mathbf{J_{s,a}}$. Rewriting (10.18) in the frequency domain we have

$$
\inf_{\Theta} \sup_{q \in L_2[0,\infty)} \frac{1}{2\pi} \int_{-\infty}^{\infty} \left[|\hat{z}(j\omega)|^2 + \sum_{k=1}^{d} \theta_k (|\hat{p}_k(j\omega)|^2 - |\hat{q}_k(j\omega)|^2) \right] d\omega.
$$

Now, introducing the slack variable $Y(\omega)$ to bound the above integrand, we can rewrite this problem as the minimization of $\frac{1}{2\pi} \int_{-\infty}^{\infty} Y(\omega) d\omega$ subject to

$$
|\hat{z}(j\omega)|^2 + \sum_{k=1}^{d} \theta_k (|\hat{p}_k(j\omega)|^2 - |\hat{q}_k(j\omega)|^2) \leq Y(\omega)
$$

for all $\hat{q}(j\omega)$ in H_2, Fourier image of $L_2[0,\infty)$. Now, since $w(t) = \delta(t)$ we have

$$
\begin{bmatrix} \hat{p}(j\omega) \\ \hat{z}(j\omega) \end{bmatrix} = \hat{M}(j\omega) \begin{bmatrix} \hat{q}(j\omega) \\ 1 \end{bmatrix},
$$

which translates the previous inequality to

$$
\begin{bmatrix} \hat{q}(j\omega) \\ 1 \end{bmatrix}^* \left(\hat{M}(j\omega)^* \begin{bmatrix} \Theta & 0 \\ 0 & I \end{bmatrix} \hat{M}(j\omega) - \begin{bmatrix} \Theta & 0 \\ 0 & Y(\omega) \end{bmatrix} \right) \begin{bmatrix} \hat{q}(j\omega) \\ 1 \end{bmatrix} \leq 0.
$$

Now we have an expression that closely resembles Problem 10.8; if $\hat{q}(j\omega)$ were allowed to be any frequency function, the inequality would reduce to (10.16) and the two problems would be equivalent. However, the constraint

$\hat{q}(j\omega) \in H_2$ that embeds some causality in Problem 10.13 will lead in general to a smaller infimum.   ∎

Examples can be given where the inequality in Proposition 10.16 is strict. Thus if we are dealing with causal uncertainty and with an impulse-based or stochastic notion of H_2 performance, $\mathbf{J_{s,a}}$ provides a tighter performance bound, and the worst-case philosophy for noise is then more pessimistic.

Next we discuss the comparison for the case of LTI uncertainty. Notice that here we have an unambiguous H_2 norm we are trying to compute, for which both approaches provide bounds.

In this regard, once again we find that removing the restriction $q \in L_2[0,\infty)$ from (10.25) will lead to the result $\mathbf{J_{f,TI}}$, but that there is a gap between the two. This would mean that if the uncertainty set is $\mathbf{\Delta}_{TI}^c$, the state space approach could in principle give a tighter bound. Notice, however, that we do not have a $\mathbf{J_{s,TI}}$ bound, only a family $\mathbf{J_{s,TI}^N}$ obtained by basis expansions of order N for the frequency varying scalings. This means that while

$$\inf_N \mathbf{J_{s,TI}^N} \leq \mathbf{J_{f,TI}},$$

we know nothing about the situation with a given, finite N. Since computational cost of LMI computation grows dramatically with state dimension, it may in practice be more tractable to use the decoupled, frequency domain computation based on Problem 10.8.

Finally, we emphasize a point which was mentioned in passing when describing each approach: they both can be significantly conservative when dealing with m-dimensional noise signals. In essence the H_2 criterion is defined in terms of *averaging* over the direction of the input noise; neither method can prevent the worst-case Δ from "lining-up" with the noise direction; as a consequence, a conservatism of up to $\sqrt{m}$ can result; examples exhibiting this can be found in the references. Removing this difficulty is a subject of active research.

In summary, we have presented two approaches to the robust H_2 performance problem, and discussed their interpretation. While other methods have been considered in the literature (see the references), these two have been highlighted since they are closest in spirit to the machinery of this course, and allow the greatest generality in the uncertainty models.

The two methods are not equivalent; one offers a tighter characterization of causal uncertainty, the other the benefit of frequency domain interpretation and computation. They both are potentially conservative when dealing with multivariable noise signals. Thus we see that robust H_2 performance has not been characterized as tightly as our earlier performance specification in terms of the L_2-induced (H_∞) norm. Nevertheless these methods allow us to combine model uncertainty and disturbance rejection in a manner which is better tuned to engineering applications. In addition to the robustness analysis problem on which we have focused here, the above conditions can also be incorporated into the heuristics for controller synthesis, thus allowing designs to be sought with more natural performance specifications.

Notes and references

The method of integral quadratic constraints has its roots in the absolute stability theory of [124, 176], in particular on the use of the S-procedure [177] for nonlinear control. However the extensive use of quadratic separation for robustness analysis appeared only very recently in the work of [101, 103, 102, 132]. In addition to the input–output viewpoint presented here, connections to state space theory such as Lyapunov functions for nonlinear analysis can be given [131]. For additional work also see [79].

The question of robustness in H_2 control goes back to the late 1970s, a time when so-called modern control theory was put to the test of robustness. In particular LQG (H_2) regulators were studied from the point of view of classical stability margins [140], but found to offer in general no guarantees [36]. This motivated efforts to reconcile LQG with classical concepts [39, 40, 153]. However, at the same time (early 1980s) the H_∞ paradigm was being put forth [184], that more easily blended with classical concepts and quickly became the centerpiece of multivariable control, since it allowed a transparent combination of performance and robustness [41].

Nevertheless, the impression has remained that the H_2 metric is more appropriate for system performance. A new opportunity for a better treatment of its robustness came in the late 1980s, with the discovery of close ties between the state space methods of H_2 and H_∞ control [38]. In particular the multi-objective H_2/H_∞ control design problem has received wide attention, a few of the many references are [15, 90, 146, 189].

Concentrating on robust H_2 performance analysis, the state space method we presented for robust H_2 performance originates in [121, 155, 189], although the presentation in terms of LMIs is more recent [78]; the extension to LTI perturbations is due to [51]. For the relationship between the impulse response and average output variance interpretations, see [8]. The frequency domain method and the worst-case white noise interpretation are developed in [117, 116]. In particular [116] contains the state space computation for $\mathbf{J_{f,a}}$. The results on characterization of $\mathbf{J_{f,TI}}$ as a test for slowly varying uncertainty, as well as the extensions to causal perturbations, are found in [115]. The example showing bound conservatism for multivariable noise is from [158]. More extensive details on these and other approaches, as well as many more references can be found in the survey chapter [118]. References that tackle the robust H_2 synthesis problem are [30, 121, 155, 179].

After two decades of research, robust control has thus developed tools to address the historic problem of robustness in LQG control. The fact that the picture is not completely tight is perhaps inevitable given the underlying coexistence of the worst-case and average case philosophies.

11
Further Topics: Synthesis

We have arrived at the final chapter of this course. As with the preceding chapter our main objective is to acquire some familiarity with two new topics, and again our treatment will be of a survey nature. The areas we will consider are linear parameter varying systems, multidimensional systems, and linear time-varying (LTV) systems. In Chapter 10 we covered new analysis techniques and problems, and our aim in this chapter is the study of additional methods and results pertaining to *synthesis*.

Up to this point in the course we have worked entirely with systems and signals of a real time variable, namely, continuous time systems; however, in this chapter we will instead consider discrete time systems. One reason for this is that the concepts of the chapter are more easily developed and understood in discrete time. This change also gives us the opportunity to reflect on how our results in earlier parts of the course translate to discrete time.

The basic state space form of a discrete time system is given below.

$$x(k + 1) = Ax(k) + Bw(k), \quad \text{with initial condition } x(0), \qquad (11.1)$$
$$z(k) = Cx(k) + Dw(k).$$

This system is described by a difference equation, and replaces the differential equation and initial condition we are accustomed to in continuous time. Thus every matrix realization (A, B, C, D) specifies both a discrete time system, and a continuous time system. We will only use *discrete time* systems in this chapter.

Before starting on the new topics we briefly digress, and define the space of sequences on which the systems discussed in this chapter will act. We use $\ell_2^n(\mathbb{N})$ to denote the space of square summable sequences mapping the non-negative integers $\mathbb{N}$ to $\mathbb{C}^n$. This is a Hilbert space with the inner product

$$\langle x, y \rangle_{\ell_2} = \sum_{k=0}^{\infty} x^*(k) y(k).$$

The space $\ell_2^n(\mathbb{N})$ is the discrete time analog of the continuous time space $L_2^n[0, \infty)$. We will usually write just ℓ_2 when the spatial dimension and argument are clear.

11.1 Linear parameter varying and multidimensional systems

So far the controllers we have aimed to synthesize have been pure state space systems. Here we will extend our state space framework to cover a broader class of systems. This extension will allow us to incorporate both *linear parameter varying* (LPV) systems and *multidimensional* systems into a common setting, with performance guarantees during synthesis.

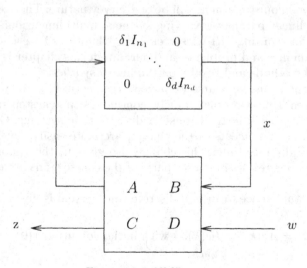

Figure 11.1. NMD system

The new set of state space systems that we introduce are shown in Figure 11.1. The picture shows the upper star product between two systems. The upper system is spatially diagonal, and the lower system is static and memoryless. Each of the blocks in the upper system is written $\delta_i I_{n_i}$, by which we mean

$$\delta_i I_{n_i} = \begin{bmatrix} \delta_i & & 0 \\ & \ddots & \\ 0 & & \delta_i \end{bmatrix} \in \mathcal{L}(\ell_2^{n_i}),$$

for some operator δ_i on $\ell_2^1(\mathbb{N})$. In words, the operator $\delta_i I_{n_i}$ is just the spatially diagonal operator formed from n_i copies of the bounded operator δ_i. For reference we call the operators δ_i *scalar operators*, because they act on scalar sequences. Here (A, B, C, D) is simply a set of real state space matrices; define n such that $A \in \mathbb{R}^{n \times n}$ and therefore $n_1 + \cdots + n_d = n$ holds. Referring to Figure 11.1, let us set

$$\Delta = \begin{bmatrix} \delta_1 I_{n_1} & & 0 \\ & \ddots & \\ 0 & & \delta_d I_{n_d} \end{bmatrix} \tag{11.2}$$

for convenient reference. Therefore we have the formal equations

$$x(k) = (\Delta A x)(k) + (\Delta B w)(k) \tag{11.3}$$
$$z(k) = C x(k) + D w(k)$$

describing the interconnection. We use $(\Delta A x)(k)$ to denote the k-th element of the sequence given by $\Delta A x$; note that since the state space matrix A has no dynamics $(A x)(k) = A x(k)$ in this notation. Here $w \in \ell_2$ and we define this system to be well-connected if $I - \Delta A$ is nonsingular. Thus there is a unique $x \in \ell_2$ which satisfies the above equations when the system is well-connected. Since the operators δ_i do not commute in general, we call these types of systems *noncommuting multidimensional systems*, or *NMD systems*, for short. NMD systems can be used to model numerous linear situations, and we now consider some examples.

Examples:

First we show that our standard state space system can be described in this setup. Let Z denote the shift operator, or delay, on ℓ_2. That is, given $x = (x(0), x(1), x(2), \ldots) \in \ell_2$ we have

$$Z x = (0, x(0), x(1), \ldots). \tag{11.4}$$

We will not distinguish between shifts that act on ℓ_2 of different spatial dimension. Therefore, given a sequence $x \in \ell_2^n$ we have

$$Z x = \begin{bmatrix} Z x_1 \\ \vdots \\ Z x_n \end{bmatrix} = \begin{bmatrix} Z & & 0 \\ & \ddots & \\ 0 & & Z \end{bmatrix} \begin{bmatrix} x_1 \\ \vdots \\ x_n \end{bmatrix}, \text{ for } x \text{ partitioned as } \begin{bmatrix} x_1 \\ \vdots \\ x_n \end{bmatrix} \in \ell_2^n.$$

Namely, Z acts independently on every scalar sequence x_i comprising the vector sequence x; so it is spatially diagonal. Based on this definition we see that the first equation in (11.1) takes the form

$$x = ZAx + ZBw,$$

for zero initial conditions. This means that by setting $\Delta = Z$ the system (11.1) is exactly of the form in (11.3).

Our next example involves varying parameters. Suppose we have a system whose state space realization depends on the real scalar parameters $\alpha_1(k), \ldots, \alpha_r(k)$ which vary with respect to the discrete time variable k; the variation with k may be unknown a priori. Let $(\tilde{A}(\alpha), \tilde{B}(\alpha), \tilde{C}(\alpha), \tilde{D}(\alpha))$ be the realization, where α signifies the dependence on the scalar parameters α_i. If the dependence of each of the matrices is rational in the parameters α_i, then it is frequently possible to convert this system to the form in (11.3) with $\delta_1 = Z$ and the multiplication-by-scalar operators $\delta_2 = \alpha_1, \delta_3 = \alpha_2, \ldots, \delta_{r+1} = \alpha_r$, for some state space realization of constant matrices (A, B, C, D). See, for instance, Exercise 8.7.

For our third example let us consider a multidimensional system in now two independent variables k_1 and k_2. The state equation for such a system follows.

$$\begin{bmatrix} \bar{x}(k_1 + 1, \, k_2) \\ \underline{x}(k_1, \, k_2 + 1) \end{bmatrix} = \begin{bmatrix} A_{11} & A_{12} \\ A_{21} & A_{22} \end{bmatrix} \begin{bmatrix} \bar{x}(k_1, \, k_2) \\ \underline{x}(k_1, \, k_2) \end{bmatrix} + \begin{bmatrix} B_1 \\ B_2 \end{bmatrix} w(k_1, \, k_2).$$

Here the independent variables can, for instance, be thought of as specifying points in a 2-dimensional grid. Now suppose that $w \in \ell_2(\mathbb{N} \times \mathbb{N})$, namely the normed space of bi-indexed sequences that are square summable. Then let Z_1 and Z_2 be the shift operators on the variables k_1 and k_2 respectively. By setting

$$\Delta = \begin{bmatrix} Z_1 & 0 \\ 0 & Z_2 \end{bmatrix},$$

this system can immediately be converted to the form in (11.3). Clearly this construction can be extended to a multidimensional system with inputs and states in $\ell_2(\mathbb{N} \times \cdots \times \mathbb{N})$. See the chapter references for more details.   □

The examples above provide motivation for the use of this model, which has a clear analogy to the uncertainty framework introduced in Chapters 8 and 9: we have replaced the unstructured blocks Δ_i of those chapters with the operators $\delta_i I$. We will say more about these ties later. However, we stress that here our view of the operators $\delta_i I$ is *not* as uncertainty, but as multidimensional variables; in the control problems to follow the controller will know these operators exactly. Thus the work in this section will be more closely connected with the H_∞ methods of Chapter 7, than our work on uncertainty. We will also make connections to the realizations and reduction work in Chapter 4.

Define the set of operators

$$\boldsymbol{\Delta} = \{\Delta \in \mathcal{L}(\ell_2) : \; \Delta \text{ is diagonal as in (11.2), and satisfies } \|\Delta\|_{\ell_2 \to \ell_2} \leq 1\}.$$

This is the set of contractive operators that have the diagonal form in (11.2). We define $\mathcal{X}$ to be the set of positive symmetric matrices in the commutant of $\boldsymbol{\Delta}$ by

$$\mathcal{X} = \{X \in \mathbb{S}^n : \ X > 0 \text{ and } X\Delta = \Delta X, \text{ for all } \Delta \in \boldsymbol{\Delta}\}.$$

Therefore every element of $\mathcal{X}$ has the block-diagonal form

$$X = \begin{bmatrix} X_1 & & 0 \\ & \ddots & \\ 0 & & X_d \end{bmatrix}, \tag{11.5}$$

where each $X_i \in \mathbb{S}^{n_i}$ and is positive. Our change in notation as compared to the commutant of an uncertainty set is made to emphasize the role of X as a generalized Lyapunov matrix, as seen below. Also, in contrast to our work of Chapter 8, only positive definite commuting matrices are considered. A final remark is that the choice to work with real symmetric, rather than complex Hermitian matrices can be shown to be inconsequential since the matrices (A, B, C, D) are real.

We are now ready to state a major result pertaining to NMD systems.

Theorem 11.1. *The following equivalences hold for the NMD system of Figure 11.1.*

(i) *The operator $I - \Delta A$ is nonsingular for all $\Delta \in \boldsymbol{\Delta}$ if and only if there exists $X \in \mathcal{X}$ such that*

$$A^*XA - X < 0.$$

(i) *The operators $I - \Delta A$ and $C(I - \Delta A)^{-1}B + D$ are nonsingular and contractive, respectively, for all $\Delta \in \boldsymbol{\Delta}$, if and only if there exists $X \in \mathcal{X}$ such that*

$$\begin{bmatrix} A & B \\ C & D \end{bmatrix}^* \begin{bmatrix} X & 0 \\ 0 & I \end{bmatrix} \begin{bmatrix} A & B \\ C & D \end{bmatrix} - \begin{bmatrix} X & 0 \\ 0 & I \end{bmatrix} < 0.$$

The first part of the theorem provides a necessary and sufficient condition for an NMD system to be well-connected, which takes the form of a generalized Lyapunov inequality. In fact an analogous result for continuous time systems is outlined in the exercises of Chapter 8, extending the full-block uncertainty case of Theorem 8.12; see Exercise 8.10. We remark as well that analogously to our work in Chapter 9, generalizations of this result are available where causality is imposed on Δ and $(I - \Delta A)^{-1}$, with the corresponding stronger implications on system stability.

The second part above incorporates an additional contractiveness condition, similarly to what was done to study performance in Chapter 8. The resulting LMI test is reminiscent of the KYP Lemma; indeed the discrete-time KYP Lemma of Proposition 8.28 takes exactly this form for the case of a single block in X. This observation about Theorem 11.1 affords the opportunity to develop exact synthesis methods for NMD systems. Synthesis is the topic of the next subsection, after which we discuss realization theory for models of this type.

11.1.1 LPV synthesis

We now look at a synthesis problem associated with NMD systems. This is commonly known as *LPV synthesis* because of the example given above using varying parameters. This problem has a direct connection with so-called *gain scheduling*, as well as synthesis for systems with both temporal and spatial independent variables. After posing the synthesis problem we will show its application to gain scheduling.

The arrangement for the synthesis problem is shown below in Figure 11.2, and corresponds to our standard synthesis problem, but now the constituent systems are NMD systems. In the setup we have an NMD system $G(\delta_1, \ldots, \delta_d)$, which is given and dependent on the scalar operators $\delta_1, \ldots, \delta_d$. The aim is to synthesize an NMD controller $K(\delta_1, \ldots, \delta_d)$ in

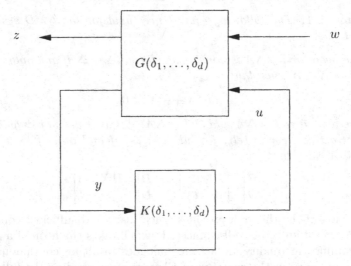

Figure 11.2. NMD synthesis configuration

terms of the same scalar operators. Using the variables shown in the figure, the equations which describe the plant are

$$x(k) = (\Delta A x)(k) + \left(\Delta \begin{bmatrix} B_1 & B_2 \end{bmatrix} \begin{bmatrix} w \\ u \end{bmatrix}\right)(k) \tag{11.6}$$

$$\begin{bmatrix} z(k) \\ y(k) \end{bmatrix} = \begin{bmatrix} C_1 \\ C_2 \end{bmatrix} x(k) + \begin{bmatrix} D_{11} & D_{12} \\ D_{21} & 0 \end{bmatrix} \begin{bmatrix} w(k) \\ u(k) \end{bmatrix},$$

for given state space matrices (A, B, C, D), where

$$\Delta = \begin{bmatrix} \delta_1 I_{n_1} & & 0 \\ & \ddots & \\ 0 & & \delta_d I_{n_d} \end{bmatrix}.$$

These are the analogous equations to those we began with in H_∞ synthesis.
Similar to this the controller equations are

$$x_K(k) = (\Delta_K A_K x_K)(k) + (\Delta_K B_K y)(k) \tag{11.7}$$
$$u(k) = C_K x_K(k) + D_K y(k),$$

with

$$\Delta_K = \begin{bmatrix} \delta_1 I_{n_{K1}} & & 0 \\ & \ddots & \\ 0 & & \delta_d I_{n_{Kd}} \end{bmatrix}.$$

We define the state dimensions of the plant and controller, respectively, to
be n and n_K, so that A is $n \times n$, the matrix A_K is $n_K \times n_K$, and therefore
the partition dimensions satisfy

$$\sum_{i=1}^{d} n_i = n \text{ and } \sum_{i=1}^{d} n_{Ki} = n_K.$$

The synthesis goal is to find suitable matrices (A_K, B_K, C_K, D_K), where
the dimension n_K can be chosen freely as can the partitioning $n_{K1}, \ldots, n_{Kd}$
respecting the condition $n_{K1} + \cdots + n_{Kd} = n_K$. However, the dimension n
and the partition $n_1, \ldots, n_d$ belonging to the plant are fixed. Our synthesis
objective is that K ensure the closed loop is both well-connected and contractive. We will make this precise soon, and show that we can easily solve
this problem using the procedure developed in Chapter 7. Before doing this
let us first look at the applicability of the NMD synthesis problem.

Example:

We return to the parameter varying setup described as an example above.
There we have a state space realization $(\tilde{A}(\alpha), \tilde{B}(\alpha), \tilde{C}(\alpha), \tilde{D}(\alpha))$, which
depends on the time-varying parameters $\alpha_1(k), \ldots, \alpha_r(k)$. Now consider a
gain scheduling scenario where these scalar parameters $\alpha_j(k)$ would not
be known a priori, but are known as each time k is reached. Thus one
can arrange for controller parameters ("gains") to be "scheduled" based
on the plant parameters $\alpha_j(k)$. Such a controller would be of the form
$K(\delta_1, \ldots, \delta_{r+1})$, where $\delta_1 = Z$ and $\delta_i = \alpha_{i-1}$, for $i \geq 2$.

Now further suppose that although they are not known ahead of time,
these parameters reside in normalized intervals, so that $\alpha_i(k) \in [-1, 1]$
always holds, and the dependence of G on the parameters takes the form

of an NMD system $G(\delta_1, \ldots, \delta_{r+1})$. In this case the search for a controller can be posed as an NMD synthesis problem.   □

Having provided some additional motivation for the setup let us return to equations governing the closed loop in Figure 11.2. The equations in (11.6) and (11.7) can be written more compactly as

$$x_{cl}(k) = (\Delta_{cl} A_{cl} x_{cl})(k) + (\Delta_{cl} B_{cl} w)(k) \qquad (11.8)$$
$$z(k) = C_{cl} x_{cl}(k) + D_{cl} w(k),$$

where the matrices $(A_{cl}, B_{cl}, C_{cl}, D_{cl})$ are appropriately defined, and

$$\Delta_{cl} = \begin{bmatrix} \Delta & 0 \\ 0 & \Delta_K \end{bmatrix}, \text{ with } x_{cl} = \begin{bmatrix} x \\ x_K \end{bmatrix}.$$

Notice that $(A_{cl}, B_{cl}, C_{cl}, D_{cl})$ have the same definitions as in the H_∞ synthesis of §7.2. We say an NMD controller K is *admissible* if for all $\delta_1, \ldots, \delta_d \in \mathcal{L}(\ell_2^1)$, with $\|\delta_i\|_{\ell_2 \to \ell_2} \leq 1$, the following two conditions are met:

- the operator $I - \Delta_{cl} A_{cl}$ is nonsingular;

- the map $w \mapsto z$ is contractive on ℓ_2.

These conditions state that a controller must satisfy a stability or well-connectedness condition, and simultaneously attenuate the input–output mapping $w \mapsto z$. They have clear analogies with the criteria we set for H_∞ synthesis.

As mentioned already this NMD synthesis problem can be solved using essentially the same methodology we employed in solving the H_∞ synthesis problem in §7.2. We will now provide a sketch of the steps in the proof. The NMD synthesis result will then be stated in Theorem 11.5.

To start we require a temporary change of basis via a permutation. Consider the form of Δ_{cl}:

$$\Delta_{cl} = \begin{bmatrix} \Delta & 0 \\ 0 & \Delta_K \end{bmatrix} = \begin{bmatrix} \begin{bmatrix} \delta_1 I_{n_1} & & 0 \\ & \ddots & \\ 0 & & \delta_d I_{n_d} \end{bmatrix} & 0 \\ 0 & \begin{bmatrix} \delta_1 I_{n_{K1}} & & 0 \\ & \ddots & \\ 0 & & \delta_d I_{n_{Kd}} \end{bmatrix} \end{bmatrix}.$$

Each scalar operator δ_i appears twice in the block structure, once in Δ and once in Δ_K. Let P be the permutation matrix that satisfies

$$P^* \Delta_{cl} P = \begin{bmatrix} \delta_1 I_{n_{cl1}} & & 0 \\ & \ddots & \\ 0 & & \delta_d I_{n_{cld}} \end{bmatrix}, \qquad (11.9)$$

where $n_{cli} = n_i + n_{Ki}$. That is, this permutation matrix simply rearranges the blocks so that each δ_i only appears in a single block. Define the positive symmetric commutant of such operators

$$\mathcal{X}_{cl} = \{X_{cl} \in \mathbb{S}^{n+n_K} : X_{cl} > 0 \text{ and } X_{cl}(P^*\Delta_{cl}P) = (P^*\Delta_{cl}P)X_{cl},$$

$$\text{for all } \delta_i \in \mathcal{L}(\ell_2) \text{ with } \Delta_{cl} \text{ defined in (11.9)}\}.$$

That is, any element of $\mathcal{X}_{cl}$ has the form

$$X_{cl} = \begin{bmatrix} X_{cl1} & & 0 \\ & \ddots & \\ 0 & & X_{cld} \end{bmatrix},$$

where each $X_{cli} \in \mathbb{S}^{n_i+n_{Ki}}$ and is positive definite.

We can now state the following result.

Proposition 11.2. *Suppose $K(\delta_1, \ldots, \delta_d)$ is a NMD controller. Then K is admissible if and only if there exists $X_{cl} \in \mathcal{X}_{cl}$ such that*

$$\begin{bmatrix} P^*A_{cl}P & P^*B_{cl} \\ C_{cl}P & D_{cl} \end{bmatrix}^* \begin{bmatrix} X_{cl} & 0 \\ 0 & I \end{bmatrix} \begin{bmatrix} P^*A_{cl}P & P^*B_{cl} \\ C_{cl}P & D_{cl} \end{bmatrix} - \begin{bmatrix} X_{cl} & 0 \\ 0 & I \end{bmatrix} < 0. \quad (11.10)$$

Proof. The argument is best visualized in terms of the configuration of Figure 11.3, which is obtained by incorporating the permutation P to the system (11.8).

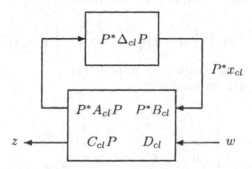

Figure 11.3. Permuted version of the closed loop NMD system.

Our controller will be admissible if and only if $I - \Delta_{cl}A_{cl}$ is nonsingular and $C_{cl}(I - \Delta_{cl}A_{cl})^{-1}\Delta_{cl}B_{cl} + D_{cl}$ is contractive, for every $\delta_1, \ldots, \delta_d$. Observe that $P^{-1} = P^*$ since P is a permutation matrix. Thus the two conditions are equivalent to saying that

$$I - (P^*\Delta_{cl}P)(P^*A_{cl}P) \quad \text{is nonsingular and}$$

$$C_{cl}P\Big(I - (P^*\Delta_{cl}P)(P^*A_{cl}P)\Big)^{-1}(P^*\Delta_{cl}P)P^*B_{cl} + D_{cl} \quad \text{is contractive.}$$

Now $P^*\Delta_{cl}P$ is of the form in (11.9), and so by invoking Theorem 11.1 we see the two latter conditions hold if and only if (11.10) has a solution. ∎

Checking the admissibility of any given controller is equivalent to an LMI feasibility problem. Our ultimate goal is to obtain synthesis conditions, so we will need to examine the block structure of X_{cl} in (11.10). Any X_{cl} in $\mathcal{X}_{cl}$ is of the block-diagonal form $\mathrm{diag}(X_{cl1}, \ldots, X_{cld})$ with blocks given by

$$X_{cli} = \begin{bmatrix} X_{1i} & X_{2i} \\ X_{2i}^* & X_{3i} \end{bmatrix},$$

and the matrices $X_{1i} \in \mathbb{S}^{n_i}$, $X_{2i} \in \mathbb{R}^{n_i \times n_{Ki}}$ and $X_{3i} \in \mathbb{S}^{n_{Ki}}$. Now consider the effect of the permutation matrix P in (11.9) on X_{cl}. Setting $X_P = P X_{cl} P^*$ it is straightforward to verify that

$$X_P = \begin{bmatrix} X_1 & X_2 \\ X_2^* & X_3 \end{bmatrix}, \tag{11.11}$$

where the constituent matrices are

$$X_1 = \mathrm{diag}(X_{11}, \ldots, X_{1d}), \quad X_2 = \mathrm{diag}(X_{21}, \ldots, X_{2d}), \text{ and}$$

$X_3 = \mathrm{diag}(X_{31}, \ldots, X_{3d})$. Now from (11.11) the matrix X_P has the same partitioned structure as the X_{cl}-matrix that appears in the H_∞ synthesis of §7.2; however, each of the matrices X_1, X_2, and X_3 is *block-diagonal*. Also, we can pre- and post-multiply the expression in (11.10) by the matrices $\mathrm{diag}(P, I)$ and $\mathrm{diag}(P^*, I)$, respectively, to arrive at

$$\begin{bmatrix} A_{cl} & B_{cl} \\ C_{cl} & D_{cl} \end{bmatrix}^* \begin{bmatrix} X_P & 0 \\ 0 & I \end{bmatrix} \begin{bmatrix} A_{cl} & B_{cl} \\ C_{cl} & D_{cl} \end{bmatrix} - \begin{bmatrix} X_P & 0 \\ 0 & I \end{bmatrix} < 0.$$

If we apply the Schur complement formula twice to the above LMI we get the following equivalent inequality.

$$\begin{bmatrix} -X_P^{-1} & A_{cl} & B_{cl} & 0 \\ A_{cl}^* & -X_P & 0 & C_{cl}^* \\ B_{cl}^* & 0 & -I & D_{cl}^* \\ 0 & C_{cl} & D_{cl} & -I \end{bmatrix} < 0. \tag{11.12}$$

That is, a controller is admissible exactly when a solution can be found to this inequality. This is the critical form for solving the synthesis problem. The left-hand side of this inequality is *affine* in the state space matrices for the controller. We can use Lemma 7.2 to show that the solution to this controller dependent LMI implies the existence of solutions to two matrix inequalities which do not explicitly or implicitly involve the controller. Another application of the Schur complement converts these inequalities into LMIs, and yields the following lemma.

Lemma 11.3. *An admissible controller exists, with partition dimensions $n_{K1}, \ldots, n_{Kd}$, if and only if there exist symmetric block-diagonal matrices X and Y such that*

(i)

$$\begin{bmatrix} N_c & 0 \\ 0 & I \end{bmatrix}^* \begin{bmatrix} AYA^* - Y & AYC_1^* & B_1 \\ C_1YA^* & C_1YC_1^* - I & D_{11} \\ B_1^* & D_{11}^* & -I \end{bmatrix} \begin{bmatrix} N_c & 0 \\ 0 & I \end{bmatrix} < 0.$$

(ii)

$$\begin{bmatrix} N_o & 0 \\ 0 & I \end{bmatrix}^* \begin{bmatrix} A^*XA - X & A^*XB_1 & C_1^* \\ B_1^*XA & B_1^*XB_1 - I & D_{11}^* \\ C_1 & D_{11} & -I \end{bmatrix} \begin{bmatrix} N_o & 0 \\ 0 & I \end{bmatrix} < 0.$$

(iii) The identities

$$PX_{cl}P^* = \begin{bmatrix} X & ? \\ ? & ? \end{bmatrix} \text{ and } PX_{cl}^{-1}P^* = \begin{bmatrix} Y & ? \\ ? & ? \end{bmatrix}$$

hold for some $X_{cl} \in \mathcal{X}_{cl}$,

where the operators N_c, N_o satisfy

$$\operatorname{Im} N_c = \operatorname{Ker} \begin{bmatrix} B_2^* & D_{12}^* \end{bmatrix} \qquad N_c^* N_c = I$$
$$\operatorname{Im} N_o = \operatorname{Ker} \begin{bmatrix} C_2 & D_{21} \end{bmatrix} \qquad N_o^* N_o = I.$$

We have concluded that an admissible controller exists exactly when there exist solutions X and Y to conditions (i)–(iii) in Lemma 11.3. Now (i) and (ii) are LMIs and are completely specified in terms of the given plant G. However, (iii) is not an LMI condition, but only depends on controller dimension. Note that X and Y must necessarily be members of the set $\mathcal{X}$ defined in (11.5) if they satisfy (iii).

So the next step is to convert (iii) to an LMI-type condition. We have the following result.

Lemma 11.4. *The block-diagonal matrices X, $Y \in \mathcal{X}$ satisfy condition (iii) in Lemma 11.3 if and only if for each $1 \leq i \leq d$, the following inequalities hold*

$$\begin{bmatrix} X_i & I \\ I & Y_i \end{bmatrix} \geq 0 \text{ and rank} \begin{bmatrix} X_i & I \\ I & Y_i \end{bmatrix} \leq n_i + n_{Ki}. \qquad (11.13)$$

Proof. By Lemma 7.9 we see that there exist positive matrices $X_{cli} \in \mathbb{S}^{n_i + n_{Ki}}$ exist such that

$$X_{cli} = \begin{bmatrix} X_i & ? \\ ? & ? \end{bmatrix} \text{ and } X_{cli}^{-1} = \begin{bmatrix} Y_i & ? \\ ? & ? \end{bmatrix}$$

for $1 \leq i \leq d$, exactly when the inequalities in (11.13) are satisfied. Clearly the block-diagonal matrix $X_{cl} \in \mathcal{X}_{cl}$. It is straightforward to see that $PX_{cl}P^*$ and $PX_{cl}^{-1}P^*$ satisfy the condition (iii) in Lemma 11.3. ∎

The first of these conditions is an LMI; however, the rank inequality is not. Note that the rank conditions are trivially met if $n_{Ki} \geq n_i$, and therefore can be eliminated if the controller order is chosen to be the same as the plant. This situation was observed in our earlier H_∞ synthesis, where we had only one variable.

We can now state the NMD system synthesis result.

Theorem 11.5. *An admissible NMD controller exists if and only if there exist matrices X, $Y \in \mathcal{X}$, which satisfy LMI conditions (i) and (ii) in Lemma 11.3 , and the LMI conditions given in (11.13), for each $k = 1, \dots, d$.*

This gives exact convex conditions for the existence of a solution to the NMD synthesis problem. Notice that if the non-LMI rank conditions in (11.13) are also achieved, then a controller exists with dimensions $n_{K1}, \dots, n_{Kd}$. Furthermore, an admissible controller exists if and only if one exists satisfying $n_{K1} = n_1, \dots, n_{Kd} = n_d$. Also if $d = 1$, namely, there is only one δ_i, this result corresponds exactly to the discrete time H_∞ synthesis problem; this is more clearly apparent by consulting Proposition 8.28. When solutions X and Y are found an explicit controller can be computed by constructing a scaling X_{cl}, and then finding a controller realization which solves the LMI given above in (11.12); this is similar to the procedure in §7.3.

11.1.2   Realization theory for multidimensional systems

Having developed some synthesis results for NMD systems, we now discuss how a number of system realization concepts can be generalized to NMD systems by utilizing an LMI approach. Most notable of these realization topics are *minimality* and the associated notions of *reducibility* and *model reduction*. Additional realization results that may be derived for NMD systems include reachability, controllability and observability conditions, and the construction of Kalman-like decomposition structures. Here we can view the δ_i either as multidimensional variables as in §11.1.1, *or* as uncertain parameters which are part of a system description.

We begin by noting that the LMI given in (11.10) has a *Lyapunov inequality* embedded in its top, left corner. As we know from Chapters 2 and 4, in standard state space systems theory Lyapunov equations play a formidable role in the development of conditions for testing the stability, minimality, controllability, and observability of a given set of realization matrices (A, B, C, D). Specifically, the following are basic results for standard state space system realizations in discrete time. They are all discrete time analogs of results from Chapters 2 and 4:

(i)   *A matrix A satisfies $\rho(A) < 1$ if and only if, for each $Q > 0$, there exists a matrix $X > 0$ satisfying*

$$A^* X A - X + Q = 0.$$

This equation is one example of a discrete time Lyapunov equation.

(ii) *Suppose $\rho(A) < 1$; then the pair (C, A) is observable if and only if there exists a positive definite solution X to the following Lyapunov equation:*

$$A^*XA - X + C^*C = 0.$$

(iii) *Suppose $\rho(A) < 1$; then the pair (A, B) is reachable if and only if there exists a positive definite solution Y to the following Lyapunov equation:*

$$AYA^* - Y + BB^* = 0.$$

The terms *observable* and *reachable* are the discrete time analogs of observability and controllability in continuous time. Observability of a realization implies that given an output sequence $y(0), \ldots, y(n-1)$, the initial state $x(0)$ can be *uniquely* determined, and reachability of a realization implies that the initial state can be transferred to any other fixed state $x \in \mathbb{C}^n$ via a finite input sequence. In this section we will not exploit these facts, here our main purpose is to emphasize the connections between the solutions to the above Lyapunov equations, namely, the gramians, and the notion of minimality for a standard state space realization. Given a realization (A, B, C, D) for a system G, we say the realization is minimal if there exists no other realization for G, for example (A_r, B_r, C_r, D_r), of lower order. That is a realization satisfying $\dim(A_r) < \dim(A)$.

As we saw in Chapters 2 and 4 a stable realization (A, B, C, D) for a system G is minimal if and only if it is both controllable and observable, that is, if and only if the gramians X and Y are positive definite. If a given realization is not reachable (or observable), the resulting associated gramian will be singular.

Returning to NMD systems, a parallel result on minimality can been shown to hold. First we need to state what we mean by equivalent realizations for NMD systems. We say that two well-connected realizations (A, B, C, D) and (A_r, B_r, C_r, D_r) are *equivalent* if the NMD systems $G(\delta_1, \ldots, \delta_d)$ and $G_r(\delta_1, \ldots, \delta_d)$ they represent satisfy

$$G(\delta_1, \ldots, \delta_d) = G_r(\delta_1, \ldots, \delta_d), \text{ for all contractive } \delta_i \in \mathcal{L}(L_2^1).$$

More explicitly the systems G and G_r are defined by

$$G(\delta_1, \ldots, \delta_d) = C(I - \Delta A)^{-1}\Delta B + D;$$
$$G_r(\delta_1, \ldots, \delta_d) = C_r(I - \Delta_r A_r)^{-1}\Delta_r B_r + D_r,$$

where

$$\Delta = \begin{bmatrix} \delta_1 I_{n_1} & & 0 \\ & \ddots & \\ 0 & & \delta_d I_{n_d} \end{bmatrix} \text{ and } \Delta_r = \begin{bmatrix} \delta_1 I_{n_{r1}} & & 0 \\ & \ddots & \\ 0 & & \delta_d I_{n_{rd}} \end{bmatrix}.$$

Thus we see that two realizations are equivalent if the operators they define on ℓ_2 are the same, for each δ_i. For multidimensional systems we define minimal as follows:

Definition 11.6. *A realization* (A, B, C, D) *for an NMD system is minimal if* $\dim(A)$ *is lowest among all equivalent realizations.*

It can be shown that any minimal NMD realization will further satisfy the condition that n_i is lowest, for each $i = 1, \ldots, d$, among all equivalent realizations.

We are now in a position to state an NMD system minimality result that is directly based on LMIs. It says a realization is minimal if and only if there exist no *singular* solutions to either of a pair of associated Lyapunov inequalities, where these solutions are restricted to be in the closure of the set $\mathcal{X}$; namely, the set of block-diagonal and positive semidefinite matrices. Note that it is assumed that the D-terms are the same in both the full and reduced systems.

Theorem 11.7. *Given a system realization* (A, B, C, D), *there exists an equivalent lower order realization* (A_r, B_r, C_r, D) *of order* n_r *if and only if there exist* $X \geq 0$ *and* $Y \geq 0$, *both in the closure* $\bar{\mathcal{X}}$, *satisfying*

(i) $AXA^* - X + BB^* \leq 0$,

(ii) $A^*YA - Y + C^*C \leq 0$,

(iii) $\lambda_{min}(XY) = 0$, *with* $rank(XY) \leq n_r$.

That is, a realization is minimal if and only if *all* solutions to the system Lyapunov inequalities are nonsingular. If a rank deficient structured gramian exists for either of the Lyapunov inequalities, then an equivalent lower dimension realization exists, and vice versa. Via the proofs for this case it is also possible to show that for a given NMD system, all minimal realizations may be found by similarity transformations in $\mathcal{X}$ and truncations. A source for a proof is provided in the chapter references.

We now have the related model reduction result; here the D terms are allowed to be different.

Theorem 11.8. *Given a system realization* (A, B, C, D) *and* $\epsilon > 0$, *there exists a reduced system realization* (A_r, B_r, C_r, D_r) *of order* n_r *such that*

$$\sup_{\|\delta_i\| \leq 1} \|G - G_r\| < \epsilon$$

if and only if there exist $X > 0$ *and* $Y > 0$, *both in the set* $\mathcal{X}$, *satisfying*

(i) $AXA^* - X + BB^* < 0$,

(ii) $A^*YA - Y + C^*C < 0$,

(iii) $\lambda_{min}(XY) = \epsilon^2$, *with* $rank(\epsilon^2 I - XY) \leq n_r$.

Although very similar in their statements, the proof of this result is significantly easier than the former minimality result, and can be obtained directly using the NMD synthesis results above in Theorem 11.5, by viewing the desired reduced dimension system in the role of the controller. As we have already noted, rank constraints as in (iii) above are not easily incorporated into convex computations. We therefore examine model reduction results along the lines of our earlier work in Chapter 4.

Balanced truncation

As with standard state space systems, an alternative for model reduction of NMD systems is that of balanced truncation. For this method, the same LMIs as just described must first be solved; that is, structured solutions must be found to the *strict* system Lyapunov inequalities. Following this, the system is balanced exactly as in the standard case of Chapter 4; since the initial matrices X, Y are in $\mathcal{X}$, then the balancing transformations will have the same block-diagonal structure. Truncating the balanced system repetitively for each δ_i leads to an additive error bound of twice the sum of the truncated singular values, much in the same way as was done in the one dimensional case. More specifically, we proceed as follows.

In order to derive the model reduction error bounds for balanced NMD systems, we partition the system matrices A, B, C and the balanced structured gramian $Y = X = \Sigma > 0$ so as to separate the subblocks which will be truncated; note that these matrices must satisfy (i)–(ii) in Theorem 11.8. That is, A, B, and C are partitioned compatibly with the block structure Δ as

$$
A = \begin{bmatrix} A_{11} & \cdots & A_{1d} \\ \vdots & \ddots & \vdots \\ A_{d1} & \cdots & A_{dd} \end{bmatrix}; \quad B = \begin{bmatrix} B_1 \\ \vdots \\ B_d \end{bmatrix}; \quad C = \begin{bmatrix} C_1 & \cdots & C_d \end{bmatrix}.
$$

We further partition each block of Σ by $\Sigma_i = \mathrm{diag}[\hat{\Sigma}_{1i}, \Sigma_{2i}]$, for $i = 1, \ldots, d$, where the realization submatrices corresponding to Σ_{2i} will be truncated. Denote

$$
\hat{\Sigma}_{1i} = \mathrm{diag}[\sigma_{i1} I_{n_{i1}}, \ldots, \sigma_{ik_i} I_{n_{ik_i}}],
$$

and

$$
\Sigma_{2i} = \mathrm{diag}[\sigma_{i(k_i+1)} I_{n_{i(k_i+1)}}, \ldots, \sigma_{it_i} I_{n_{it_i}}], \quad k_i \leq t_i.
$$

We then truncate both Σ_{2i} and the corresponding parameter matrices; for example, we truncate

$$
A_{11} = \begin{bmatrix} \hat{A}_{11} & A_{11_{12}} \\ A_{11_{21}} & A_{11_{22}} \end{bmatrix}, \quad B_1 = \begin{bmatrix} \hat{B}_1 \\ B_{1_2} \end{bmatrix}, \quad \text{and } C_1 = \begin{bmatrix} \hat{C}_1 & C_{1_2} \end{bmatrix}
$$

to $\hat{A}_{11}$, $\hat{B}_1$ and $\hat{C}_1$. Partitioning and truncating each A_{ij}, B_j and C_i, for $i, j = 1, \ldots, d$, in a similar fashion, results in the truncated system realization

$$
\begin{bmatrix} A_r & B_r \\ C_r & D \end{bmatrix} = \begin{bmatrix} \hat{A}_{11} & \cdots & \hat{A}_{1d} & \hat{B}_1 \\ \vdots & \ddots & \vdots & \vdots \\ \hat{A}_{d1} & \cdots & \hat{A}_d & \hat{B}_d \\ \hat{C}_1 & \cdots & \hat{C}_d & D \end{bmatrix},
$$

with an associated set structure of $\Delta_r = \{\mathrm{diag}[\delta_1 I_{r_1}, \ldots, \delta_d I_{r_d}] : \delta_i \in \mathcal{L}(l_2)\}$ where $r_i = \sum_{j=1}^{k_i} n_{ij}$. Note that Δ_r is constructed with the same uncertainty variables δ_i as is Δ, thus the $\Delta \in \Delta$ and $\Delta_r \in \Delta_r$ are not independent.

As in the one dimensional case, truncating a balanced well-connected NMD system realization results in a lower dimension realization which is balanced and well-connected, which is easily seen by considering the system Lyapunov inequalities. We now state the balanced truncation model reduction error bound theorem for multidimensional systems.

Theorem 11.9. *Suppose (A_r, B_r, C_r, D) is the reduced model obtained from the balanced well-connected system (A, B, C, D). Then*

$$
\sup_{\|\delta_i\| \le 1} \|G(\delta_1, \ldots, \delta_d) - G_r(\delta_1, \ldots, \delta_d)\| \le 2 \sum_{i=1}^d \sum_{j=k_i+1}^{t_i} \sigma_{ij}. \tag{11.14}
$$

This result states that the reduced order system is guaranteed to satisfy an error bound very much similar to the one we saw for standard systems in Chapter 4. Notice that this result requires the use of Lyapunov inequalities rather than equations. Most importantly, unlike the reduction result of Theorem 11.8, it can be computed directly using LMIs.

We now leave NMD systems and turn to the new topic of time-varying state space systems.

11.2 A framework for time-varying systems: Synthesis and analysis

In this section we will consider a class of linear time-varying systems in discrete time. The standard way of describing such a system is using the state space system equations

$$
\begin{aligned}
x(k+1) &= A(k)x(k) + B(k)w(k) \\
z(k) &= C(k)x(k) + D(k)w(k),
\end{aligned} \tag{11.15}
$$

for $w(k)$ an input sequence, where $A(k)$, $B(k)$, $C(k)$, and $D(k)$ are bounded matrix sequences which are given a priori. Here, we will assume the initial condition of the system is $x(0) = 0$.

Such system models arise in a number of applications. One way they arise is when a nonlinear system is linearized along a trajectory; they can arise in multirate control or optimal filter design, where the system is usually periodic; also they occur when plants and subsystem components are naturally time-varying.

Our main objective is to present a framework for LTV systems that is consistent with the techniques we have learned so far, and that will allow us to pursue synthesis and analysis. Many of the standard state space methods we have learned for LTI systems can be applied directly to LTV systems using the machinery we now introduce. We illustrate this by focusing on controller synthesis for the ℓ_2-induced norm (i.e., an H_∞ synthesis for LTV systems). In short the technique we now pursue makes a strong connection with state space and LMI methods so that the transition from time invariant to time-varying analysis appears straightforward.

Although our main objective is a synthesis method for such systems, our general goals will be broader, and we will introduce machinery which can be applied to a variety of problems involving LTV systems. In §11.2.1 and §11.2.2 our focus will be introducing tools for working with LTV systems. Following this we illustrate the use of this framework by considering analysis of LTV systems. Having established these results we state the LTV synthesis conditions in §11.2.4. The results we derive will be operator inequalities when our initial systems are general LTV systems. If the system has the specific structure of being periodic or of finite time horizon, then the general operator conditions become structured LMIs. In both cases the conditions we obtain will be convex.

Before beginning our investigation we bring in a tool which is extensively used below, the *infinite matrix* representation of an ℓ_2 operator. This representation stems from the observation that the time-domain space ℓ_2 has a canonical basis, just as the finite dimensional space $\mathbb{C}^n$ does. Therefore, given that M is an operator on ℓ_2, we can regard the equation $y = Mu$ as the infinite matrix equation

$$\begin{bmatrix} y(0) \\ y(1) \\ y(2) \\ \vdots \end{bmatrix} = \begin{bmatrix} M(0,0) & M(0,1) & M(0,2) & \cdots \\ M(1,0) & M(1,1) & M(1,2) & \cdots \\ M(2,0) & M(2,1) & M(2,2) & \cdots \\ \vdots & \vdots & \vdots & \vdots & \ddots \end{bmatrix} \begin{bmatrix} u(0) \\ u(1) \\ u(2) \\ \vdots \end{bmatrix},$$

where the sequences u and y are in ℓ_2. Here each entry $M(k,i)$ is a matrix, and the above equation states that each value $y(k)$ of the output sequence can be expressed as

$$y(k) = \sum_{i=0}^{\infty} M(k,i)u(i).$$

In the sequel we will often not distinguish between an operator and its infinite matrix representation.

Example:

Consider the unilateral shift operator Z on ℓ_2 from (11.4). Then Z is represented by the infinite matrix

$$\begin{bmatrix} 0 & 0 & \cdots \\ I & 0 & 0 & \cdots \\ 0 & I & 0 & 0 & \cdots \\ & 0 & I & 0 & 0 & \cdots \\ & & \ddots & \ddots & \ddots & \ddots \end{bmatrix}.$$

$\square$

Let us now discuss infinite matrices for *partitioned* operators between Cartesian products of ℓ_2. Given an operator $M : \ell_2^{m_1} \times \ell_2^{m_2} \to \ell_2^{p_1} \times \ell_2^{p_2}$, we can express it as

$$M = \begin{bmatrix} M_{11} & M_{12} \\ M_{21} & M_{22} \end{bmatrix},$$

where each M_{ij} is in $\mathcal{L}(\ell_2)$. Here the matrix representation is being used for spatial partitioning; in addition, each operator M_{ij} can be viewed as an infinite matrix indexed by time, which would make the above a 2×2 matrix of infinite matrices. Alternatively, we can think of M as an operator between the spaces $\ell_2^{m_1+m_2}$ and $\ell_2^{p_1+p_2}$, with an associated (single) infinite matrix; this change amounts in essence to permuting rows and columns in the previous representation. Such permutations will be discussed in more detail below, for a specific class of operators which is now introduced.

11.2.1 *Memoryless time-varying operators*

We now define a class of operators which is key to LTV systems analysis.

Definition 11.10. *A bounded operator Q mapping ℓ_2^m to ℓ_2^n is memoryless if there exists a sequence of matrices $Q(k)$ in $\mathbb{C}^{m \times n}$ such that, for all w, z, if $z = Qw$ then $z(k) = Q(k)w(k)$. That is, Q has the block-diagonal infinite matrix representation*

$$\begin{bmatrix} Q(0) & & & 0 \\ & Q(1) & & \\ & & Q(2) & \\ 0 & & & \ddots \end{bmatrix}.$$

Further, if $Q(k) \in \mathbb{C}^{m \times n}$ is a uniformly bounded sequence of matrices we use $Q = diag(Q(0), Q(1), \ldots)$ to denote the memoryless operator associated with the sequence $Q(k)$.

We now return to the study of a partitioned operator

$$A = \begin{bmatrix} F & G \\ R & S \end{bmatrix} : \ell_2^{m_1} \times \ell_2^{m_2} \to \ell_2^{p_1} \times \ell_2^{p_2},$$

focusing on the case where F, G, R, and S are memoryless operators. Then we define the following notation:

$$\begin{bmatrix}\!\begin{bmatrix} F & G \\ R & S \end{bmatrix}\!\end{bmatrix} := \mathrm{diag}\left(\begin{bmatrix} F(0) & G(0) \\ R(0) & S(0) \end{bmatrix}, \begin{bmatrix} F(1) & G(1) \\ R(1) & S(1) \end{bmatrix}, \cdots \right),$$

which we call the *diagonal realization* of A. This is clearly a diagonal operator formed from A, and maps $\ell_2^{m_1+m_2}$ to $\ell_2^{p_1+p_2}$. Implicit in the definition of $[\![A]\!]$ is the underlying block structure of the partitioned operator A. Clearly, for any given operator A of this particular structure, $[\![A]\!]$ is simply an equivalent version of A with the rows and columns interchanged appropriately so that

$$\begin{bmatrix}\!\begin{bmatrix} F & G \\ R & S \end{bmatrix}\!\end{bmatrix}(k) = \begin{bmatrix} F(k) & G(k) \\ R(k) & S(k) \end{bmatrix}.$$

Hence, there exist suitably defined permutation operators, which we shall denote by $P_l(A)$ and $P_r(A)$, such that $P_l(A)AP_r(A) = [\![A]\!]$ or equivalently

$$P_l\left(\begin{bmatrix} F & G \\ R & S \end{bmatrix}\right)\begin{bmatrix} F & G \\ R & S \end{bmatrix}P_r\left(\begin{bmatrix} F & G \\ R & S \end{bmatrix}\right) = \begin{bmatrix}\!\begin{bmatrix} F & G \\ R & S \end{bmatrix}\!\end{bmatrix}.$$

For any operator A whose elements are memoryless operators

$$P_l(A)P_l(A)^* = P_l(A)^*P_l(A) = I,$$
$$P_r(A)P_r(A)^* = P_r(A)^*P_r(A) = I.$$

and if A is self-adjoint, then $P_l(A) = P_r(A)^*$. For a concrete example, consider $\begin{bmatrix} F & G \end{bmatrix}$. Then

$$P_l\left(\begin{bmatrix} F & G \end{bmatrix}\right) = I, \qquad P_r\left(\begin{bmatrix} F & G \end{bmatrix}\right) = \begin{bmatrix} E \\ Z^*EZ \end{bmatrix},$$

where

$$E = \begin{bmatrix} 1 & 0 & 0 & & & \\ 0 & 0 & 1 & 0 & & \\ 0 & 0 & 0 & 0 & 1 & 0 \\ \vdots & & & & & \ddots \end{bmatrix}.$$

The following is immediate.

Proposition 11.11. *For any real number β, and any partitioned operator A consisting of elements which are memoryless, $A < \beta I$ holds if and only if $[\![A]\!] < \beta I$. That is, positivity is preserved under permutation.*

Two further useful facts for the above permutations are the following.

Proposition 11.12.

(i) *Suppose that $A : \ell_2^{m_1} \times \ell_2^{m_2} \to \ell_2^{p_1} \times \ell_2^{p_2}$, and $B : \ell_2^{m_1} \times \ell_2^{m_2} \to \ell_2^{p_1} \times \ell_2^{p_2}$, are partitioned operators consisting of memoryless elements. Then*

$$[\![A + B]\!] = [\![A]\!] + [\![B]\!].$$

(ii) *Suppose that $A : \ell_2^{m_1} \times \ell_2^{m_2} \to \ell_2^{p_1} \times \ell_2^{p_2}$, and $C : \ell_2^{q_1} \times \ell_2^{q_2} \to \ell_2^{m_1} \times \ell_2^{m_2}$, are partitioned operators, each of which consists of elements which are memoryless. Then the product $AC : \ell_2^{q_1} \times \ell_2^{q_2} \to \ell_2^{p_1} \times \ell_2^{p_2}$ is also partitioned into memoryless components and*

$$[\![AC]\!] = [\![A]\!]\,[\![C]\!].$$

Proof. Part (i) is obvious.

Part (ii) is simple to see, since $P_r(A)$, the right permutation of A, depends only on the column dimensions m_1, m_2 of the blocks in A. Since C has blocks with the same row dimensions, $P_l(C) = P_r(A)^*$, and hence

$$\begin{aligned}
[\![A]\!]\,[\![C]\!] &= P_l(A)AP_r(A)P_l(C)CP_r(C) \\
&= P_l(A)ACP_r(C) \\
&= P_l(AC)ACP_r(AC) \\
&= [\![AC]\!].
\end{aligned}$$

∎

The diagonal realization of a partitioned operator, and the notation $[\![\cdot]\!]$ will allow us to easily manipulate expressions involving diagonal operators. We now introduce another new tool.

11.2.2   The system function

We start by writing the system (11.15) in operator form. Using the previously defined notation, clearly $A(k)$, $B(k)$, $C(k)$, and $D(k)$ in (11.15) define memoryless operators. Using the shift operator Z, we can rewrite equation (11.15) as

$$\begin{aligned}
x &= ZAx + ZBw \\
z &= Cx + Dw.
\end{aligned}$$

The question of whether this set of equations is well-defined, that is, whether or not there exists an $x \in \ell_2$ such that they are satisfied, is one of *stability* of the system. If the equations are well-defined, then we can write

$$G = C(I - ZA)^{-1}ZB + D, \tag{11.16}$$

and $z = Gw$. These equations are clearly well-defined if $1 \notin \mathrm{spec}(ZA)$. The next result states that this condition is equivalent to a standard notion of stability of LTV systems, that is, exponential stability.

Definition 11.13. *The system G is exponentially stable if, when $w = 0$, there exist constants $c > 0$ and $0 < \lambda < 1$ such that, for each $k_0 \geq 0$ and any initial condition x_{k_0}, the inequality $|x(k)| \leq c\lambda^{(k-k_0)}|x_{k_0}|$ holds for all $k \geq k_0$.*

Proposition 11.14. *Suppose $A(k)$ is a bounded sequence of matrices and $A = diag(A(0), A(1), \cdots)$. Then the difference equation $x(k + 1) = A(k)x(k)$ is exponentially stable if and only if $1 \notin spec(ZA)$.*

This says that exponential stability is equivalent to ℓ_2 stability of the system $x(k + 1) = A(k)x(k) + v(k)$; versions of this result can be found in any standard reference on Lyapunov theory. See the chapter references. Thus the system is stable if and only if $1 \notin spec(ZA)$; we will work with this latter condition.

Throughout the sequel we will refer to the memoryless operators A, B, C, and D, and the operator G they define, without formal reference to their definitions in (11.15) and (11.16).

We now consider the properties of operators of the form of Equation (11.16). Formally, this equation looks very much like the frequency domain description of a discrete-time time invariant system. It is well known that for such systems, one can replace the shift operator Z with a complex number, and then the induced norm of the system is given by the maximum norm of this transfer function over the unit ball in the complex plane.

We will show that, for linear *time-varying* systems, very similar statements can be made. Indeed, the induced norm of a linear time-varying system can be analyzed by computing the maximum norm of an operator-valued function over a complex ball. However, in this context, we will use a bounded sequence $\lambda_k \in \mathbb{C}$ of complex numbers as our notion of frequency. Robust control techniques to date have been primarily developed for LTI systems; the system function derived here provides an important and direct link between LTI and LTV systems, making the techniques of robust control available for LTV systems.

Given such a sequence, we will make use of two associated memoryless operators on ℓ_2. These are

$$\Lambda = \begin{bmatrix} \lambda_0 I & & & 0 \\ & \lambda_1 I & & \\ & & \lambda_2 I & \\ 0 & & & \ddots \end{bmatrix} \quad \text{and} \quad \Omega = \begin{bmatrix} \lambda_0 I & & & 0 \\ & \lambda_0\lambda_1 I & & \\ & & \lambda_0\lambda_1\lambda_2 I & \\ 0 & & & \ddots \end{bmatrix}.$$

$$\tag{11.17}$$

It is easily verified that

$$\Omega Z = \Lambda Z \Omega. \tag{11.18}$$

Also note that if each element of the sequence λ_k is on the unit circle $\mathbb{T}$ then Ω is invertible in $\mathcal{L}(\ell_2)$. Using the definition of Λ we define the *system function* of the operator G by

$$\hat{G}(\Lambda) := C(I - \Lambda Z A)^{-1}\Lambda Z B + D,$$

when the inverse is defined. We can now state the main result of this section, which provides an LTV parallel to standard z-transform theory for LTI systems.

Theorem 11.15. *Suppose* $1 \notin spec(ZA)$. *Then*

$$\|C(I - ZA)^{-1}ZB + D\| = \sup_{\lambda_k \in \mathbb{D}} \|\hat{G}(\Lambda)\|,$$

where Λ *depends on* λ_k *as in (11.17).*

This theorem says that the induced ℓ_2 norm of the system G, which equals $\|C(I - ZA)^{-1}ZB + D\|$, is given by the maximum of the norm $\|\hat{G}(\Lambda)\|$, when the λ_k are chosen in the unit disk. This result looks similar to the well-known result for transfer functions of time invariant systems, and points the way to extending time invariant techniques to time-varying systems.

   In particular, we will use this result to derive a time-varying version of the KYP Lemma, characterizing those systems which are contractive. However, first we must prove a preliminary result.

Lemma 11.16. *Suppose* $1 \notin spec(ZA)$. *Then given any sequence* λ_k *in the unit circle* $\mathbb{T}$, *the operator* $I - \Lambda Z A$ *is invertible and we have*

$$\|C(I - ZA)^{-1}ZB + D\| = \|\hat{G}(\Lambda)\|.$$

Proof. Fix a sequence $\lambda_k \in \mathbb{T}$ and define the operator Ω as in (11.17). Now notice that both Ω and Ω^{-1} are isometries and therefore

$$\|C(I - ZA)^{-1}ZB + D\| = \|\Omega\{C(I - ZA)^{-1}ZB + D\}\Omega^{-1}\|.$$

To complete the proof consider the operator on the right-hand side above

$$\Omega\{C(I - ZA)^{-1}ZB + D\}\Omega^{-1} = C\Omega(I - ZA)^{-1}Z\Omega^{-1}B + D$$
$$= C\Omega(I - ZA)^{-1}\Omega^{-1}\Lambda ZB + D$$
$$= C(I - \Lambda ZA)^{-1}\Lambda ZB + D = \hat{G}(\Lambda),$$

where we have used the fact that Ω commutes with A, B, C, and D, and the relationship described by Equation (11.18).   ■

This lemma states that it is possible to scale the system matrices A and B by any complex sequence on the unit circle without affecting the norm of the system. Note that this can equivalently be thought of as scaling Z, the shift operator. The next lemma describes the effect of the operator Λ on the spectrum of ZA.

Lemma 11.17. *Suppose that* λ_k *is a sequence in the closed unit disc* $\bar{\mathbb{D}}$ *and define* Λ *as in (11.17):*

   (i) If $\mu \notin spec(ZA)$, *then* $\mu \notin spec(\Lambda ZA)$.

(ii) If the sequence λ_k is further restricted to be in $\mathbb{T}$, then $spec(ZA) = spec(\Lambda ZA)$.

Proof. First note that without loss of generality we may work with $\mu = 1$ in (i), and therefore will show that $1 \notin spec(ZA)$ implies that $1 \notin spec(\Lambda ZA)$.

We begin proving (i) by invoking Proposition 11.14 to see that, since $1 \notin spec(ZA)$, the difference equation $x(k+1) = A(k)x(k)$ is exponentially stable. Each λ_k satisfies $|\lambda_k| \le 1$ and so

$$x(k+1) = \lambda_{k+1} A(k)x(k)$$

is also exponentially stable; this is easily verified by applying the above equation recursively. Again use Proposition 11.14 to conclude that $1 \notin spec(ZQ)$ where Q is the memoryless operator corresponding to $Q(k) = \lambda_{k+1} A(k)$. It is routine to verify that $ZQ = \Lambda ZA$.

Part (ii) is immediate by applying (11.18) to see that $\Omega ZA\Omega^{-1} = \Lambda ZA$. ∎

Note that in particular (i) and (ii) imply the that the spectrum of ZA is an entire disc centered at zero[1]; to see this, set $\Lambda = \lambda I$ and let λ be in $\bar{\mathbb{D}}$. We can now prove the theorem.

Proof of Theorem 11.15. For convenience define $\gamma := \|\hat{G}(I)\|$, which is equal to $\|G\|$ by definition. Suppose contrary to the theorem that there exists a sequence $\lambda_k \in \bar{\mathbb{D}}$ such that $\|\hat{G}(\Lambda)\| > \gamma$. Then there exist elements $x, y \in \ell_2$ satisfying $\|x\| = \|y\| = 1$ and

$$|\langle y, \hat{G}(\Lambda)w\rangle| > \gamma.$$

Without loss of generality we may assume that w and y have *finite* support, which we denote by n.

Now it is routine to verify that $\hat{G}(\Lambda)$ is lower triangular and has the representation

$$\hat{G}(\Lambda) = \begin{bmatrix} D(0) & & & & 0 \\ \lambda_1 T_{10} & D(1) & & & \\ \lambda_2\lambda_1 T_{20} & \lambda_2 T_{21} & D(2) & & \\ \lambda_3\lambda_2\lambda_1 T_{30} & \vdots & & \ddots & \\ \vdots & & & & \end{bmatrix}, \qquad (11.19)$$

where $T_{kl} = C(k)A(k-1)\cdots A(l+1)B(l)$. Therefore, recalling that w and y have finite support, the inner product

$$\langle y, \hat{G}(\Lambda)w\rangle = p(\lambda_1, \ldots, \lambda_n),$$

---

[1]Operators of the form ZA are commonly known as weighted shifts.

where $p(\cdot, \ldots, \cdot)$ is some multinomial. Multinomials satisfy a maximum principle[2]; specifically p satisfies

$$\max_{\mu_k \in \bar{\mathbb{D}}} |p(\mu_1, \ldots, \mu_n)| = \max_{\mu_k \in \mathbb{T}} |p(\mu_1, \ldots, \mu_n)|.$$

Thus there exist numbers $\lambda'_1, \ldots, \lambda'_n$ on the unit circle $\mathbb{T}$ so that

$$|p(\lambda'_1, \ldots, \lambda'_n)| \geq |p(\lambda_1, \ldots, \lambda_n)| > \gamma. \tag{11.20}$$

Let Λ' be the operator, of form (11.17), that corresponds to the sequence $\{1, \lambda'_1, \ldots, \lambda'_n, 1, \ldots\}$. Observe that by Lemma 11.16 we have $\|\hat{G}(I)\| = \|\hat{G}(\Lambda')\|$. Also note that $\hat{G}(\Lambda')$ has the same lower triangular form as $\hat{G}(\Lambda)$ in (11.19) and therefore

$$\langle y, \hat{G}(\Lambda')w \rangle_2 = p(\lambda'_1, \ldots, \lambda'_n).$$

Thus by (11.20) the inequality $|\langle y, \hat{G}(\Lambda')w \rangle| > \gamma$ holds.

Now certainly $\|\hat{G}(\Lambda')\| \geq |\langle y, \hat{G}(\Lambda')w \rangle|$ and hence $\|\hat{G}(\Lambda')\| > \gamma$; also recall that $\|\hat{G}(I)\| = \|\hat{G}(\Lambda')\|$. But this is a contradiction since by definition $\gamma = \|\hat{G}(I)\|$.   ∎

In the sequel we primarily work with the system function when $\Lambda = \lambda I$, where λ is a complex scalar. Observe by defining the notation

$$\hat{G}(\lambda) := C(I - \lambda ZA)^{-1}\lambda ZB + D,$$

this specialized function $\hat{G}(\lambda)$ looks and acts very much like the transfer function of an LTI system, and therefore plays an instrumental role in our viewpoint in the next section.

11.2.3 Evaluating the ℓ_2 induced norm

The previous section showed that the induced norm of a linear time-varying system was given by the maximum of an operator norm over a complex ball. In this section, our primary goal is to show that this can be recast into a convex condition on the system matrices. We will see that the results derived appear very similar to those we have already seen for time invariant systems, and indeed the methodology parallels it closely.

To start we state the following technical lemma.

Lemma 11.18. *The following conditions are equivalent:*

(i) $\sup_{\lambda \in \bar{\mathbb{D}}} \|C(I - \lambda ZA)^{-1}\lambda ZB + D\| < 1$ *and* $\rho(ZA) < 1$.

(ii) *There exists* $\bar{X} \in \mathcal{L}(\ell_2)$, *which is self-adjoint and* $\bar{X} > 0$, *such that*

$$\begin{bmatrix} ZA & ZB \\ C & D \end{bmatrix}^* \begin{bmatrix} \bar{X} & 0 \\ 0 & I \end{bmatrix} \begin{bmatrix} ZA & ZB \\ C & D \end{bmatrix} - \begin{bmatrix} \bar{X} & 0 \\ 0 & I \end{bmatrix} < 0. \tag{11.21}$$

---

[2]See the reference provided, or the Chapter 8 exercises.

This is an operator version the matrix KYP Lemma. It does *not* depend on the structure of A, B, C, or D, or the presence of the operator Z. For a proof, see the references at the end of the chapter.

For comparison, the corresponding standard result for linear time invariant discrete time systems is now stated from Proposition 8.28. Given a system G with transfer function $\hat{G}(z) := C_0(I - zA_0)^{-1}zB_0 + D_0$, then $\rho(A_0) < 1$ and the H_∞ norm of G is less than 1 if and only if there exists a matrix $X_0 > 0$ such that

$$\begin{bmatrix} A_0 & B_0 \\ C_0 & D_0 \end{bmatrix}^* \begin{bmatrix} X_0 & 0 \\ 0 & I \end{bmatrix} \begin{bmatrix} A_0 & B_0 \\ C_0 & D_0 \end{bmatrix} - \begin{bmatrix} X_0 & 0 \\ 0 & I \end{bmatrix} < 0.$$

Thus the above LTV result looks very similar to the time invariant result just stated.

In Lemma 11.18 the variable $\bar{X}$ is self-adjoint and positive definite but has otherwise no specific structure. We will now see that for the specific class of LTV systems under consideration, the result can be improved, and the variable chosen to be *memoryless*. We will denote by $\mathcal{X}$ the class of memoryless, positive definite operators, with infinite matrix representation

$$X = \begin{bmatrix} X(0) & & & 0 \\ & X(1) & & \\ & & X(2) & \\ 0 & & & \ddots \end{bmatrix} > 0. \qquad (11.22)$$

Here the block dimension (spatial dimension of the operator) is the same as that of the operator A.

Before stating the result below, it useful to gain some intuition from the perspective of *commutant* sets. In Lemma 11.18(i) the supremum is taken over a single λI operator; the positive commutant of such class is indeed the set involved in condition (11.21). However, for the class of LTV systems we know from Theorem 11.15 that the supremum can be extended to memoryless time-varying operators of the form $\Lambda = \text{diag}(\lambda_0 I, \lambda_1 I, \cdots)$. The positive commutant of such set is the above set $\mathcal{X}$, so it appears natural that this set should be involved in the KYP-type characterization. This is precisely the main result, which we now state.

Theorem 11.19. *The following conditions are equivalent:*

(i) $\|C(I - ZA)^{-1}ZB + D\| < 1$ *and* $1 \notin spec(ZA)$.

(ii) There exists $X \in \mathcal{X}$ *such that*

$$\begin{bmatrix} ZA & ZB \\ C & D \end{bmatrix}^* \begin{bmatrix} X & 0 \\ 0 & I \end{bmatrix} \begin{bmatrix} ZA & ZB \\ C & D \end{bmatrix} - \begin{bmatrix} X & 0 \\ 0 & I \end{bmatrix} < 0. \qquad (11.23)$$

Formally, the result is the same as that for the linear time invariant case, but the operators ZA and ZB replace the usual A-matrix and B-matrix, and X is memoryless. We shall see in the sequel that this is a general property of this formalism, and that this gives a simple way to construct and to understand the relationship between time invariant and time varying systems.

Proof. We start by invoking Theorem 11.15 and Lemma 11.17 with $\Lambda := \lambda I$: condition (i) above is equivalent to condition (i) in Lemma 11.18. Therefore it suffices to show that (ii) above is equivalent to (ii) in Lemma 11.18. Also, a solution $X \in \mathcal{X}$ to (11.23) immediately satisfies (ii) in Lemma 11.18 with $\bar{X} := X$.

It only remains to show that a solution $\bar{X}$ to (11.21) implies that there exists $X \in \mathcal{X}$ satisfying (11.23), which we now demonstrate. Suppose $\bar{X} \in \mathcal{L}(\ell_2)$ is self-adjoint, and satisfies both $\bar{X} > 0$ and (11.23). Our goal is to construct $X \in \mathcal{X}$ from $\bar{X}$ and show that it has the desired properties.

Define the operator $E(k) = [\underbrace{0 \cdots 0}_{k \text{ zeros}} I \, 0 \cdots]^*$, for $k \geq 0$, mapping $\mathbb{C}^n \to \ell_2$, which then satisfies

$$E(k)^* A = [0 \, \cdots \, 0 \, A(k) \, 0 \cdots].$$

Observe that $E(k)^* E(k) = I$. Using $E(k)$ define X to be the memoryless operator $\ell_2 \to \ell_2$ corresponding to the sequence defined by

$$X(k) = E(k)^* \bar{X} E(k), \quad \text{for each } k \geq 0.$$

Thus, X is a memoryless operator, whose elements are the blocks on the diagonal of $\bar{X}$. Clearly X is self-adjoint and satisfies $X > 0$ because $\bar{X}$ has these properties. This proves $X \in \mathcal{X}$.

To complete the proof we must now demonstrate that X satisfies (11.23). Grouping Z in (11.23) with X we apply Proposition 11.11 to see that (11.23) holds if and only if the permuted inequality

$$\left[\begin{bmatrix} A & B \\ C & D \end{bmatrix}^* \begin{bmatrix} Z^* X Z & 0 \\ 0 & I \end{bmatrix} \begin{bmatrix} A & B \\ C & D \end{bmatrix} - \begin{bmatrix} X & 0 \\ 0 & I \end{bmatrix} \right] < 0$$

holds. Now we can apply Proposition 11.12 to show that the above is tantamount to

$$\begin{bmatrix} A & B \\ C & D \end{bmatrix}^* \begin{bmatrix} Z^* X Z & 0 \\ 0 & I \end{bmatrix} \begin{bmatrix} A & B \\ C & D \end{bmatrix} - \begin{bmatrix} X & 0 \\ 0 & I \end{bmatrix} < 0. \tag{11.24}$$

We will now show that this inequality is satisfied.

Observe that, for each $k \geq 0$, the following holds: [3]

$$E(k)^* C = [0 \, \cdots \, 0 \, C(k) \, 0 \cdots].$$

---

[3] Here we do not distinguish between versions of $E(k)$ that differ only in the spatial dimension of the identity block.

Now using the fact $E(k)^* E(k) = I$ it is routine to verify the important property that

$$\begin{bmatrix} A & B \\ C & D \end{bmatrix} \begin{bmatrix} E(k) & 0 \\ 0 & E(k) \end{bmatrix} = \begin{bmatrix} E(k) & 0 \\ 0 & E(k) \end{bmatrix} \begin{bmatrix} A & B \\ C & D \end{bmatrix} (k) \quad \text{holds,} \quad (11.25)$$

for each $k \geq 0$.

Since $\bar{X}$ by assumption satisfies (11.21) there exists a $\beta > 0$ such that

$$\begin{bmatrix} A & B \\ C & D \end{bmatrix}^* \begin{bmatrix} Z^*\bar{X}Z & 0 \\ 0 & I \end{bmatrix} \begin{bmatrix} A & B \\ C & D \end{bmatrix} - \begin{bmatrix} \bar{X} & 0 \\ 0 & I \end{bmatrix} < -\beta I.$$

Pre- and post-multiply this by $\mathrm{diag}(E(k), E(k))^*$ and $\mathrm{diag}(E(k), E(k))$, respectively, and use (11.25) to get that the matrix inequality that

$$\begin{bmatrix} A & B \\ C & D \end{bmatrix}(k)^* \begin{bmatrix} E(k)^* & 0 \\ 0 & E(k)^* \end{bmatrix} \begin{bmatrix} Z^*\bar{X}Z & 0 \\ 0 & I \end{bmatrix} \begin{bmatrix} E(k) & 0 \\ 0 & E(k) \end{bmatrix} \begin{bmatrix} A & B \\ C & D \end{bmatrix}(k)$$

$$- \begin{bmatrix} E(k)^* & 0 \\ 0 & E(k)^* \end{bmatrix} \begin{bmatrix} \bar{X} & 0 \\ 0 & I \end{bmatrix} \begin{bmatrix} E(k) & 0 \\ 0 & E(k) \end{bmatrix} < -\beta I$$

holds, for every $k \geq 0$. Finally use the definition of X to see that this last inequality is exactly

$$\begin{bmatrix} A & B \\ C & D \end{bmatrix}^* (k) \begin{bmatrix} Z^*XZ & 0 \\ 0 & I \end{bmatrix} (k) \begin{bmatrix} A & B \\ C & D \end{bmatrix} (k) - \begin{bmatrix} X & 0 \\ 0 & I \end{bmatrix} (k) < -\beta I, \quad (11.26)$$

for each $k \geq 0$. This immediately implies that inequality (11.24) is satisfied. ∎

The following corollary relates the infinite dimensional linear matrix inequality to the pointwise properties of the system matrices.

Corollary 11.20. *The following conditions are equivalent:*

(i) $\|C(I - ZA)^{-1}ZB + D\| < 1$ *and* $1 \notin spec(ZA)$.

(ii) *There exists a sequence of matrices* $X(k) > 0$, *bounded above and below, such that the matrices*

$$\begin{bmatrix} A(k) & B(k) \\ C(k) & D(k) \end{bmatrix}^* \begin{bmatrix} X(k+1) & 0 \\ 0 & I \end{bmatrix} \begin{bmatrix} A(k) & B(k) \\ C(k) & D(k) \end{bmatrix} - \begin{bmatrix} X(k) & 0 \\ 0 & I \end{bmatrix}$$

are uniformly negative definite.

Proof. The result follows immediately from Equation (11.26) in the proof of Theorem 11.19 using the fact that $(Z^*XZ)(k) = X(k+1)$. ∎

In this section we have developed an analysis condition for evaluating the induced norm of an LTV system. In this framework the condition looks formally equivalent to the KYP Lemma for LTI systems.

11.2.4 LTV synthesis

The previous two sections have developed a framework for dealing with
LTV systems and provided analysis results of the form we require to apply the approach of Chapter 7 to solve the LTV synthesis problem. The
synthesis problem here is, given a discrete linear time-varying system, we
would like to find a controller such that the closed loop is contractive. In
the results of the previous section we saw that, using the framework developed, it was possible to perform the analysis for the time-varying case by
following directly the methods for the time-invariant case.

Let the LTV system G be defined by the following state space equations

$$x(k+1) = A(k)x(k) + B_1(k)w(k) + B_2(k)u(k), \qquad x(0) = 0,$$
$$z(k) = C_1(k)x(k) + D_{11}(k)w(k) + D_{12}(k)u(k), \qquad (11.27)$$
$$y(k) = C_2(k)x(k) + D_{21}(k)w(k).$$

We make the physical and technical assumption that the matrices
A, B, C, D are uniformly bounded functions of time. The only restrictions
on this system are that the direct feed-through term $D_{22} = 0$.

We suppose this system is being controlled by an LTV controller K
characterized by

$$x_K(k+1) = A_K(k)x_K(k) + B_K(k)y(k)$$
$$u(k) = C_K(k)x_K(k) + D_K(k)y(k). \qquad (11.28)$$

Here we use n to denote the number of states of G and n_K is the number
of states of K. The connection of G and K is shown in Figure 11.4.

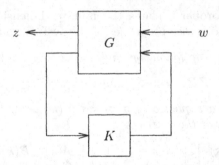

Figure 11.4. Closed-loop system

We write the realization of the closed loop system as

$$x_{cl}(k+1) = A_{cl}(k)x_{cl}(k) + B_{cl}(k)w(k)$$
$$z(k) = C_{cl}(k)x_{cl}(k) + D_{cl}(k)w(k), \qquad (11.29)$$

where $x_{cl}(k)$ contains the combined states of G and K, and $A_{cl}(k)$, $B_{cl}(k)$,
$C_{cl}(k)$ and $D_{cl}(k)$ are appropriately defined.

We are only interested in controllers K that both stabilize G and provide acceptable performance as measured by the induced norm of the map $w \mapsto z$. The following definition expresses our synthesis goal. A controller K is an *admissible synthesis* for G in Figure 11.4, if

- the spectral condition $1 \notin \operatorname{spec}(ZA_{cl})$ holds;

- the closed-loop performance $\|w \mapsto z\|_{\ell_2 \to \ell_2} < 1$ is achieved.

Hence, recalling Proposition 11.14 we are requiring the closed-loop system defined by Equations (11.29) be exponentially stable, in addition to being strictly contractive. We have the following theorem, which is written entirely in terms of memoryless operators and the shift.

Theorem 11.21. *There exists an admissible synthesis K for G, with state dimension $n_K \geq n$, if and only if there exist memoryless operators $Y > 0$ and $X > 0$ satisfying*

(i) $$\begin{bmatrix} N_c & 0 \\ 0 & I \end{bmatrix}^* \begin{bmatrix} ZAYA^*Z^* - Y & ZAYC_1^* & ZB_1 \\ C_1YA^*Z^* & C_1YC_1^* - I & D_{11} \\ B_1^*Z^* & D_{11}^* & -I \end{bmatrix} \begin{bmatrix} N_c & 0 \\ 0 & I \end{bmatrix} < 0,$$

(ii) $$\begin{bmatrix} N_o & 0 \\ 0 & I \end{bmatrix}^* \begin{bmatrix} A^*Z^*XZA - X & A^*Z^*XZB_1 & C_1^* \\ B_1^*Z^*XZA & B_1^*Z^*XZB_1 - I & D_{11}^* \\ C_1 & D_{11} & -I \end{bmatrix} \begin{bmatrix} N_o & 0 \\ 0 & I \end{bmatrix} < 0,$$

(iii) $$\begin{bmatrix} Y & I \\ I & X \end{bmatrix} \geq 0,$$

where the operators N_c, N_o satisfy

$$\operatorname{Im} N_c = \operatorname{Ker} \begin{bmatrix} B_2^*Z^* & D_{12}^* \end{bmatrix} \qquad\qquad N_c^*N_c = I$$
$$\operatorname{Im} N_o = \operatorname{Ker} \begin{bmatrix} C_2 & D_{21} \end{bmatrix} \qquad\qquad N_o^*N_o = I$$

This makes the correspondence with the time invariant case clear. Recalling that discrete-time H_∞ synthesis is a special case of Theorem 11.5, we can formally replace in those formulae the A-matrix by ZA and the B-matrix by ZB to arrive at the above conditions.

The above synthesis theorem can be derived routinely by using the machinery developed in §11.2.1 and §11.2.2, and the KYP Lemma of §11.2.3. This is done simply by following the methodology used in the time invariant case in Chapter 7. Further, this solution has the important property of being convex.

11.2.5 Periodic systems and finite dimensional conditions

The synthesis condition stated in Theorem 11.21 is in general infinite dimensional, as is the analysis condition of Theorem 11.19. However, there are two important cases in which these results reduce to finite dimensional

convex problems. The first is when one is only interested in behavior on
the finite horizon. In this case the matrix sequences $A(k)$, $B(k)$, $C(k)$, and
$D(k)$ would be chosen to be zero for $k \geq N$ the length of the horizon.
Thus the associated analysis and synthesis inequalities immediately reduce
to finite dimensional conditions. The second major case that reduces occurs when the system G is periodic, and explaining this connection is the
purpose of this section.

An operator P on ℓ_2 is said to be q-periodic if

$$Z^q P = P Z^q;$$

namely, it commutes with q shifts. Throughout the sequel we fix $q \geq 1$ to
be some integer.

Before stating the next result we require some additional notation. Suppose Q is a q-periodic memoryless operator, then we define $\tilde{Q}$ to be the
first period truncation of Q, namely,

$$\tilde{Q} := \begin{bmatrix} Q(0) & & 0 \\ & \ddots & \\ 0 & & Q(q-1) \end{bmatrix},$$

which is a matrix. Also define the cyclic shift matrix $\tilde{Z}$, for $q \geq 2$, by

$$\tilde{Z} = \begin{bmatrix} 0 & \cdots & 0 & I \\ I & \ddots & & 0 \\ & \ddots & & \vdots \\ & & I & 0 \end{bmatrix}, \text{ so } \tilde{Z}^* \tilde{Q} \tilde{Z} = \begin{bmatrix} Q(1) & & & 0 \\ & \ddots & & \\ & & Q(q-1) & \\ 0 & & & Q(0) \end{bmatrix}.$$

For $q = 1$ set $\tilde{Z} = I$. Also define the truncation of the set $\mathcal{X}$, defined in
(11.22), by

$$\tilde{\mathcal{X}} := \{\tilde{X} : X \in \mathcal{X}\}.$$

Using these new definitions, we have the following theorem, which is a
periodic version of Theorem 11.19.

Theorem 11.22. *Suppose A, B, C, and D are q-periodic operators. The
following conditions are equivalent:*

(i) $\|C(I - ZA)^{-1} ZB + D\| < 1$ and $1 \notin spec(ZA)$.

(ii) There exists a matrix $\tilde{X} \in \tilde{\mathcal{X}}$ such that

$$\begin{bmatrix} \tilde{Z}\tilde{A} & \tilde{Z}\tilde{B} \\ \tilde{C} & \tilde{D} \end{bmatrix}^* \begin{bmatrix} \tilde{X} & 0 \\ 0 & I \end{bmatrix} \begin{bmatrix} \tilde{Z}\tilde{A} & \tilde{Z}\tilde{B} \\ \tilde{C} & \tilde{D} \end{bmatrix} - \begin{bmatrix} \tilde{X} & 0 \\ 0 & I \end{bmatrix} < 0. \qquad (11.30)$$

Thus this result gives an LMI condition to determine the ℓ_2 induced norm
of a periodic system of the form in (11.15). Notice that the statement of
this theorem simply involves replacing all the objects in Theorem 11.19

with their "wave" equivalent; for instance, A now appears as $\tilde{A}$. We have the following synthesis theorem, which mirrors this pattern exactly.

Theorem 11.23. *Suppose G has a q-periodic realization. Then an admissible controller of order $n_K \geq n$ exists if and only if there exist solutions to the inequalities of Theorem 11.21, where A, B, C, D, X, Y, and Z, are replaced by the block-matrices defined by $\tilde{A}$, $\tilde{B}$, $\tilde{C}$, $\tilde{D}$, $\tilde{X}$, $\tilde{Y}$, and $\tilde{Z}$.*

Given this correspondence, one wonders whether a finite dimensional system function can be defined. The answer is yes. Define

$$\tilde{G}(\tilde{\Lambda}) := \tilde{C}(I - \tilde{\Lambda}\tilde{Z}\tilde{A})^{-1}\tilde{\Lambda}\tilde{Z}\tilde{B} + \tilde{D},$$

where the matrix $\tilde{\Lambda}$ is defined by

$$\tilde{\Lambda} = \begin{bmatrix} \lambda_0 & & 0 \\ & \ddots & \\ & & \lambda_{q-1} \end{bmatrix},$$

and the λ_k are complex scalars.

This brings us to the end of our quick look at LTV systems. The framework presented here reduces general time-varying problems to solutions in terms of *structured* operator inequalities; these inequalities follow from the standard LTI problems we have studied in earlier chapters. We have explicitly illustrated this by applying the new tools to deriving a KYP Lemma for LTV systems, and then provided the corresponding synthesis results without proof. These general results become bona fide LMI problems when periodic or finite time horizon systems are being considered. In summary the main feature of the framework is that it makes direct ties to the standard LMI techniques for time invariant systems, and thus many results and derivations become formally equivalent.

Notes and references

The work in §11.1 on NMD systems is based on related ideas and concepts from [96, 112]. The first of these papers studies LPV control for stabilization only, whereas the second includes performance. References [11, 12] focus on minimality and model reduction of NMD systems, respectively, and provide the basis for our presentation here on realization and model reduction. Theorem 11.1, which states when the above results are exact, is precisely the scalar-operator-times-identity version of Theorem 8.12 in discrete time; see Exercise 8.10 for a starting point to prove it. We remark that Theorem 11.1 remains true when δ_1 is fixed to be the shift Z.

A continuous time version of the LPV synthesis is derived in [5]. Also [175] provides a new approach which admits rate constraints on the varying parameters.

See [9], and the references therein, for work on multidimensional systems containing both temporal and spatial independent variables in the context of robust control. Also [31] where multidimensional systems are approached directly using NMD systems. For earlier related research see [19, 85] and the references therein.

The presentation and technical machinery for LTV systems developed in §11.2 are based on [46]. For the Lyapunov theory required on exponential stability in this section, see, for example, [88, 167]. In Theorem 11.15 a maximum principle for multinomials is required, and can be found in [128]; this monograph also provides many useful details on complex functions of several variables. A proof of Lemma 11.18 can be found in [178]. The papers [6, 77] use techniques closely related to the one presented here, and are based on that in [3]; see also [50]. Problems involving time-varying systems have been extensively studied and there is a large literature. For references in a Riccati equation context see, for instance, [10, 67] and the references therein.

Appendix A
Some Basic Measure Theory

In this appendix we will make precise some of the terminology used in Chapter 3. Specifically we target the terms "for almost every" and "essential supremum." These terms are introduced because many mathematical properties of functions (of a real variable) hold in nearly all instances, but fail on very small subsets of the real line; for this reason it is worthwhile to specify rigorously what we mean by small sets. What we present here is a very brief treatment aimed only at providing basic understanding of these terms and the required concepts from measure theory; for further reading see the references provided at the end of this appendix.

A.1 Sets of zero measure

We will confine ourselves to considering subsets of the real line $\mathbb{R}$, with the specific goal of classifying the smallest of these. To do this our first task is to investigate some of the basic notions of size or *measure*[1] of sets. Suppose we have two subsets S_1 and S_2 of $\mathbb{R}$, and $S_2 \subset S_1$. Then clearly S_2 is no larger than S_1 and our usual intuition about measuring size must preserve this. The notion we aim to make precise is, when will we say that these sets are the same size. Clearly this question distills down to

When is the difference $S_1 \backslash S_2$ small?

---

[1] We will restrict ourselves to Lebesgue measure.

Said yet another way, when do we say that the size of the difference between the sets is insignificant? Thus in order to answer this question we need to define precisely what we mean by insignificant in size. We will introduce the idea of a set having zero size, or in mathematical language having *zero measure*.

The first thing we require is a definition for the size of an open interval (a, b) in $\mathbb{R}$. We define the size of this set to be $b - a$, which is exactly our usual notion of length. Generalizing this, suppose we have a collection of n *disjoint* intervals

$$(a_1, b_1), (a_2, b_2), \ldots, (a_n, b_n),$$

and define the associated set

$$\mathcal{G} = (a_1, b_1) \cup (a_2, b_2) \cup \ldots \cup (a_n, b_n)$$

to be their union. Since the intervals are disjoint, our usual intuition about size would dictate that the size of $\mathcal{G}$ should be additively based on these interval lengths. We therefore accordingly define the size of this set to be

$$\sum_{k=1}^{n} (b_k - a_k) \ = \ \text{size of } \mathcal{G}.$$

With these definitions in place, consider a subset $\mathcal{S}$ which is contained in such a union:

$$\mathcal{S} \subset \cup_{k=1}^{n} (a_k, b_k).$$

Then if $\mathcal{S}$ has some size associated with it we would want this size to satisfy

$$\text{size of } \mathcal{S} \ \leq \ \sum_{k=1}^{n} (b_k - a_k).$$

Notice that this bound will remain true regardless of whether or not these intervals are disjoint; we therefore proceed assuming that they are not necessarily disjoint. Let us generalize this idea to a countable number of intervals. Suppose that

$$\mathcal{S} \subset \cup_{k=1}^{\infty} (a_k, b_k).$$

Then if we have a size associated with $\mathcal{S}$, we would naturally conclude that

$$\text{size of } \mathcal{S} \ \leq \ \sum_{k=1}^{\infty} (b_k - a_k) \quad \text{must hold.}$$

If the series on the right converges we have an upper bound on the possible size, or *measure*, of the set $\mathcal{S}$. Of course if the series diverges then the above inequality gives us no information about the set $\mathcal{S}$.

Having done a little exploration above about measuring sets, we are now ready to define a set of *zero measure*.

Definition A.1. *A subset* $S \subset \mathbb{R}$ *has* zero measure *if, for every* $\epsilon > 0$, *there exists a countable family of intervals* (a_k, b_k) *such that the following conditions hold:*

(a) *the set* S *is a subset of the union* $\cup_{k=1}^{\infty}(a_k, b_k)$;

(b) *the sum* $\sum_{k=1}^{\infty}(b_k - a_k) < \epsilon$.

Given our discussion above, this definition can be interpreted as follows: a set S has zero measure if the upper bound on its size can be made as small as desired. To see how this definition applies we consider two simple examples.

Examples:

First we consider the simplest nonempty sets in $\mathbb{R}$, those containing one element; let $S = \{t\}$ be such a set. For $\epsilon > 0$ this set is contained in the interval $(t - \frac{\epsilon}{2}, t + \frac{\epsilon}{2})$, whose length is ϵ. Thus directly from the definition this set has zero measure. More intuitively this construction says that

$$\text{size } S \leq \epsilon.$$

Therefore S has zero size, since the above inequality holds for any $\epsilon > 0$. Using the same argument it is not difficult to show that any set composed of a finite number of points $\{t_1, \ldots, t_n\}$ has zero measure.

Let us now turn to the set of natural numbers $\mathbb{N} = \{1, 2, 3, \ldots\}$. This set contains a countably infinite number of points; yet we will now see that it too has zero measure. Set ϵ to be any number satisfying $0 < \epsilon < 1$, and define the intervals

$$(a_k, b_k) = \left(k - \frac{(1-\epsilon)\epsilon^k}{2}, k + \frac{(1-\epsilon)\epsilon^k}{2}\right),$$

for each $k \in \mathbb{N}$. Since $k \in (a_k, b_k)$, for each $k > 0$, we see that

$$\mathbb{N} \subset \cup_{k=1}^{\infty}(a_k, b_k).$$

The length of each interval (a_k, b_k) is $(1 - \epsilon)\epsilon^k$, and so

$$\sum_{k=1}^{\infty}(b_k - a_k) = \sum_{k=1}^{\infty}(1 - \epsilon)\epsilon^k = \epsilon,$$

where we have used the geometric series formula. From our definition above we conclude that $\mathbb{N}$ has zero measure; its size is clearly smaller than any ϵ. We leave as an exercise the extension of this example to show that the integers $\mathbb{Z}$ are also a subset of $\mathbb{R}$ that has zero measure. Similarly it is possible to show that any countable subset of the real line has measure zero. In particular the set of rational numbers $\mathbb{Q}$ is of zero measure; this fact is perhaps surprising at first glance since the rationals are so densely distributed on the real line. The examples we have given here of zero measure sets all have a finite or countable number of elements; not all sets of zero measure are countable, but constructing these other cases is more involved.   □

A.2   Terminology

Having introduced the definition of a set of zero measure, we can explain the meaning of the term "for almost every." Suppose that $P(t)$ is a logical condition which depends on the real variable t. Then recall that a statement

$$\text{"For every } t \in \mathbb{R} \text{ the condition } P(t) \text{ holds"}$$

means that, for any chosen value $t_0 \in \mathbb{R}$, the condition $P(t_0)$ is true. Then we define the following terminology.

Definition A.2. *Given a logical condition* $P(t)$, *which depends on the real variable* t, *the expression "For almost every* $t \in \mathbb{R}$ *the condition* $P(t)$ *holds" means that the set*

$$S = \{t_0 \in \mathbb{R} : \ P(t_0) \ \text{is false}\} \quad \text{has zero measure.}$$

This definition states that "for almost every" means that the condition $P(t)$ can fail for some values of t, provided that it only fails on a very small set of points. Put more precisely, the set S of points where the condition $P(t)$ is false has zero measure. Notice that this means that "for every" implies for "for almost every" but the converse is not true; namely, the former is the stronger condition. To see the further implications of this terminology, we consider some examples.

Examples:

Consider the function $f(t) = \sin^2 \pi t$. This function does *not* satisfy $f(t) > 0$, for all $t \in \mathbb{R}$, since the positivity condition fails when t is an integer. Since we know $\mathbb{Z}$ is a set of measure zero, it follows that

$$f(t) > 0, \text{ for almost all } t \in \mathbb{R}.$$

For the purpose of another example consider the function

$$d(t) = \begin{cases} 1, & \text{for } t \in \mathbb{Q}; \\ 0, & \text{for } t \notin \mathbb{Q}. \end{cases}$$

Then we see that $d(t) = 0$, for almost all t. Further given any function $g(t)$, it follows from the properties of d that that $(g\,d)(t) = 0$, for almost all t.   □

So far we have assumed that $P(t)$ is defined on $\mathbb{R}$; however, it not uncommon for logical conditions to depend on subsets of the real line, and we therefore extend our above definition. If D is a subset of $\mathbb{R}$, then "For almost every $t \in D$ the condition $P(t)$ holds" is defined to mean that the set $S = \{t_0 \in D : \ P(t_0) \text{ is false}\}$ is of zero measure.

We can now turn to the definition of the *essential supremum* of a function.

Definition A.3. *Suppose that $\mathcal{D}$ is a subset of $\mathbb{R}$, and the function $f : \mathbb{R} \to \mathbb{R}$. The essential supremum of the function over $\mathcal{D}$ is defined by*

$$\operatorname*{ess\,sup}_{t \in \mathcal{D}} f(t) = \inf\{\beta \in \mathbb{R} : \quad f(t) < \beta, \text{ for almost every } t \in \mathcal{D}\}.$$

In other words a function is never greater than its essential supremum, except on a set of measure zero, and the essential supremum is the smallest number that has this property. Thus we immediately see that the essential supremum of a function can never be greater than the supremum. The basic property which makes the essential supremum useful is that it ignores values of the function that are only approached on a set of zero measure. Again we look at some concrete examples to make this definition clear.

Examples:

Define the function $h : [0, \infty) \to \mathbb{R}$ by

$$h(t) = \begin{cases} e^{-t}, & \text{for } t > 0; \\ 2, & \text{for } t = 0. \end{cases}$$

Then according the definitions of supremum and essential supremum we have

$$\sup_{t \in [0, \infty)} h(t) = 2 \quad \text{and} \quad \operatorname*{ess\,sup}_{t \in [0, \infty)} h(t) = 1.$$

The distinguishing property here is that the supremum of 2 is only approached (in fact achieved) at one point, namely $t = 0$, whereas the function is otherwise less than one, so the essential supremum can be no greater than one. However, for any value of $\beta < 1$, the set of points for which $h(t) \geq \beta$ is never of zero measure for it always contains an interval. Thus the essential supremum is indeed one.

Recall the function $d(t)$ just defined above. It satisfies

$$\sup_{t \in \mathbb{R}} d(t) = 1 \quad \text{and} \quad \operatorname*{ess\,sup}_{t \in \mathbb{R}} d(t) = 0.$$

To see this simply realize that $d(t)$ is only near the value one on a set of zero measure, namely, the rational numbers; otherwise it is always equal to zero. In fact, given any function $g(t)$ we have that $\operatorname{ess\,sup}(g\,d)(t) = 0$.

Finally we leave as an exercise the verification of the fact that if $f(t)$ is a continuous function, then its supremum is equal to its essential supremum.

$\square$

A.3 Comments on norms and L_p spaces

To end this appendix we discuss how sets of measure zero play a role in defining the elements in an $L_p(-\infty, \infty)$ space. We begin by focusing on L_∞ and an example.

Example:

Let $f(t)$ be the function that is zero at every time, and then clearly $\|f\|_\infty = 0$. Also define the function g by

$$g(t) = \left\{ \begin{array}{ll} 1, & t \in \mathbb{Z} \\ 0, & t \notin \mathbb{Z}. \end{array} \right.$$

From the above discussion of the essential supremum we know that $\|g\|_\infty = 0$. Thus we have that f and g are functions in L_∞ which both have norm zero. In fact it is clear that we can define many different functions with zero infinity norm.

$\square$

This example seems to indicate that $\|\cdot\|_\infty$ is not a norm, since it violates the requirement that only one element can have zero norm. What is needed to reconcile this dichotomy is a reinterpretation of what we mean by an element of L_∞:

> Functions that differ only on a set of measure zero are considered to represent the same element.

Thus in our example above f and g both represent the zero element in L_∞. Furthermore if h and w are L_∞ functions, and satisfy $\|h - w\|_\infty = 0$, then they represent the same element. Thus strictly speaking the elements of L_∞ are not functions but instead sets of functions, where each set contains functions that are equivalent.

We now generalize to L_p spaces, for $1 \le p < \infty$. Recall that the norm is defined by

$$\|h\|_p = \left(\int_{-\infty}^{\infty} |h(t)|_p^p \, dt \right)^{\frac{1}{p}}.$$

It is a fact that if h is zero everywhere except on a set of zero measure, then

$$\|h\|_p = 0.$$

That is function values on a measure zero set do not contribute to the integral.[2] Thus we see that the two example functions f and g given above have zero norm in every L_p space, and as a result we cannot rigorously

---

[2]This is based on Lebesgue integration theory.

regard them as distinct. So just as in L_∞ we regard any functions that differ only on a set of measure zero as representing the same element in L_p. In doing this all the mappings $\| \cdot \|_p$ indeed define norms.

To conclude we emphasize that for our purposes in this course the distinction between functions and elements of an L_p space is not crucial, and elements of L_p spaces can be viewed as functions without compromising understanding.

Notes and references

The subject of measure theory is vast, and we refer the reader to [170] for more details at an introductory level, and to [70] for a more advanced treatment of the subject.

Appendix B
Proofs of Strict Separation

This appendix presents two technical proofs which were omitted in Chapters 8 and 9, part of the argument to establish necessity of scaled small gain conditions for robustness. Specifically we will prove two propositions which concerned the *strict* separation of the sets ∇ and Π in $\mathbb{R}^d$. These were defined as

$$\Pi = \{(r_1, \ldots, r_d) \in \mathbb{R}^d : \ r_k \geq 0, \text{ for each } k = 1, \ldots, d\};$$

$$\nabla = \{(\phi_1(q), \ldots, \phi_d(q)) \in \mathbb{R}^d : \ q \in L_2 \text{ satisfying } \|q\| = 1\}.$$

We recall that $\phi_k(q) = \|E_k M q\|^2 - \|E_k q\|^2$, M is the nominal LTI system under consideration, and the projection matrices E_k break up signals in components conformably with the uncertainty structure $\Delta = \text{diag}(\Delta_1, \ldots, \Delta_d)$.

In what follows, $L_2[a, b]$ denotes the subspace of functions in $L_2[0, \infty)$ with support in the interval $[a, b]$, and $P_{[a,b]} : L_2[0, \infty) \to L_2[a, b]$ is the natural projection. We now state the first pending result.

Proposition B.1 (Proposition 8.9, Chapter 8). *Suppose* (M, Δ_a) *is robustly well-connected. Then the sets* Π *and* ∇ *are strictly separated, i.e.,*

$$D(\Pi, \nabla) := \inf_{r \in \Pi, y \in \nabla} |r - y| > 0.$$

We will give a proof by contrapositive, based on the following key lemma.

Lemma B.2. *Suppose* $D(\nabla, \Pi) = 0$. *Given any* $\epsilon > 0$ *and any* $t_0 \geq 0$ *the following conditions can be satisfied:*

1. *There exists a closed interval $[t_0, t_1]$, and two functions $p, q \in L_2[t_0, t_1]$, with $\|q\| = 1$, such that*

$$\|E_k p\| \geq \|E_k q\|, \text{ for each } k = 1, \ldots, d. \tag{B.1}$$

$$\epsilon^2 > \|(I - P_{[t_0, t_1]})Mq\| \tag{B.2}$$

$$\epsilon\sqrt{d} = \|p - P_{[t_0, t_1]}Mq\|. \tag{B.3}$$

2. *With the above choice of $[t_0, t_1]$ and q, there exists an operator $\Delta = diag(\Delta_1, \ldots, \Delta_d)$ in $\mathcal{L}(L_2[t_0, t_1]) \cap \mathbf{\Delta}_a$, such that $\|\Delta\| \leq 1$ and*

$$\| (I - \Delta P_{[t_0, t_1]}M) q \| \leq \epsilon\sqrt{d}. \tag{B.4}$$

Proof. Fix $\epsilon > 0$ and $t_0 \geq 0$. By hypothesis, there exists $q \in L_2$, $\|q\| = 1$, satisfying $\phi_k(q) > -\epsilon^2$ for each $k = 1, \ldots, d$. This amounts to

$$\epsilon^2 + \|E_k Mq\|^2 > \|E_k q\|^2, \text{ for each } k = 1, \ldots, d.$$

Now clearly if the support of q is truncated to a sufficiently long interval, and q is rescaled to have unit norm, the above inequality will still be satisfied by continuity of the norm. Also since $Mq \in L_2$, by possibly enlarging this truncation interval we can obtain $[t_0, t_1]$ satisfying (B.2), and also

$$\epsilon^2 + \|E_k P_{[t_0, t_1]}Mq\|^2 > \|E_k q\|^2, \text{ for each } k = 1, \ldots, d.$$

Next choose $\eta \in L_2[t_0, t_1]$ such that $E_k \eta$ has norm ϵ and is orthogonal to $E_k P_{[t_0, t_1]}Mq$, for each $k = 1, \ldots, d$. Then define

$$p = P_{[t_0, t_1]}Mq + \eta.$$

Now $\|\eta\| = \epsilon\sqrt{d}$ so (B.3) follows, and also

$$\|E_k p\|^2 = \epsilon^2 + \|E_k P_{[t_0, t_1]}Mq\|^2 > \|E_k q\|^2, \text{ for every } k = 1, \ldots, d,$$

which proves (B.1) and completes Part 1.

For Part 2, we start from (B.1) and invoke Lemma 8.4, Chapter 8 (notice that it holds in any L_2 space), to construct a contractive, block diagonal Δ satisfying $\Delta p = q$. Then

$$\left(I - \Delta P_{[t_0, t_1]}M\right) q = \Delta \left(p - P_{[t_0, t_1]}Mq\right),$$

so (B.4) follows from (B.3). ∎

Proof. (Proposition B.1) The argument is by contrapositive: we assume that $D(\nabla, \Pi) = 0$, the objective is to construct a perturbation $\Delta \in \mathbf{\Delta}_a$ such that $I - \Delta M$ is singular.

Fix any positive sequence $\epsilon_n \to 0$ as n tends to ∞. For each n, we construct $q^{(n)}$ and $\Delta^{(n)}$ as in Lemma B.2. Since their supports can be shifted arbitrarily, we choose them to be of the form $[t_n, t_{n+1}]$, with $t_0 = 0$, so that these intervals form a complete partition of $[0, \infty)$. Now we can combine the $\Delta^{(n)} \in \mathcal{L}(L_2[t_n, t_{n+1}]) \cap \mathbf{\Delta}_a$ to construct a single $\Delta \in$

$\mathcal{L}(L_2[0, \infty))$, defined by

$$\Delta = \sum_{n=1}^{\infty} \Delta^{(n)} P_{[t_n, t_{n+1}]}. \tag{B.5}$$

Descriptively, this operator breaks up a signal u into its components in the time partition $[t_n, t_{n+1}]$, applies $\Delta^{(n)}$ to each "piece" $P_{[t_n, t_{n+1}]}u$, and puts the resulting pieces back together. It is easy to see that $\|\Delta\| \leq 1$, since all the $\Delta^{(n)}$ are contractive. Furthermore Δ inherits the block-diagonal spatial structure so $\Delta \in \mathbf{\Delta}_a$.

Now apply Δ to the signal $Mq^{(n)}$ for a fixed n. We can write

$$\Delta Mq^{(n)} = \Delta \left\{ P_{[t_n, t_{n+1}]} + (I - P_{[t_n, t_{n+1}]}) \right\} Mq^{(n)}$$
$$= \Delta^{(n)} P_{[t_n, t_{n+1}]} Mq^{(n)} + \Delta(I - P_{[t_n, t_{n+1}]}) Mq^{(n)},$$

Applying the triangle inequality this leads to

$$\| (I - \Delta M) q^{(n)} \|$$
$$\leq \left\| \left(I - \Delta^{(n)} P_{[t_n, t_{n+1}]} M \right) q^{(n)} \right\| + \| (I - P_{[t_n, t_{n+1}]}) Mq^{(n)} \|$$
$$\leq \epsilon_n \sqrt{d} + \epsilon_n^2,$$

where we have used (B.4) and (B.2), respectively. Now we let $n \to \infty$ to see that the right-hand side tends to zero, and thus so does the left-hand side. Therefore $I - \Delta M$ cannot have a bounded inverse since for each n we know by definition that $\|q^{(n)}\| = 1$. This contradicts robust well-connectedness.   ∎

We turn now to our second result, which states that if we restrict ourselves to the causal operators in $\mathbf{\Delta}_a$, our first result still holds.

Proposition B.3 (Proposition 9.7, Chapter 9). *Suppose that the uncertain system $(M, \mathbf{\Delta}_a^c)$ is robustly stable. Then $D(\Pi, \nabla) > 0$.*

As compared to Proposition B.1, the hypothesis has now changed to state that $I - \Delta M$ has a causal, bounded inverse, for every causal $\Delta \in \mathbf{\Delta}_a$.

This means that we would already have a proof by contradiction if the Δ we constructed in the previous proposition were *causal*. Looking more closely, we see that the issue is the causality of each term $\Delta^{(n)} P_{[t_n, t_{n+1}]}$; unfortunately, the basic construction of $\Delta^{(n)}$ mapping $p^{(n)}$ to $q^{(n)}$ in Lemma B.2 cannot guarantee causality inside the interval $[t_n, t_{n+1}]$. Obtaining the desired causality requires a more refined argument.

Lemma B.4. *Suppose $D(\nabla, \Pi) = 0$. Given $\epsilon = \frac{1}{\sqrt{n}} > 0$ and $t_0 \geq 0$ there exist:*

  (i) *an interval $[t_0, \tilde{t}_1]$;*

  (ii) *a signal $\tilde{q} \in L_2[t_0, \tilde{t}_1]$, $\|\tilde{q}\| = 1$;*

(iii) a contractive operator $\tilde{\Delta}$ in $\mathcal{L}(L_2[t_0, \tilde{t}_1]) \cap \boldsymbol{\Delta}_a$, with $\tilde{\Delta} P_{[t_0, \tilde{t}_1]}$ causal, satisfying

$$\| (I - P_{[t_0, \tilde{t}_1]}) M \tilde{q} \| \leq \frac{1}{\sqrt{n}} \tag{B.6}$$

$$\left\| \left(I - \tilde{\Delta} P_{[t_0, \tilde{t}_1]} M \right) \tilde{q} \right\| \leq \frac{\kappa}{\sqrt{n}} \tag{B.7}$$

for some constant κ.

Before embarking on the proof of this lemma, we observe that it suffices to prove Proposition B.3. In fact, we can repeat the construction of (B.5) and obtain Δ, which is now causal and makes $I - \Delta M$ singular. The latter fact is established by using (B.6) and (B.7) instead of (B.2) and (B.4) , for $\epsilon_n = 1/\sqrt{n}$.

Therefore we concentrate our efforts in proving Lemma B.4.

Proof. We first invoke Lemma B.2 to construct an interval $[t_0, t_1]$, functions p, q and an operator Δ with the stated properties. For simplicity, we will take $t_0 = 0$ from now on; it is clear that everything can be shifted appropriately. Also we denote $h = t_1$.

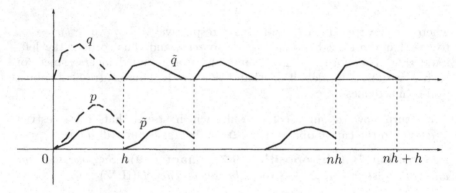

Figure B.1. Signals q and p (dashed); $\tilde{q}$ and $\tilde{p}$ (solid)

An illustration of the functions q and p is given by the broken line in Figure B.1. Notice that in this picture p appears to have greater norm than q, but this "energy" appears later in time; this would preclude a *causal*, contractive Δ from mapping p to q.

To get around this difficulty, we introduce a periodic repetition ($n = 1/\epsilon^2$ times) of the signals p and q, defining

$$\tilde{q} = \frac{1}{\sqrt{n}} \sum_{i=1}^{n} \tilde{S}_{ih} q, \qquad \tilde{p} = \frac{1}{\sqrt{n}} \sum_{i=0}^{n-1} S_{ih} p,$$

where S_τ denotes time shift as usual. The signals are sketched in Figure B.1. Notice that there is an extra delay for $\tilde{q}$; this is done deliberately so that $\tilde{p}$ anticipates its energy to $\tilde{q}$. Also, the normalizing factor is added to ensure $\|\tilde{q}\| = 1$. Both signals are supported in $[0, \tilde{t}_1]$ where $\tilde{t}_1 = (n+1)h$.

Now we introduce the operator

$$\tilde{\Delta} = \sum_{i=0}^{n-1} S_{(i+1)h} \Delta S_{-ih} P_{[ih,(i+1)h]},$$

Notice that each term of the above sum truncates a signal to $[ih, (i+1)h]$, shifts back to the interval $[0, h]$, applies Δ we had obtained from Lemma B.2, and shifts it again forward to $[(i+1)h, (i+2)h]$ (i.e., by one extra interval of h). A little thought will convince the reader that

- $\tilde{\Delta}$ maps $L_2[0, (n+1)h]$ to itself;

- Since Δ is contractive, so is $\tilde{\Delta}$;

- Since $\Delta p = q$, then $\tilde{\Delta} \tilde{p} = \tilde{q}$.

We claim that $\tilde{\Delta}$ is causal. By definition, this means that $P_T \tilde{\Delta} P_T = P_T \tilde{\Delta}$ for all T, where P_T denotes truncation to $[0, T]$; the only non-trivial case here is when $T \in [0, (n+1)h]$. In particular assume that

$$i_0 h < T \le (i_0 + 1)h,$$

for some integer i_0 between 0 and n. First observe that

$$P_T \tilde{\Delta} = P_T \sum_{i=0}^{i_0-1} S_{(i+1)h} \Delta S_{-ih} P_{[ih,(i+1)h]}, \qquad (B.8)$$

since the remaining terms in the sum $\tilde{\Delta}$ have their image supported in $[(i_0 + 1)h, \infty)$. For the terms in (B.8) we have $(i+1)h \le i_0 h < T$ so

$$P_{[ih,(i+1)h]} P_T = P_{[ih,(i+1)h]}.$$

Therefore multiplying (B.8) on the right by P_T is inconsequential, i.e. $P_T \tilde{\Delta} P_T = P_T \tilde{\Delta}$.

It only remains to show that the given $\tilde{\Delta}$ and $\tilde{q}$ satisfy (B.6) and (B.7). We first write

$$\|(I - P_{[0,\tilde{t}_1]})M\tilde{q}\| \le \frac{1}{\sqrt{n}} \sum_{i=1}^{n} \|(I - P_{[0,(n+1)h]})S_{ih}Mq\|$$

$$\le \frac{1}{\sqrt{n}} \sum_{i=1}^{n} \|(I - P_{[ih,(i+1)h]})S_{ih}Mq\|$$

$$= \frac{1}{\sqrt{n}} \sum_{i=1}^{n} \|S_{ih}(I - P_{[0,h]})Mq\|$$

$$= \sqrt{n} \, \|(I - P_{[0,h]})Mq\| \le \frac{1}{\sqrt{n}}. \qquad (B.9)$$

The first step relies on the time invariance of M, and the last bound follows from (B.2), since $\epsilon^2 = 1/n$. This proves (B.6).

To prove (B.7) it suffices to show that for some constant κ,

$$\|\tilde{p} - P_{[0,\tilde{t}_1]}M\tilde{q}\| \leq \frac{\kappa}{\sqrt{n}}; \tag{B.10}$$

in this case contractiveness of $\tilde{\Delta}$ gives (B.7) because

$$\tilde{\Delta}\left(\tilde{p} - P_{[0,\tilde{t}_1]}M\tilde{q}\right) = \left(I - \tilde{\Delta}P_{[0,\tilde{t}_1]}M\right)\tilde{q}.$$

We thus focus on (B.10); this bound is broken in two parts, the first is

$$\left\|P_{[0,\tilde{t}_1]}M\tilde{q} - \frac{1}{\sqrt{n}}\sum_{i=1}^{n}S_{ih}P_{[0,h]}Mq\right\| \leq \frac{1}{\sqrt{n}}, \tag{B.11}$$

and its derivation is left as an exercise since it is almost identical to (B.9). The second quantity to bound is

$$\left\|\tilde{p} - \frac{1}{\sqrt{n}}\sum_{i=1}^{n}S_{ih}P_{[0,h]}Mq\right\| =$$

$$\frac{1}{\sqrt{n}}\left\|p - S_{nh}P_{[0,h]}Mq + \sum_{i=1}^{n-1}S_{ih}(p - P_{[0,h]}Mq)\right\|. \tag{B.12}$$

Notice that we have isolated two terms inside the norm sign on the right-hand side, since the sums we are comparing have slightly different index ranges; these two terms have a bounded norm, however, since $\|q\| = 1$, M is a bounded operator, and p is close to Mq because of (B.3). As for the last sum, the terms have disjoint support so

$$\left\|\sum_{i=1}^{n-1}S_{ih}(p - P_{[0,h]}Mq)\right\|^2 = \sum_{i=1}^{n-1}\|p - P_{[0,h]}Mq\|^2 \leq (n-1)\epsilon^2 d \leq d,$$

where we invoked (B.3). This means that the right-hand side of (B.12) is bounded by some constant times $1/\sqrt{n}$. Now combining this with (B.11) we have (B.10), concluding the proof.   ∎

Notes and references

The preceding proofs follow the ideas in [148], which in particular proposed the periodic repetition method to construct a causal, destabilizing Δ.

Appendix C
μ-Simple Structures

This appendix is devoted to the proof of Theorem 8.27, which characterizes the uncertainty structures for which the structured singular value is equal to its upper bound, that is,

$$\mu(M, \boldsymbol{\Delta}_{s,f}) = \inf_{\Theta \in \boldsymbol{P\Theta}_{s,f}} \sigma(\Theta^{\frac{1}{2}} M \Theta^{-\frac{1}{2}}).$$

We will focus on showing that the condition $2s + f \leq 3$ is sufficient for the above; the references can be consulted for counterexamples in the remaining cases.

Clearly because of the scalability of μ and its upper bound it suffices to show that

$$\mu(M, \boldsymbol{\Delta}_{s,f}) < 1 \text{ implies } \inf_{\Theta \in \boldsymbol{P\Theta}_{s,f}} \bar{\sigma}(\Theta^{\frac{1}{2}} M \Theta^{-\frac{1}{2}}) < 1.$$

This step is fairly technical and found in very few places in the literature; our treatment here is based on the common language of quadratic forms, developed in Chapter 8.

We recall the definition of the sets

$$\nabla_{s,f} := \{(\Phi_1(q), \ldots, \Phi_s(q), \phi_{s+1}(q), \ldots, \phi_{s+f}(q)) : q \in \mathbb{C}^m, |q| = 1\}$$
$$\Pi_{s,f} := \{(R_1, \ldots, R_s, r_{s+1}, \ldots, r_{s+f}), R_k = R_k^* \geq 0, \ r_k \geq 0\},$$

where the quadratic functions

$$\Phi_k(q) = E_k M q q^* M^* E_k^* - E_k q q^* E_k^*,$$
$$\phi_k(q) = q^* M^* E_k^* E_k M q - q^* E_k^* E_k q,$$

were used to characterize the uncertainty structure. In particular, we showed in Propositions 8.25 and 8.26 that

- $\mu(M, \boldsymbol{\Delta}_{s,f}) < 1$ if and only if $\nabla_{s,f}$ and $\Pi_{s,f}$ are disjoint.

- $\inf_{\Theta \in P\Theta_{s,f}} \bar{\sigma}(\Theta^{\frac{1}{2}} M \Theta^{-\frac{1}{2}}) < 1$ if and only if $\mathrm{co}(\nabla_{s,f})$ and $\Pi_{s,f}$ are disjoint.

Thus our problem reduces to establishing that when $2s + f \leq 3$, if $\nabla_{s,f}$ and $\Pi_{s,f}$ are disjoint, then $\mathrm{co}(\nabla_{s,f})$ and $\Pi_{s,f}$ are also disjoint. Unfortunately this *cannot* be established by invoking convexity of $\nabla_{s,f}$, which in general does not hold; thus a specialized argument is required in each case.

We will concentrate our efforts in the "extreme" cases $(s,f) = (1,1)$ and $(s,f) = (0,3)$. These suffice to cover all cases since if the bound is exact for a certain structure, it must also be exact with fewer uncertainty blocks of each type. This can be shown by starting with a smaller problem, e.g., $(s,f) = (0,2)$, then defining an augmented problem with an extra uncertainty block which is "inactive," i.e., the added blocks of the M matrix are zero. Then the result for the larger structure can be invoked; we leave details to the reader.

The two key cases are covered, respectively, in Sections C.1 and C.2.

C.1   The case of $\boldsymbol{\Delta}_{1,1}$

Let us start our proof by writing the partition

$$M = \begin{bmatrix} M_{11} & M_{12} \\ M_{21} & M_{11} \end{bmatrix}$$

in correspondence with the two blocks in $\boldsymbol{\Delta}_{1,1}$. A first observation is that if $\rho(M_{11}) \geq 1$, then there exists a complex scalar δ satisfying $|\delta| \leq 1$, such that

$$\begin{bmatrix} I & 0 \\ 0 & I \end{bmatrix} - \begin{bmatrix} M_{11} & M_{12} \\ M_{21} & M_{11} \end{bmatrix} \begin{bmatrix} \delta I & 0 \\ 0 & 0 \end{bmatrix} \quad \text{is singular.}$$

Now the matrix on the right is a member of $\boldsymbol{\Delta}_{1,1}$ and has a maximum singular value of at most one, and therefore we see that $\mu(M, \boldsymbol{\Delta}_{1,1}) \geq 1$. This means that if $\mu(M, \boldsymbol{\Delta}_{1,1}) < 1$, then necessarily $\rho(M_{11}) < 1$ is satisfied. We will therefore assume as we proceed that the latter condition holds.

We now recast our problem in terms of the sets $\nabla_{1,1}$ and $\Pi_{1,1}$, which are subsets of $\mathbb{V} = \mathbb{S}^{m_1} \times \mathbb{R}$. It will also be convenient to introduce the subset of $\Pi_{1,1}$ given by

$$\Upsilon_{1,1} := \{(0,r) : 0 \in \mathbb{S}^{m_1}, r \in \mathbb{R}, r \geq 0\}.$$

We can state the main result:

Theorem C.1. *Suppose that* $\rho(M_{11}) < 1$. *The following are equivalent:*

(a) $co(\nabla_{1,1})$ and $\Pi_{1,1}$ are disjoint;

(b) $\nabla_{1,1}$ and $\Pi_{1,1}$ are disjoint;

(c) $\nabla_{1,1}$ and $\Upsilon_{1,1}$ are disjoint;

(d) $co(\nabla_{1,1})$ and $\Upsilon_{1,1}$ are disjoint.

Clearly what we are after is the equivalence of (a) and (b), which by Propositions 8.25 and 8.26 implies the equality of the structured singular value and its upper bound. The other two steps will be convenient for the proof. An important comment is that the result is true even though the set $\nabla_{1,1}$ is not in general convex.

Let us now examine these conditions. Condition (a) obviously implies all the others; also (c) is immediately implied by all the other conditions. Therefore to prove the theorem it is therefore sufficient to show that (c) implies (a). We do this in two steps. First we show that (d) implies (a) in Lemma C.2 below; and then finally the most challenging part, that (c) implies (d), is proved in Lemma C.6 of the sequel. Having made these observations we are ready to begin proving the theorem, which starts with the following lemma.

Lemma C.2. *Suppose that $\rho(M_{11}) < 1$. If $co(\nabla_{1,1})$ and $\Upsilon_{1,1}$ are disjoint, then $co(\nabla_{1,1})$ and $\Pi_{1,1}$ are disjoint.*

Proof. Start by noting that $co(\nabla_{1,1})$ and $\Upsilon_{1,1}$ are disjoint convex sets in $\mathbb{V}$, with $co(\nabla_{1,1})$ compact and $\Upsilon_{1,1}$ closed. Hence they are strictly separated by a hyperplane; namely, there exists a symmetric matrix Θ and a real number θ such that

$$\text{Tr}(\Theta\Phi_1(q)) + \theta\phi_2(q) < \alpha \leq \theta r \quad \text{for every } q, \ |q| = 1, \text{ and every } r \geq 0.$$

It follows that $\theta \geq 0$, since $\alpha \leq \theta r$ for all positive numbers r; and therefore that we can choose $\alpha = 0$ in the above separation. Now analogously to the proof of Proposition 8.26 the first inequality can be rewritten as

$$M^* \begin{bmatrix} \Theta & 0 \\ 0 & \theta I \end{bmatrix} M - \begin{bmatrix} \Theta & 0 \\ 0 & \theta I \end{bmatrix} < 0.$$

Using the partition for M, the top-left block of this matrix inequality is

$$M_{11}^* \Theta M_{11} + \theta M_{21}^* M_{21} - \Theta < 0,$$

and therefore

$$M_{11}^* \Theta M_{11} - \Theta < 0.$$

This is a discrete time Lyapunov inequality, so using the hypothesis $\rho(M_{11}) < 1$ we conclude that $\Theta > 0$. Now this implies that

$$\text{Tr}(\Theta R) + \theta r \geq 0$$

for every $(R, r) \in \Pi_{1,1}$, and therefore the hyperplane strictly separates $\mathrm{co}(\nabla_{1,1})$ and $\Pi_{1,1}$.

■

Thus we have now shown above that (d) does indeed imply (a) in Theorem C.1. It only remains for us to demonstrate that (c) implies (d). This is the key step in proving the theorem and will require a little preliminary work. The first step is to obtain a more convenient characterization of $\mathrm{co}(\nabla_{1,1})$, which will allow us to bring some matrix theory to bear on our problem.

By definition $\Phi \in \nabla_{1,1}$ means there exists a vector q of unit norm such that

$$\Phi = \begin{bmatrix} \Phi_1(q) \\ \phi_2(q) \end{bmatrix}.$$

Consider a convex combination of two points Φ_q and Φ_v in $\nabla_{1,1}$.

$$\lambda\Phi_q + (1-\lambda)\Phi_v = \begin{bmatrix} \lambda\Phi_1(q) + (1-\lambda)\Phi_1(v) \\ \lambda\phi_2(q) + (1-\lambda)\phi_2(v) \end{bmatrix}.$$

The following is readily obtained using the transposition property of the matrix trace:

$$\lambda\Phi_1(q) + (1-\lambda)\Phi_1(v) = E_1 M W M^* E_1^* - E_1 W E_1^*,$$
$$\lambda\phi_2(q) + (1-\lambda)\phi_2(v) = \mathrm{Tr}\{W(M^* E_2^* E_2 M - E_2^* E_2)\},$$

where $W = \lambda q q^* + (1-\lambda)vv^*$. Given a symmetric matrix V we define the extended notation

$$\Phi_1(V) = E_1 M V M^* E_1^* - E_1 V E_1^*$$
$$\phi_2(V) = \mathrm{Tr}\{V(M^* E_2^* E_2 M - E_2^* E_2)\}.$$

Thus the above equations can be written compactly as

$$\lambda\Phi_1(q) + (1-\lambda)\Phi_1(v) = \Phi_1(W)$$
$$\lambda\phi_2(q) + (1-\lambda)\phi_2(v) = \phi_2(W).$$

This leads to the following parametrization of the convex hull of $\nabla_{1,1}$.

Proposition C.3. *The convex hull* $\mathrm{co}(\nabla_{1,1})$ *is equal to the set of points*

$$\{(\Phi_1(W), \phi_2(W)) : \text{ for some } r \geq 1,$$
$$W = \sum_{i=1}^{r} \lambda_i q_i q_i^*, \ |q_i| = 1, \ \lambda_i \geq 0 \text{ and } \sum_{i=1}^{r} \lambda_i = 1\}.$$

We remark that in the above parametrization, W is always positive semidefinite and nonzero. Next we prove two important technical lemmas.

Lemma C.4. *Suppose P and Q are matrices of the same dimensions such that*

$$PP^* = QQ^* .$$

Then there exists a matrix U satisfying

$$P = QU \quad and \quad UU^* = I .$$

Proof. Start by taking the singular value decomposition

$$PP^* = QQ^* = V\Sigma^2 V^* .$$

Clearly this means the singular value decompositions of P and Q are

$$P = V\Sigma U_1^* \quad and \quad Q = V\Sigma U_2^* .$$

Now set $U = U_2 U_1^*$.

∎

We use the lemma just proved in demonstrating the next result, which is the key to our final step.

Lemma C.5. *Suppose P and Q are both $n \times r$ matrices. If $PWP^* - QWQ^* = 0$, for some symmetric matrix $W \geq 0$, then*

$$W = \sum_{i=1}^{r} w_i w_i^* \quad for \ some \ vectors \ w_i ,$$

such that for each $1 \leq i \leq r$ we have

$$0 = Pw_i w_i^* P^* - Qw_i w_i^* Q^* .$$

Proof. By Lemma C.4 there exists a unitary matrix U such that

$$PW^{\frac{1}{2}} = QW^{\frac{1}{2}}U .$$

Since U is unitary, there exists scalars r_i on the unit circle, and orthonormal vectors q_i such that

$$U = \sum_{i=1}^{r} r_i q_i q_i^* \quad and \quad \sum_{i=1}^{r} q_i q_i^* = I .$$

Thus we have for each i that

$$PW^{\frac{1}{2}}q_i = QW^{\frac{1}{2}}Uq_i = r_i QW^{\frac{1}{2}}q_i .$$

Thus we set $w_i = W^{\frac{1}{2}}q_i$ to obtain the desired result. ∎

We now complete our proof of Theorem C.1, and establish that (c) implies (d) in the following lemma.

Lemma C.6. *If the sets $\nabla_{1,1}$ and $\Upsilon_{1,1}$ are disjoint, then $\mathrm{co}(\nabla_{1,1})$ and $\Upsilon_{1,1}$ are disjoint.*

Proof. We prove the result using the contrapositive. Suppose that $\mathrm{co}(\nabla_{1,1})$ and $\Upsilon_{1,1}$ intersect. Then by Proposition C.3 there exists a nonzero positive semidefinite matrix W such that

$$\Phi_1(W) = 0 \quad \text{and} \quad \phi_2(W) \geq 0 .$$

By definition this means

$$E_1 M W M^* E_1^* - E_1 W E_1^* = 0 \quad \text{and} \quad \mathrm{Tr}\{W(M^* E_2^* E_2 M - E_2^* E_2)\} \geq 0$$

are both satisfied.

Focusing on the former equality we see that by Lemma C.5 there exist vectors so that $W = \sum_{i=1}^{r} w_i w_i^*$ and for each i the following holds.

$$E_1 M w_i w_i^* M^* E_1^* - E_1 w_i w_i^* E_1^* = 0 . \qquad (C.1)$$

Now looking at the inequality

$$\mathrm{Tr}\{W(M^* E_2^* E_2 M - E_2^* E_2)\} \geq 0 ,$$

we substitute for W to get

$$\sum_{i=1}^{r} w_i^* \{M^* E_2^* E_2 M - E_2^* E_2\} w_i \geq 0 .$$

Thus we see that there must exist a nonzero w_{i_0} such that

$$w_{i_0}^* \{M^* E_2^* E_2 M - E_2^* E_2\} w_{i_0} \geq 0 .$$

Also by (C.1) we see that

$$E_1 M w_{i_0} w_{i_0}^* M^* E_1^* - E_1 w_{i_0} w_{i_0}^* E_1^* = 0 .$$

Set $q = \frac{w_{i_0}}{|w_{i_0}|}$ and then we have that $\Phi_1(q) = 0$ and $\phi_2(q) \geq 0$ both hold, where $|q| = 1$. This directly implies that $\nabla_{1,1}$ and $\Upsilon_{1,1}$ intersect, which completes our contrapositive argument.   ∎

With the above lemma proved we have completely proved Theorem C.1.

C.2   The case of $\Delta_{0,3}$

We begin by reviewing the definition of the set $\nabla_{0,3}$, namely,

$$\nabla_{0,3} = \{(q^* H_1 q,\ q^* H_2 q,\ q^* H_3 q) : q \in \mathbb{C}^m, |q| = 1\} \subset \mathbb{R}^3,$$

where

$$H_k := M_k^* E_k^* E_k M_k - E_k^* E_k \in \mathbb{H}^m \quad \text{for } k = 1, 2, 3.$$

In fact this structure for the H_k will be irrelevant to us from now on.
The proof hinges around the following key lemma.

Lemma C.7. *Given two distinct points x and y in $\nabla_{0,3}$, there exists an ellipsoid $\mathcal{E}$ in $\mathbb{R}^3$ which contains both points and is a subset of $\nabla_{0,3}$.*

By an ellipsoid we mean here the image through an affine mapping of the unit sphere (with no interior)

$$\mathcal{S} = \{x \in \mathbb{R}^3 : x_1^2 + x_2^2 + x_3^2 = 1\}.$$

In other words, $\mathcal{E} = \{v_0 + Tx : x \in \mathcal{S}\}$ for some fixed $v_0 \in \mathbb{R}^3$, $T \in \mathbb{R}^{3\times 3}$.

Proof. Let

$$x = (q_x^* H_1 q_x, \; q_x^* H_2 q_x, \; q_x^* H_3 q_x),$$
$$y = (q_y^* H_1 q_y, \; q_y^* H_2 q_y, \; q_y^* H_3 q_y),$$

where $q_x, q_y \in \mathbb{C}^m$ and $|q_x| = |q_y| = 1$. Since $x \neq y$, it follows that the vectors q_x and q_y must be linearly independent, and thus the matrix

$$Q := \begin{bmatrix} q_x & q_y \end{bmatrix}^* \begin{bmatrix} q_x & q_y \end{bmatrix} > 0.$$

Now consider the two by two matrices

$$\tilde{H}_k := Q^{-\frac{1}{2}} \begin{bmatrix} q_x & q_y \end{bmatrix}^* H_k \begin{bmatrix} q_x & q_y \end{bmatrix} Q^{-\frac{1}{2}}, \quad k = 1, 2, 3,$$

and define the set

$$\mathcal{E} := \{ (\eta^* \tilde{H}_1 \eta, \; \eta^* \tilde{H}_2 \eta, \; \eta^* \tilde{H}_3 \eta) : \eta \in \mathbb{C}^2, \; |\eta| = 1 \}.$$

We have the following properties:

- $\mathcal{E} \subset \nabla_{0,3}$. In fact $\eta^* \tilde{H}_k \eta = q^* H_k q$ for $q = \begin{bmatrix} q_x & q_y \end{bmatrix} Q^{-\frac{1}{2}} \eta$, and if $|\eta| = 1$ it follows from the definition of Q that

$$\left| \begin{bmatrix} q_x & q_y \end{bmatrix} Q^{-\frac{1}{2}} \eta \right|^2 = \eta^* Q^{-\frac{1}{2}} \begin{bmatrix} q_x & q_y \end{bmatrix}^* \begin{bmatrix} q_x & q_y \end{bmatrix} Q^{-\frac{1}{2}} \eta = 1.$$

- $x, y \in \mathcal{E}$. Taking

$$\eta_x = Q^{\frac{1}{2}} \begin{bmatrix} 1 \\ 0 \end{bmatrix}$$

we have $\begin{bmatrix} q_x & q_y \end{bmatrix} Q^{-\frac{1}{2}} \eta_x = q_x$, and also

$$\eta_x^* \eta_x = \begin{bmatrix} 1 \\ 0 \end{bmatrix}^* Q \begin{bmatrix} 1 \\ 0 \end{bmatrix} = q_x^* q_x = 1.$$

An analogous construction holds for y.

- $\mathcal{E}$ is an ellipsoid. To see this, we first parametrize the generating η's by

$$\eta = \begin{bmatrix} r_1 \\ r_2 e^{j\varphi} \end{bmatrix},$$

where $r_1 \geq 0$, $r_2 \geq 0$, $r_1^2 + r_2^2 = 1$, and $\varphi \in [0, 2\pi)$. Notice that we have made the first component real and positive; this restriction does not change the set $\mathcal{E}$ since a complex factor of unit magnitude applied to η does not affect the value of the quadratic forms $\eta^* \tilde{H}_k \eta$. We can also parametrize the valid r_1, r_2 and write

$$\eta = \begin{bmatrix} \cos(\frac{\theta}{2}) \\ \sin(\frac{\theta}{2}) e^{j\varphi} \end{bmatrix}, \quad \theta \in [0, \pi], \quad \varphi \in [0, 2\pi).$$

Now setting

$$\tilde{H}_k = \begin{bmatrix} a_k & b_k \\ b_k^* & c_k \end{bmatrix},$$

we have

$$\eta^* \tilde{H}_k \eta = a_k \cos^2 \left(\frac{\theta}{2} \right) + c_k \sin^2 \left(\frac{\theta}{2} \right) + 2 \sin \left(\frac{\theta}{2} \right) \cos \left(\frac{\theta}{2} \right) \operatorname{Re}(b_k e^{i\varphi}).$$

Employing some trigonometric identities and some further manipulations, the latter is rewritten as

$$\eta^* \tilde{H}_k \eta = \frac{a_k + c_k}{2} + \begin{bmatrix} \dfrac{a_k - c_k}{2} & \operatorname{Re}(b_k) & -\operatorname{Im}(b_k) \end{bmatrix} \begin{bmatrix} \cos(\theta) \\ \sin(\theta)\cos(\varphi) \\ \sin(\theta)\sin(\varphi) \end{bmatrix}.$$

Collecting the components for $k = 1, 2, 3$, we arrive at the formula

$$v = v_0 + T \begin{bmatrix} \cos(\theta) \\ \sin(\theta)\cos(\varphi) \\ \sin(\theta)\sin(\varphi) \end{bmatrix},$$

where $v_0 \in \mathbb{R}^3$ and $T \in \mathbb{R}^{3\times 3}$ are fixed, and θ and φ vary, respectively, over $[0, \pi]$ and $[0, 2\pi)$. Now we recognize that the above vector varies precisely over the unit sphere in $\mathbb{R}^3$ (the parametrization corresponds to the standard spherical coordinates). Thus $\mathcal{E}$ is an ellipsoid as claimed.

∎

The above lemma does not imply that the set $\nabla_{0,3}$ is convex; indeed such an ellipsoid can have "holes" in it. However, it is geometrically clear that if the segment between two points intersects the positive orthant $\Pi_{0,3}$, the same happens with any ellipsoid going through these two points; this is the direction we will follow to establish that $\operatorname{co}(\nabla_{0,3}) \cap \Pi_{0,3}$ nonempty implies $\nabla_{0,3} \cap \Pi_{0,3}$ nonempty.

However, the difficulty is that not all points in $\operatorname{co}(\nabla_{0,3})$ lie in a segment between two points in $\nabla_{0,3}$: convex combinations of more than two points are in general required. The question of how many points are actually required is answered by a classical result from convex analysis; see the references for a proof.

Lemma C.8 (Carathéodory). *Let $\mathcal{K} \subset \mathcal{V}$, where $\mathcal{V}$ is a d dimensional real vector space. Every point in $\mathrm{co}(\mathcal{K})$ is a convex combination of at most $d+1$ points in $\mathcal{K}$.*

We will require the following minor refinement of the above statement.

Corollary C.9. *If $\mathcal{K} \subset \mathcal{V}$ is compact, then every point in the boundary of $\mathrm{co}(\mathcal{K})$ is a convex combination of at most d points in $\mathcal{K}$.*

Proof. The Carathéodory result implies that for every $v \in \mathrm{co}(\mathcal{K})$, there exists a finite convex hull of the form

$$\mathrm{co}\{v_1,\ldots,v_{d+1}\} = \left\{ \sum_{k=1}^{d+1} \alpha_k v_k : \alpha_k \geq 0, \sum_{k=1}^{d+1} \alpha_k = 1 \right\}$$

with vertices $v_k \in \mathcal{K}$, which contains v.

If the v_k are in a lower dimensional hyperplane, then d points will suffice to generate v by invoking the same result. Otherwise, every point in $\mathrm{co}\{v_1,\ldots,v_{d+1}\}$ which is generated by $\alpha_k > 0$ for every k will be *interior* to $\mathrm{co}\{v_1,\ldots,v_{d+1}\} \subset \mathrm{co}(\mathcal{K})$. Therefore for points v in the *boundary* of $\mathrm{co}(\mathcal{K})$, one of the α_k's must be 0 and a convex combination of d points will suffice. ∎

Equipped with these tools, we are now ready to tackle the main result.

Theorem C.10. *If $\mathrm{co}(\nabla_{0,3}) \cap \Pi_{0,3}$ is nonempty, then $\nabla_{0,3} \cap \Pi_{0,3}$ is nonempty.*

Proof. By hypothesis there exists a point $v \in \mathrm{co}(\nabla_{0,3}) \cap \Pi_{0,3}$; since $\mathrm{co}(\nabla_{0,3})$ is compact, such a point can be chosen from its boundary. Since we are in the space $\mathbb{R}^3$, Corollary C.9 implies that there exist *three* points x, y, z in $\nabla_{0,3}$ such that

$$v = \alpha x + \beta y + \gamma z \quad \in \Pi_{0,3},$$

with α, β, γ non-negative and $\alpha + \beta + \gamma = 1$. Geometrically, the triangle $\mathrm{co}\{x,y,z\}$ intersects the positive orthant at some point v.
Claim: v lies in a segment between *two* points in $\nabla_{0,3}$.

This is obvious if x, y, z are aligned or if any of α, β, γ is 0. We thus focus on the remaining case, where the triangle $\mathrm{co}\{x,y,z\}$ is non-degenerate and v is interior to it, as illustrated in Figure C.1.
We first write

$$v = \alpha x + \beta y + \gamma z = \alpha x + (\beta + \gamma)\frac{1}{\beta+\gamma}(\beta y + \gamma z) = \alpha x + (\beta+\gamma)w,$$

where the constructed w lies in the segment $L(y,z)$. Now consider the ellipsoid $\mathcal{E} \subset \nabla_{0,3}$ through y and z, obtained from Lemma C.7. If it degenerates to 1 or 2 dimensions, then $w \in \mathcal{E} \subset \nabla_{0,3}$ and the claim is proved. If not, w must lie inside the ellipsoid $\mathcal{E}$. The half-line starting at x, through w must "exit" the ellipsoid at a point $u \in \mathcal{E} \subset \nabla_{0,3}$ such that w is in the segment

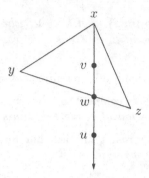

Figure C.1. Illustration of the proof

$L(x, u)$. Therefore $v \in L(x, w) \subset L(x, u)$. Since $x, u \in \nabla_{0,3}$, we have proved the claim.

Now to finish the proof, we have found two points in $\nabla_{0,3}$ such that the segment between them intersects $\Pi_{0,3}$. The corresponding ellipsoid $\mathcal{E} \subset \nabla_{0,3}$ between these points must clearly also intersect $\Pi_{0,3}$. Therefore $\nabla_{0,3} \cap \Pi_{0,3}$ is non-empty.   ∎

We remark that the above argument depends strongly on the dimensionality of the space; an extension to, e.g., four dimensions would require a construction of four-dimensional ellipsoids going through *three* points, extending Lemma C.7. In fact such extension is not possible and the result is not true for block structures of the form $\Delta_{0,f}$ with $f \geq 4$.

Notes and references

The above results are basically from [37] and [113] in the structured singular value literature, although the presentation is different, in particular in regard to the definitions of the ∇ sets. Also parallel results have been obtained in the Russian literature in terms of the S-procedure [55, 177].

Our proof for the case of $\Delta_{1,1}$ follows that of [129] where the focus is the KYP Lemma, and our proof of the $\Delta_{0,3}$ case is from [115]. A reference for the Carathéodory theorem is [134].

Notation

Symbol	Meaning	Page
$\mathbb{R}^n$	n-dimensional real vector space	21
$\mathbb{C}^n$	n-dimensional complex vector space	21
$\mathbb{R}^{m \times n}$	space of real m by n matrices.	21
$\mathbb{C}^{m \times n}$	space of complex m by n matrices.	21
$\mathbb{H}^n$	complex n by n Hermitian matrices.	22
$\mathbb{S}^n$	real n by n symmetric matrices.	22
A'	transpose of a matrix.	21
Q^*	adjoint of a matrix or operator Q	22,115
$\mathrm{span}\{v_1, \ldots, v_m\}$	subspace spanned by vectors $v_1, \ldots, v_m$.	23
e_i	canonical basis vectors in $\mathbb{F}^n$	25
E_{ir}	canonical basis matrices in $\mathbb{F}^{m \times n}$	25
E_k	block matrix $\begin{bmatrix} 0 & \cdots & 0 & I & 0 \cdots & 0 \end{bmatrix}$	250
$\mathrm{Im} A$	image subspace of a linear mapping	27
$\mathrm{Ker} A$	kernel subspace of a linear mapping	28
$\mathrm{rank} A$	rank of a mapping or matrix A	28
I	identity mapping or matrix	28
$\dim(\mathcal{S})$	dimension of a subspace $\mathcal{S}$	24
$\mathcal{S}^\perp$	orthogonal complement of the subspace $\mathcal{S}$	67,132
$\bar{\mathcal{Q}}$	closure of the set $\mathcal{Q}$	33
$\mathrm{co}(\mathcal{Q})$	convex hull of a set $\mathcal{Q}$	35
$\mathrm{Tr}(A)$	trace of the square matrix A	38
$\det(A)$	determinant of the square matrix A	40
$\mathrm{eig}(A)$	set of eigenvalues of the matrix A	40
$\lambda_{max}(A)$	maximum eigenvalue of the matrix A	151
$\lambda_{min}(A)$	minimum eigenvalue of the matrix A	168
$\bar{\sigma}(A)$	maximum singular value of the matrix A	48
$\underline{\sigma}(A)$	minimum singular value of the matrix A	56
$\rho(Q)$	spectral radius of a matrix or operator	56,112
$Q \geq 0$	matrix or operator is positive semidefinite	45,116
$Q > 0$	matrix or operator is positive definite	45,116
$Q^{\frac{1}{2}}$	square root of a matrix or operator	46,116
e^M	exponential of the matrix M	60
$\mathcal{R}_t$	set of reachable states at time t	63
$\mathcal{C}_{AB}$	controllability subspace	64
W_t	controllability gramian at time t	64
$\mathcal{N}_{CA}$	unobservable subspace	81
$\hat{f}(s)$	Laplace transform of the function $f(t)$.	89
$\hat{G}(s)$	transfer function	90
RP	set of real rational proper functions	92
$\left[\begin{array}{c|c} A & B \\ \hline C & D \end{array}\right]$	transfer function of state space system (A, B, C, D)	90

Symbol	Meaning	Page
$\|v\|_p$	p-norm of a vector $v \in \mathbb{C}^n$, $1 \le p \le \infty$	103
$\|v\| = \|v\|_2$	Euclidean norm of a vector $v \in \mathbb{C}^n$	43
$\|v\|_p$	L_p-norm of a function $v(t)$, $1 \le p \le \infty$	103
$\|v\| = \|v\|_2$	L_2-norm of a function $v(t)$	103
$\langle u, v \rangle$	inner product in L_2 space	104
$\langle \hat{F}, \hat{P} \rangle_2$	inner product in the matrix space RH_2	196
$\|\hat{G}\|_\infty$	infinity norm of a transfer function	124
$L_p(-\infty, \infty)$	functions $v : \mathbb{R} \to \mathbb{C}^n$ with $\|v\|_p < \infty$	103
$L_p[0, \infty)$	L_p functions supported on $[0, \infty)$.	103
$L_p[a, b]$	L_p functions supported on $[a, b]$.	389
L_{2e}	extended L_2 space	285
ℓ_2	space of square summable sequences	132
$\hat{L}_2(j\mathbb{R})$	frequency domain L_2 space	117
$\hat{L}_\infty(j\mathbb{R})$	frequency response function space	124
H_2	Hardy space on $\bar{\mathbb{C}}^+$, subspace of $\hat{L}_2(j\mathbb{R})$	120
$H_2^\perp$	Hardy space on $\bar{\mathbb{C}}^-$, subspace of $\hat{L}_2(j\mathbb{R})$	122
H_∞	Hardy space on $\bar{\mathbb{C}}^+$, subspace of $\hat{L}_\infty(j\mathbb{R})$	129
$R\hat{L}_2$	rational functions in $\hat{L}_2(j\mathbb{R})$	123
RH_2	rational functions in H_2, vector case	123
	matrix case	195
$RH_2^\perp$	rational functions in $H_2^\perp$	123
$R\hat{L}_\infty(j\mathbb{R})$	rational functions in $\hat{L}_\infty(j\mathbb{R})$	130
RH_∞	rational functions in H_∞	130
Φ	Fourier transform $\Phi : L_2(-\infty, \infty) \to \hat{L}_2$	118
Λ	Laplace transform $\Lambda : L_2[0, \infty) \to H_2$	119
Λ_-	Laplace transform $\Lambda_- : L_2(-\infty, 0] \to H_2^\perp$	122
$\mathcal{L}(\mathcal{V}, \mathcal{Z})$	space of bounded operators between normed spaces $\mathcal{V}$ and $\mathcal{Z}$.	107
$\mathcal{L}(\mathcal{V})$	same as $\mathcal{L}(\mathcal{V}, \mathcal{V})$	107
$\mathrm{spec}(M)$	spectrum of an operator M	112
$M_{\hat{G}}$	multiplication operator defined by $\hat{G}$	125
S_τ	shift operator	125
P_τ	truncation operator to interval $(-\infty, \tau]$	126
Γ_G	Hankel operator	150
Ψ_o	observability operator	140
Y_o	observability gramian	141
Ψ_c	controllability operator	143
X_c	controllability gramian	144
$\hat{U}^\sim(s)$	para-Hermitian conjugate of $\hat{U}$	160
(A_K, B_K, C_K, D_K)	controller state space matrices	174
$(A_{cl}, B_{cl}, C_{cl}, D_{cl})$	closed loop state space matrices	224

Symbol	Meaning	Page
$\underline{S}(G, K)$	lower star product	196
$\mathrm{Ric}(H)$	stabilizing solution of Riccati equation	
	defined by Hamiltonian matrix H	200
N_P	full rank matrix such that $\mathrm{Ker}\,P = \mathrm{Im}N_P$	219
$\bar{S}(M, \Delta)$	upper star product	241
$\mathcal{R}$	uncertainty relation, general	238
$\mathcal{R}_a$	arbitrary structured	245
$\mathcal{R}_{TI}$	LTI structured	259
$\mathrm{diag}(\Delta_1, \ldots, \Delta_d)$	block diagonal matrix or operator	245
Δ	uncertainty set, general	240
Δ_a	arbitrary structured	245
Δ_{TI}	LTI structured	259
Δ^c	causal uncertainty set,	290
Δ_a^c	arbitrary structured	291
Δ_{TI}^c	LTI structured	324
$\Delta_{s,f}$	block-structured matrix perturbations	262
$C\Delta,$	cone of structured perturbations	256
$C\Delta_{s,f}$	cone of block-structured matrices	262
Θ_Δ	commutant, general uncertainty set	258
Θ_a	of set Δ_a	247
Θ_{TI}	of set Δ_a	261
$\Theta_{s,f}$	of set $\Delta_{s,f}$	267
$P\Theta_\Delta$	positive set, associated with Θ	258
$P\Theta_a$	associated with Θ_a	247
$P\Theta_{TI}$	associated with Θ_{TI}	301
$P\Theta_{s,f}$	associated with $\Theta_{s,f}$	267
$\mu(M, \Delta)$	structured singular value of M with	
	respect to Δ	256
$\phi_k(q)$	quadratic form on L_2 or $\mathbb{C}^n$	250
$\Phi_k(q)$	matrix valued quadratic function on $\mathbb{C}^n$	269
$\nabla, \nabla_{s,f}$	image set of quadratic forms	251
$\Pi_{s,f}$	positive cone in commutant matrix space.	269
$\psi \begin{pmatrix} p \\ q \end{pmatrix}$	quadratic form on $L_2 \times L_2$	311
$\mathcal{C}_\psi^+, \mathcal{C}_\psi^{-\epsilon}$	cones of signals defined by ψ	312
$\mathbf{J_{f,a}}, \mathbf{J_{f,TI}}$	frequency domain robust H_2 bounds	325-7
$\mathbf{J_{s,a}}, \mathbf{J_{s,TI}^N}$	state space robust H_2 bounds	330-3
Z	shift operator on ℓ_2.	341
$G(\delta_1, \ldots, \delta_d)$	noncommuting multidimensional	
	(NMD) system	344
$\mathcal{X}$	commutant set, NMD system	343

References

[1] V.M. Adamjan, D.Z. Arov, and M.G. Krein. Infinite block Hankel matrices and related extension problems. *American Mathematical Society Translations*, 111:133–156, 1978.

[2] U.M. Al-Saggaf and G.F. Franklin. An error bound for a discrete reduced order model of a linear multivariable system. *IEEE Transactions on Automatic Control*, 32:815–819, 1987.

[3] D. Alpay, P. Dewilde, and H. Dym. Lossless inverse scattering and reproducing kernels for upper triangular operators. *Operator Theory: Advances and Applications, Birkäuser*, 47:61–135, 1992.

[4] B. D. O. Anderson and J. B. Moore. *Optimal Control: Linear Quadratic Methods*. Prentice Hall, 1990.

[5] P. Apkarian and P. Gahinet. A convex characterization of gain-scheduled H_∞ controllers. *IEEE Transactions on Automatic Control*, 40:853–864, 1995.

[6] J.A. Ball, I. Gohberg, and M.A. Kaashoek. Nevanlinna-Pick interpolation for time-varying input-output maps: the discrete case. *Operator Theory: Advances and Applications, Birkäuser*, 56:1–51, 1992.

[7] B. Bamieh and M. Dahleh. On robust stability with structured time-invariant perturbations. *Systems and Control Letters*, 21:103–108, 1993.

[8] B. Bamieh and J.B. Pearson Jr. The H_2 problem for sampled-data systems. *Systems and Control Letters*, 19:1–12, 1992.

[9] B. Bamieh, F. Paganini, and M. Dahleh. Optimal control of distributed arrays with spatial invariance. In *Robustness in Identification and Control; editors A. Garulli, A. Tesi, A. Vicino*. Springer, 1999.

[10] T. Basar and P. Bernhard. H_∞-Optimal Control and Related Mini-Max Design Problems: A Dynamic Game Approach. Birkhäuser, 1991.

[11] C.L. Beck and J. C. Doyle. A necessary and sufficient minimality condition for uncertain systems. IEEE Transactions on Automatic Control, 44:1802–1813, 1999.

[12] C.L. Beck, J.C. Doyle, and K. Glover. Model reduction of multi-dimensional and uncertain systems. IEEE Transactions on Automatic Control, 41:1466–1477, 1996.

[13] V. Belevitch. Classical Network Theory. Holden-Day, 1968.

[14] H. Bercovici, C. Foias, and A. Tannenbaum. Structured interpolation theory. Operator Theory Advances and Applications, 47:195–220, 1990.

[15] D.S. Bernstein and W. H. Haddad. LQG control with an H_∞ performance bound: A Riccati equation approach. IEEE Transactions on Automatic Control, 34:293–305, 1989.

[16] J. Bernussou, P.L.D. Peres, and J.C. Geromel. A linear programming oriented procedure for quadratic stabilization of uncertain systems. Systems and Control Letters, 13:65–72, 1989.

[17] S. Bochner and K. Chandrasekharan. Fourier Transforms. Princeton University Press, 1949.

[18] B. Bollobas. Linear Analysis. Cambridge University Press, 1990.

[19] N. K. Bose. Multidimensional Systems : Theory and Applications. IEEE Press; distributor Wiley, 1979.

[20] S. Boyd and C. Barratt. Linear Controller Design: Limits of Performance. Prentice Hall, 1991.

[21] S.P. Boyd, L. El Ghaoui, E. Feron, and V. Balakrishnan. Linear Matrix Inequalities in System and Control Theory. Society for Industrial and Applied Mathematics, 1994.

[22] R. Braatz, P. Young, J. C. Doyle, and M. Morari. Computational complexity of μ calculation. IEEE Transactions on Automatic Control, 39:1000–1002, 1994.

[23] R.W. Brockett. Finite Dimensional Linear Systems. Wiley, 1970.

[24] H. Chapellat and M. Dahleh. Analysis of time-varying control strategies for optimal disturbance rejection and robustness. IEEE Transactions on Automatic Control, 37:1734–1745, 1992.

[25] C.T. Chen. Linear System Theory and Design. Holt, Rhinehart, and Winston, 1984.

[26] B. R. Copeland and M. G. Safonov. A generalized eigenproblem solution for singular H_2 and H_∞ problems. In Control and Dynamic Systems; editor C. T. Leondes. Academic Press, 1992.

[27] R.F. Curtain and H.J. Zwart. An Introduction to Infinite Dimensional Systems Theory. Springer, 1995.

[28] M. A. Dahleh and I. J. Diaz-Bobillo. Control of Uncertain Systems: a Linear Programming Approach. Prentice Hall, 1995.

[29] R. D'Andrea. H_∞ optimization with spatial constraints. In Proc. IEEE Conference on Decision and Control, 1995.

[30] R. D'Andrea. Generalized ℓ_2 synthesis. *IEEE Transactions on Automatic Control*, 44:1145–1156, 1999.

[31] R. D'Andrea, G.E. Dullerud, and S. Lall. Convex L_2 synthesis for multidimensional systems. In *Proc. IEEE Conference on Decision and Control*, 1998.

[32] K.R. Davidson. *Nest Algebras*. Longman Scientific and Technical, 1988.

[33] C.A. Desoer, R.W. Liu, J. Murray, and R. Saeks. Feedback system design: the fractional representation approach. *IEEE Transactions on Automatic Control*, 25:399–412, 1980.

[34] C.A. Desoer and M. Vidyasagar. *Feedback Systems: Input-Output Properties*. Academic Press, 1975.

[35] P. Van Dooren. A generalized eigenvalue approach for solving Riccati equations. *SIAM Journal of Scientific and Statistical Computing*, 2:121–135, 1981.

[36] J. C. Doyle. Guaranteed margins for LQG regulators. *IEEE Transactions on Automatic Control*, 23:756–757, 1978.

[37] J. C. Doyle. Analysis of feedback systems with structured uncertainty. *IEE Proceedings*, 129:242–250, 1982.

[38] J. C. Doyle, K. Glover, P. Khargonekar, and B. A. Francis. State-Space solutions to standard H_2 and H_∞ control problems. *IEEE Transactions on Automatic Control*, 34:831–847, 1989.

[39] J. C. Doyle and G. Stein. Robustness with observers. *IEEE Transactions on Automatic Control*, 24:607–611, 1979.

[40] J. C. Doyle and G. Stein. Multivariable feedback design: Concepts for a classical/modern synthesis. *IEEE Transactions on Automatic Control*, 26:4–16, 1981.

[41] J. C. Doyle, J . E. Wall, and G. Stein. Performance and robustness analysis for structured uncertainty. In *Proc. IEEE Conference on Decision and Control*, 1982.

[42] J.C. Doyle. Lecture notes in advances in multivariable control. *ONR / Honeywell Workshop*, 1984.

[43] J.C. Doyle. Structured uncertainty in control system design. In *Proc. IEEE Conference on Decision and Control*, 1985.

[44] J.C. Doyle, B.A. Francis, and A. Tannenbaum. *Feedback Control Theory*. Macmillan, 1992.

[45] G.E. Dullerud. *Control of Uncertain Sampled-Data Systems*. Birkhäuser, 1995.

[46] G.E. Dullerud and S.G. Lall. A new approach to analysis and synthesis of time-varying systems. *IEEE Transactions on Automatic Control*, 44:1486–1497, 1999.

[47] P.L. Duren. *Theory of H_p Spaces*. Academic Press, 1970.

[48] D.F. Enns. Model reduction with balanced realizations: an error bound and frequency weighted generalization. In *Proc. IEEE Conference on Decision and Control*, 1984.

[49] M. Fan, A. Tits, and J. C. Doyle. Robustness in the presence of mixed parametric uncertainty and unmodeled dynamics. *IEEE Transactions on Automatic Control*, 36:25–38, 1991.

[50] A. Feintuch. *Robust Control Theory in Hilbert Space*. Springer, 1998.

[51] E. Feron. Analysis of robust H_2 performance using multiplier theory. *SIAM Journal of Control and Optimization*, 35:160–177, 1997.

[52] E. Feron, V. Balakrishnan, S. Boyd, and L. El Ghaoui. Numerical methods for H_2 related problems. In *Proc. American Control Conference*, 1992.

[53] C. Foias and A.E. Frazho. *The Commutant Lifting Approach to Interpolation Problems*. Birkhauser, 1990.

[54] C. Foias, H. Ozbay, and A. Tannenbaum. *Robust Control of Infinite Dimensional Systems*. Springer, 1996.

[55] A. Fradkov and V. A. Yakubovich. The S-procedure and duality theorems for nonconvex problems of quadratic programming. *Vestnik Leningrad University*, 31:81–87, 1973. In Russian.

[56] B.A. Francis. *A Course in H_∞ Control Theory*. Springer, 1987.

[57] B.A. Francis. Notes on introductory state space systems. 1997. Personal communication.

[58] P. Gahinet and P. Apkarian. A Linear Matrix Inequality approach to H_∞ control. *International Journal of Robust and Nonlinear Control*, 4:421–448, 1994.

[59] J.B. Garnett. *Bounded Analytic Functions*. Academic Press, 1981.

[60] T. Georgiou and M. Smith. Optimal robustness in the gap metric. *IEEE Transactions on Automatic Control*, 35:673–686, 1990.

[61] L. El Ghaoui and S. Niculescu (eds.). *Recent Advances on LMI Methods in Control*. SIAM, 1999.

[62] E. Gilbert. Controllability and observability in multivariable control systems. *SIAM Journal of Control*, 1:128–151, 1963.

[63] K. Glover. All optimal Hankel-norm approximations of linear multivariable systems and their L_∞ error bounds. *International Journal of Control*, 39:1115–1193, 1984.

[64] K. Glover. A tutorial on model reduction. In *From Data to Model; editor J.C. Willems*. Springer, 1989.

[65] K. Glover and D. McFarlane. Robust stabilization of normalized coprime factor plant descriptions with H_∞-bounded uncertainty. *IEEE Transactions on Automatic Control*, 34:821–830, 1989.

[66] G.H. Golub and C.F. Van Loan. *Matrix Computations*. The Johns Hopkins University Press, 1996.

[67] M. Green and D.J.N. Limebeer. *Linear Robust Control*. Prentice Hall, 1995.

[68] W.H. Greub. *Linear Algebra*. Springer, 1981.

[69] J. Guckenheimer and P. Holmes. *Nonlinear oscillations, dynamical systems and bifurcations of vector fields*. Springer, 1986.

[70] P.R. Halmos. *Measure Theory*. Springer, 1974.

[71] P.R. Halmos. *A Hilbert Space Problem Book*. Springer, 1982.

[72] M.L.J. Hautus. Controllability and observability conditions of linear autonomous systems. In *Proc. Kon. Ned. Akad. Wetensch. Ser. A.*, 1969.

[73] D. Hinrichsen and A.J. Pritchard. An improved error estimate for reduced order models of discrete time systems. *IEEE Transactions on Automatic Control*, 35:317–320, 1990.

[74] K. Hoffman. *Banach Spaces of Analytic Functions*. Prentice-Hall, 1962.

[75] R.A. Horn and C.R. Johnson. *Matrix Analysis*. Cambridge University Press, 1991.

[76] R.A. Horn and C.R. Johnson. *Topics in Matrix Analysis*. Cambridge University Press, 1995.

[77] P.A. Iglesias. An entropy formula for time-varying discrete-time control systems. *SIAM Journal of Control and Optimization*, 34:1691–1706, 1996.

[78] T. Iwasaki. Robust performance analysis for systems with norm-bounded time-varying structured uncertainty. In *Proc. American Control Conference*, 1994.

[79] U. Jonsson and A. Rantzer. Optimization of integral quadratic constraints. In *Recent Advances on LMI Methods in Control; editors L. El Ghaoui, S. Niculescu*. SIAM, 1999.

[80] T. Kailath. *Linear Systems*. Prentice-Hall, 1980.

[81] R.E. Kalman. A new approach to linear filtering and prediction theory. *ASME Transactions, Series D: Journal of Basic Engineering*, 82:35–45, 1960.

[82] R.E. Kalman. On the general theory of control systems. In *Proc. IFAC World Congress*, 1960.

[83] R.E. Kalman. Mathematical descriptions of linear systems. *SIAM Journal of Control*, 1:152–192, 1963.

[84] R.E. Kalman and R.S. Bucy. New results in linear filtering and prediction theory. *ASME Transactions, Series D: Journal of Basic Engineering*, 83:95–108, 1960.

[85] F.W. Kamen. Stabilization of linear spatially-distributed continuous time and discrete-time systems. *in Multdimensional Systems Theory; editor N.K. Bose*, Kluwer, 1985.

[86] D. Kavranoglu and M. Bettayeb. Characterization of the solution to the optimal H_∞ model reduction problem. *Systems and Control Letters*, 20:99–107, 1993.

[87] C. Kenig and P. Tomas. Maximal operators defined by Fourier multipliers. *Studia Math.*, 68:79–83, 1980.

[88] H.K. Khalil. *Nonlinear Systems*. Prentice-Hall, 1996.

[89] M. Khammash and J. B. Pearson. Performance robustness of discrete-time systems with structured uncertainty. *IEEE Transactions on Automatic Control*, 36:398–412, 1991.

[90] P. Khargonekar and M. Rotea. Mixed H_2/H_∞ control: A convex optimization approach. *IEEE Transactions on Automatic Control*, 36:824–837, 1991.

[91] P.P. Khargonekar and E. Sontag. On the relation between stable matrix fraction factorizations and regulable realizations of linear systems over rings. *IEEE Transactions on Automatic Control*, 27:627–638, 1982.

[92] V. Kucera. *Discrete Linear Control: the Polynomial Approach*. Wiley, 1979.

[93] H. Kwakernaak and R. Sivan. *Linear Optimal Control Systems*. Wiley-Interscience, 1972.

[94] P. Lancaster and L. Rodman. *Algebraic Riccati Equations*. Clarendon Press, 1995.

[95] A.J. Laub, M.T. Heath, C.C. Page, and R.C. Ward. Computation of balancing transformations and other applications of simultaneous diagonalization algorithms. *IEEE Transactions on Automatic Control*, 32:115–122, 1987.

[96] W. M. Lu, K. Zhou, and J. C. Doyle. Stabilization of LFT systems. In *Proc. IEEE Conference on Decision and Control*, 1991.

[97] D.G. Luenberger. Observing the state of a linear system. *IEEE Transactions on Military Electronics*, 8:74–80, 1964.

[98] D.G. Luenberger. *Optimization by Vector Space Methods*. Wiley, 1969.

[99] A.I. Lur'e. *Some Nonlinear Problems in the Theory of Automatic Control*. Her Majesty's Stationary Office, 1957. In Russian 1951.

[100] A.I. Lur'e and V.N. Postnikov. On the theory of stability of controlled systems. *Prikladnaya Matematika i Mekhanika*, 8:246–248, 1944. In Russian.

[101] A. Megretski. Necessary and sufficient conditions of stability: A multi-loop generalization of the circle criterion. *IEEE Transactions on Automatic Control*, 38:753–756, 1993.

[102] A. Megretski and A. Rantzer. System analysis via integral quadratic constraints. *IEEE Transactions on Automatic Control*, 42:819–830, 1997.

[103] A. Megretski and S. Treil. Power distribution inequalities in optimization and robustness of uncertain systems. *Jour. Math. Sys., Est. & Control*, 3:301–319, 1993.

[104] M. Mesbahi, M.G. Safonov, and G.P. Papavassilopoulos. Bilinearity and complementarity in robust control. In *Recent Advances on LMI Methods in Control; editors L. El Ghaoui, S. Niculescu*. SIAM, 1999.

[105] Y. Meyer. *Wavelets and Operators*. Cambridge University Press, 1992. In French 1990.

[106] B.C. Moore. Principal component analysis in linear systems: controllability, observablity, and model reduction. *IEEE Transactions on Automatic Control*, 26:17–32, 1981.

[107] M. Morari and E. Zafiriou. *Robust Process Control*. Prentice Hall, 1989.

[108] Y. Nesterov and A. Nemirovskii. *Interior-Point Polynomial Algorithms in Convex Programming*. Society for Industrial and Applied Mathematics, 1994.

[109] C.N. Nett, C.A. Jacobson, and M.J. Balas. A connection between state space and doubly coprime fractional representations. *IEEE Transactions on Automatic Control*, 29:831–834, 1984.

[110] J. G. Owen and G. Zames. Duality theory for robust disturbance attenuation. *Automatica*, 29:695–705, 1993.

[111] A. Packard. *What's New With μ: Structured Uncertainty in Multivariable Control.* PhD thesis, University of California, Berkeley, 1988.

[112] A. Packard. Gain scheduling via linear fractional transformations. *Systems and Control Letters*, 22:79–92, 1994.

[113] A. Packard and J. C. Doyle. The complex structured singular value. *Automatica*, 29:71–109, 1993.

[114] A. Packard, K. Zhou, P. Pandey, J. Leonhardson, and G. Balas. Optimal, constant I/O similarity scaling for full-information and state-feedback control problems. *Systems and Control Letters*, 19:271–280, 1992.

[115] F. Paganini. *Sets and Constraints in the Analysis of Uncertain Systems.* PhD thesis, California Institute of Technology, 1996.

[116] F. Paganini. Convex methods for robust H_2 analysis of continuous time systems. *IEEE Transactions on Automatic Control*, 44:239–252, 1999.

[117] F. Paganini. Frequency domain conditions for robust H_2 performance. *IEEE Transactions on Automatic Control*, 44:38–49, 1999.

[118] F. Paganini and E. Feron. LMI methods for robust H_2 analysis: a survey with comparisons. In *Recent Advances on LMI Methods in Control; editors L. El Ghaoui and S. Niculescu.* SIAM, 1999.

[119] J.R. Partington. *An Introduction to Hankel Operators.* Cambridge University Press, 1988.

[120] L. Pernebo and L.M. Silverman. Model reduction by balanced state space representations. *IEEE Transactions on Automatic Control*, 27:382–387, 1982.

[121] I. Petersen, D. McFarlane, and M. Rotea. Optimal guaranteed cost control of discrete-time uncertain linear systems. In *Proc. IFAC World Congress*, 1993.

[122] J.W. Polderman and J.C. Willems. *Introduction to Mathematical Systems Theory: A Behavioral Approach.* Springer, 1998.

[123] K. Poolla and A. Tikku. Robust performance against time-varying structured perturbations. *IEEE Transactions on Automatic Control*, 40:1589–1602, 1995.

[124] V. M. Popov. On the absolute stability of nonlinear systems of automatic control. *Automation and Remote Control*, 22:857–875, 1961.

[125] V.M. Popov. The solution of a new stability problem for controlled system. *Automation and Remote Control*, 24:1–23, 1963.

[126] V.M. Popov. *Hyperstability of Control Systems.* Springer, 1973. In Romanian 1966.

[127] S.C. Power. *Hankel Operators on Hilbert Space.* Pitman, 1981.

[128] R.M. Range. *Holomorphic Functions and Integrable Representations in Several Complex Variables.* Springer, 1986.

[129] A. Rantzer. On the Kalman-Yakubovich-Popov lemma. *Systems and Control Letters*, 28:7–10, 1996.

[130] A. Rantzer and A. Megretski. A convex parameterization of robustly stabilizing controllers. *IEEE Transactions on Automatic Control*, 39:1802–1808, 1994.

[131] A. Rantzer and A. Megretski. Tutorial on integral quadratic constraints. 1999. Preprint.

[132] A. Rantzer and A. Megretskii. System analysis via integral quadratic constraints. In *Proc. IEEE Conference on Decision and Control*, 1994.

[133] R.M. Redheffer. On a certain linear fractional transformation. *J. Math. and Physics*, 39:269–286, 1960.

[134] R.T. Rockafellar. *Convex Analysis*. Princeton University Press, 1997.

[135] M. Rosenblum and J. Rovnyak. *Hardy Classes and Operator Theory*. Oxford University Press, 1985.

[136] W. Rudin. *Real and Complex Analysis*. McGraw-Hill, 1987.

[137] M. G. Safonov. Tight bounds on the response of multivariable systems with component uncertainty. In *Proc. Allerton Conf.*, 1978.

[138] M. G. Safonov and M. Athans. A multiloop generalization of the circle criterion for stability margin analysis. *IEEE Transactions on Automatic Control*, 26:415–422, 1981.

[139] M. G. Safonov and R.Y. Chiang. Real/complex K_m synthesis without curve fitting. In *Control and Dynamic Systems; editor C. T. Leondes*. Academic Press, 1993.

[140] M.G. Safonov. *Stability and Robustness of Multivariable Feedback Systems*. MIT Press, 1980.

[141] M.G. Safonov and R.Y. Chiang. A Schur method for balanced-truncation model reduction. *IEEE Transactions on Automatic Control*, 34:729–733, 1989.

[142] M. Sampei, T. Mita, and M. Nakamichi. An algebraic approach to H_∞ output feedback control problems. *Systems and Control Letters*, 14:13–24, 1990.

[143] R.S. Sanchez-Pena and M. Sznaier. *Robust Systems Theory and Applications*. Wiley, 1998.

[144] I.W. Sandberg. An observation concerning the application of the contraction mapping fixed-point theorem and a result concerning the norm-boundedness of solutions of nonlinear functional equations. *Bell Systems Technical Journal*, 44:809–1812, 1965.

[145] C. Scherer. H_∞ optimization without assumptions on finite or infinite zeros. *SIAM Journal of Control and Optimization*, 30:143–166, 1992.

[146] C. Scherer, P. Gahinet, and M. Chilali. Multiobjective output-feedback control via LMI-optimization. *IEEE Transactions on Automatic Control*, 42:896–911, 1997.

[147] J.S. Shamma. The necessity of the small-gain theorem for time-varying and nonlinear systems. *IEEE Transactions on Automatic Control*, 36:1138–1147, 1991.

[148] J.S. Shamma. Robust stability with time varying structured uncertainty. *IEEE Transactions on Automatic Control*, 39:714–724, 1994.

[149] R.E. Skelton, T. Iwasaki, and K.M. Grigoriadis. *A Unified Algebraic Approach to Linear Control Design.* Taylor and Francis, 1998.

[150] S. Skogestad and I. Postlethwaite. *Multivariable Feedback Control : Analysis and Design.* Wiley, 1996.

[151] E.D. Sontag. *Mathematical Control Theory.* Springer, 1998.

[152] M.W. Spong and D.J. Block. The pendubot: A mechatronic system for control research and education. In *Proc. IEEE Conference on Decision and Control*, 1995.

[153] G. Stein and M. Athans. The LQG/LTR procedure for multivariable feedback control design. *IEEE Transactions on Automatic Control*, 32:105–114, 1987.

[154] A. A. Stoorvogel. *The H_∞ Control Problem: A State Space Approach.* Prentice Hall, 1992.

[155] A. A. Stoorvogel. The robust H_2 control problem: A worst-case design. *IEEE Transactions on Automatic Control*, 38:1358–1370, 1993.

[156] G. Strang. *Linear Algebra and its Applications.* Academic Press, 1980.

[157] B. Sz.-Nagy and C. Foias. *Harmonic Analysis of Operators on Hilbert Space.* North-Holland, 1970.

[158] M. Sznaier and J. Tierno. Is set modeling of white noise a good tool for robust H_2 analysis? In *Proc. IEEE Conference on Decision and Control*, 1998.

[159] A. L. Tits and M. K. H. Fan. On the small-μ theorem. *Automatica*, 31:1199–1201, 1995.

[160] O. Toker and H. Ozbay. On the NP-hardness of the purely complex μ computation, analysis/synthesis, and some related problems in multidimensional systems. In *Proc. American Control Conference*, 1995.

[161] S. Treil. The gap between the complex structured singular value and its upper bound is infinite. 1999. Preprint.

[162] B. van Keulen. H_∞ *Control for Distributed Parameter Systems: A State-Space Approach.* Birkhauser, 1993.

[163] L. Vandenberghe and S. Boyd. Semidefinite programming. *SIAM Review*, 38:49–95, 1996.

[164] M. Vidyasagar. Input-output stability of a broad class of linear time invariant multvariable feedback systems. *SIAM Journal of Control*, 10:203–209, 1972.

[165] M. Vidyasagar. The graph metric for unstable plants and robustness estimates for feedback stability. *IEEE Transactions on Automatic Control*, 29:403–418, 1984.

[166] M. Vidyasagar. *Control System Synthesis: A Factorization Approach.* MIT Press, 1985.

[167] M. Vidyasagar. *Nonlinear Systems Analysis.* Prentice-Hall, 1993.

[168] G. Vinnicombe. Frequency domain uncertainty and the graph topology. *IEEE Transactions on Automatic Control*, 38:1371–1383, 1993.

[169] G. Weiss. Representation of shift-invariant operators on L^2 by H^∞ transfer functions: an elementary proof, a generalization to L^p, and a counterexample for L^∞. *Math. of Control, Signals, and Systems*, 4:193–203, 1991.

[170] R.L. Wheeden and A. Zygmund. *Measure and Integral*. Marcel Dekker, 1977.

[171] J. C. Willems. Least squares stationary optimal control and the algebraic Riccati equation. *IEEE Transactions on Automatic Control*, 16:621–634, 1971.

[172] J.C. Willems. *The Analysis of Feedback Systems*. MIT Press, 1971.

[173] W.M. Wonham. On pole assignment in multi-input controllable linear systems. *IEEE Transactions on Automatic Control*, 12:660–665, 1967.

[174] W.M. Wonham. *Linear Multivariable Control*. Springer, 1985.

[175] F. Wu, A. Packard, and G. Becker. Induced L_2-norm control for LPV systems with bounded parameter variation rates. *International Journal of Robust and Nonlinear Control*, 6:983–998, 1996.

[176] V. A. Yakubovich. Frequency conditions for the absolute stability of control systems with several nonlinear or linear nonstationary units. *Automat. Telemech.*, pages 5–30, 1967.

[177] V. A. Yakubovich. S-procedure in nonlinear control theory. *Vestnik Leningrad Univ.*, pages 62–77, 1971. In Russian. English translation in Vestnik Leningrad Univ. Math., (1977), 73-93.

[178] V.A. Yakubovich. A frequency theorem for the case in which the state and control spaces are Hilbert spaces with an application to some problems of synthesis of optimal controls. *Sibirskii Mat. Zh.*, 15:639–668, 1975. English translation in Siberian Mathematics Journal.

[179] K.Y. Yang, S. R. Hall, and E. Feron. Robust H_2 control. In *Recent Advances on LMI Methods in Control; editors L. El Ghaoui, S. Niculescu.* SIAM, 1999.

[180] D.C. Youla, H.A. Jabr, and J.J. Bongiorno Jr. Modern Wiener-Hopf design of optimal controllers: part II. *IEEE Transactions on Automatic Control*, 21:319–338, 1976.

[181] N. Young. *An Introduction to Hilbert Space*. Cambridge University Press, 1988.

[182] P. M. Young. *Robustness with Parametric and Dynamic Uncertainty*. PhD thesis, California Institute of Technology, 1993.

[183] G. Zames. On the input-output stability of time varying nonlinear feedback systems. Part I: Conditions derived using concepts of loop gain, conicity, and positivity. *IEEE Transactions on Automatic Control*, 11:228–238, 1966.

[184] G. Zames. Feedback and optimal sensitivity: Model reference transformations, multiplicative seminorms, and approximate inverses. *IEEE Transactions on Automatic Control*, 26:301–320, 1981.

[185] G. Zames and A.K. El-Sakkary. Unstable systems and feedback: the gap metric. In *Proc. Allerton Conf.*, 1980.

[186] G. Zames and J. G. Owen. Duality theory for MIMO robust disturbance rejection. *IEEE Transactions on Automatic Control*, 38:743–52., 1993.

[187] K. Zhou and J.C. Doyle. *Essentials of Robust Control*. Prentice Hall, 1998.

[188] K. Zhou, J.C. Doyle, and K. Glover. *Robust and Optimal Control*. Prentice Hall, 1996.

[189] K. Zhou, K. Glover, B. Bodenheimer, and J.C. Doyle. Mixed H_2 and H_∞ performance objectives I: Robust performance analysis. *IEEE Transactions on Automatic Control*, 39:1564–1574, 1994.

Index